Random Graphs

The book is devoted to the study of classical combinatorial structures, such as random graphs, permutations, and systems of random linear equations in finite fields. The author shows how the application of the generalized scheme of allocation in the study of random graphs and permutations reduces the combinatorial problems to classical problems of probability theory on the summation of independent random variables. He concentrates on recent research by Russian mathematicians, including a discussion of equations containing an unknown permutation. This is the first English-language presentation of techniques for analyzing systems of random linear equations in finite fields.

These results will interest specialists in combinatorics and probability theory and will also be useful in applied areas of probabilistic combinatorics, such as communication theory, cryptology, and mathematical genetics.

V. F. Kolchin is a leading researcher at the Steklov Institute and a professor at the Moscow Institute of Electronics and Mathematics (MIEM). He has written four books and many papers in the area of probabilistic combinatorics. His papers have been published mainly in the Russian journals *Theory of Probability and Its Applications*, *Mathematical Notes*, and *Discrete Mathematics*, and in the international journal *Random Structures and Algorithms*.

Random Graphs

ENCYCLOPEDIA OF MATHEMATICS AND ITS APPLICATIONS

Random Graphs

V. F. KOLCHIN

Steklov Mathematical Institute, Moscow

CAMBRIDGE
UNIVERSITY PRESS

CAMBRIDGE UNIVERSITY PRESS
Cambridge, New York, Melbourne, Madrid, Cape Town, Singapore, São Paulo, Delhi

Cambridge University Press
The Edinburgh Building, Cambridge CB2 8RU, UK

Published in the United States of America by Cambridge University Press, New York

www.cambridge.org
Information on this title: www.cambridge.org/9780521119689

First published 1999
This digitally printed version 2009

A catalogue record for this publication is available from the British Library

Library of Congress Cataloguing in Publication data
Kolchin, V. F. (Valentin Fedorovich)
Random graphs / V. F. Kolchin
p. cm. – (Encyclopedia of mathematics and its applications; v. 53)
Includes bibliographical references and index.
ISBN 0 521 44081 5 hardback
1. Random graphs. I. Title. II. Series.
QA166.17.K65 1999 98-24390
511´.5 – dc20 CIP

ISBN 978-0-521-44081-3 hardback
ISBN 978-0-521-11968-9 paperback

CONTENTS

PREFACE

Combinatorics played an important role in the development of probability theory and the two have continued to be closely related. Now probability theory, by offering new approaches to problems of discrete mathematics, is beginning to repay its debt to combinatorics. Among these new approaches, the methods of asymptotic analysis, which have been well developed in probability theory, can be used to solve certain complicated combinatorial problems.

If the uniform distribution is defined on the set of combinatorial structures in question, then the numerical characteristics of the structures can be regarded as random variables and analyzed by probabilistic methods. By using the probabilistic approach, we restrict our attention to "typical" structures that constitute the bulk of the set, excluding the small fraction with exceptional properties.

The probabilistic approach that is now widely used in combinatorics was first formulated by V. L. Goncharov, who applied it to S_n, the set of all permutations of degree n, and to the runs in random (0,1)-sequences. S. N. Bernstein, N. V. Smirnov, and V. E. Stepanov were among those who developed probabilistic combinatorics in Russia, building on the famous Russian school of probability founded by A. A. Markov, P. L. Lyapunov, A. Ya. Khinchin, and A. N. Kolmogorov.

This book is based on results obtained primarily by Russian mathematicians and presents results on random graphs, systems of random linear equations in GF(2), random permutations, and some simple equations involving permutations.

Selecting material for the book was a difficult job. Of course, this book is not a complete treatment of the topics mentioned. Some results (and their proofs) did not seem ready for inclusion in a book, and there may be relevant results that have escaped the author's attention.

There is a large body of literature on random graphs, and it is not possible to review it here. Among the probabilistic tools that have been used to analyze random structures are the method of moments, Poisson and Gaussian approximations, generating functions using the saddle-point method, Tauberian-type theorems, analysis

of singularities, and martingale theory. In the past two decades, a method called
the generalized scheme of allocation has been widely used in probabilistic combinatorics. It is so named because of its connection with the problem of assigning
n objects randomly to N cells. Let $\eta_1, \ldots, \eta_N$ be random variables that are, for
example, the sizes of components of a graph. If there are independent random
variables $\xi_1, \ldots, \xi_N$ so that the joint distribution of $\eta_1, \ldots, \eta_N$ for any integers
$k_1, \ldots, k_N$ can be written as

$$\mathbf{P}\{\eta_1 = k_1, \ldots, \eta_N = k_N\} = \mathbf{P}\{\xi_1 = k_1, \ldots, \xi_N = k_N \mid \xi_1 + \cdots + \xi_N = n\},$$

where n is a positive integer, then we say that $\eta_1, \ldots, \eta_N$ satisfy the generalized
scheme of allocation with parameters n and N and independent random variables
$\xi_1, \ldots, \xi_N$.

Graph evolution is the random process of sequentially adding new edges to a
graph. For many classes of random graphs with n labeled vertices and T edges, the
parameter $\theta = 2T/n$ plays a role of time in the process; various graph properties
often change abruptly at the critical point $\theta = 1$. Graph evolution is the most
fascinating object in the theory of random graphs, and it appears that it is well
suited to the generalized scheme. We will show that applying generalized schemes
makes it possible to analyze random graphs at different stages of their evolution
and to obtain limit distributions in those cases in which only properties similar to
the law of large numbers have been proved.

The theory of random equations in finite fields is shared by probability, combinatorics, and algebra. In this book, we will consider systems of linear equations in
GF(2) with random coefficients. The matrix of such a system corresponds to a random graph or hypergraph; therefore, results on random graphs help to study these
systems. We are sure that this application alone justifies developing the theory of
random graphs.

The theory of random permutations is a well-developed branch of probabilistic combinatorics. Although Goncharov has investigated the cycle structure of a
random permutation in great detail, there is still great interest in this area. We will
fully describe the asymptotic behavior of $\mathbf{P}\{v_n = k\}$ for the total number v_n of
cycles in a random permutation for all possible behaviors of the parameters n and
$k = k(n)$ as $n \to \infty$. We will also give some of the asymptotic results for the
number of solutions of the equation $X^d = e$, where an unknown $X \in S_n$, d is a
fixed positive integer, and e is the identity of the group S_n.

Although the generalized scheme of allocation cannot be applied to nonequiprobable graphs, we present some results in this situation by using the method
of moments. The statistical applications of nonequiprobable graphs call for the
development of regular methods of analyzing these structures.

The book consists of five chapters. Chapter 1 describes the generalized scheme
of allocation and its applications to a random forest of nonrooted trees, a random

graph consisting of unicyclic components, and a random graph with a mixture of trees and unicyclic components. In Chapter 2, these results are applied to the study of the evolution of random graphs. Chapter 3 is devoted to systems of random linear equations in GF(2). Much of this branch of probabilistic combinatorics is the work of Russian mathematicians; this is the first English-language presentation of many of the results. Random permutations are considered in Chapter 4, and Chapter 5 contains some results on permutation equations of the form $X^d = e$.

Most results presented in this book derive from work done over the past fifteen years; notes and references can be found in the last section of each chapter. (It is, of course, impossible to give a complete list in each particular area.) In addition to articles used in the text, the summary sections of all chapters include references to papers on related topics, especially those in which the same results were obtained by other methods.

We assume that the reader is familiar with basic combinatorics. This book should be accessible to those who have completed standard courses of mathematical analysis and probability theory. Section 1.1 includes a list of pertinent results from probability.

This book continues in the tradition of *Random Mappings* [78] and differs from other treatments of random graphs in the systematic use of the generalized scheme of allocation. We hope that the chapter on systems of random linear equations in GF(2) will be of interest to a broad audience. I wish to express my sincere appreciation to G.-C. Rota, who encouraged me to write this book for the *Encyclopedia of Mathematics* series, even though there are already several excellent books on random graphs.

My greatest concern is writing the book in English. I am indebted to the editors who have brought the text to an acceptable form. It is apparent that no amount of editing can erase the heavy Russian accent of my written English, so my special thanks go to those readers who will not be deterred by the language of the book.

I greatly appreciate the support I received from my colleagues at the Steklov Mathematical Institute while I wrote this book.

1

The generalized scheme of allocation and the components of random graphs

1.1. The probabilistic approach to enumerative combinatorial problems

The solution to enumerative combinatorial problems consists in finding an exact or approximate expression for the number of combinatorial objects possessing the property under investigation. In this book, the probabilistic approach to enumerative combinatorial problems is adopted.

The fundamental notion of probability theory is the probability space $(\Omega, \mathcal{A}, \mathbf{P})$, where Ω is a set of arbitrary elements, $\mathcal{A}$ is a set of subsets of Ω forming a σ-algebra of events with the operations of union and intersection of sets, and $\mathbf{P}$ is a nonnegative countably additive function defined for each event $A \in \mathcal{A}$ so that $\mathbf{P}(\Omega) = 1$. The set Ω is called the space of elementary events and $\mathbf{P}$ is a probability. A random variable is a real-valued measurable function $\xi = \xi(\omega)$ defined for all $\omega \in \Omega$.

Suppose Ω consists of finitely many elements. Then the probability $\mathbf{P}$ is defined on all subsets of Ω if it is defined for each elementary event $\omega \in \Omega$. In this case, any real-valued function $\xi = \xi(\omega)$ on such a space of elementary events is a random variable.

Instead of a real-valued function, one may consider a function $f(\omega)$ taking values from some set Y of arbitrary elements. Such a function $f(\omega)$ may be considered a generalization of a random variable and is called a random element of the set Y.

In studying combinatorial objects, we consider probability spaces that have a natural combinatorial interpretation: For the space of elementary events Ω, we take the set of combinatorial objects under investigation and assign the same probability to all the elements of the set. In this case, numerical characteristics of combinatorial objects of Ω become random variables. The term "random element of the set Ω" is usually used for the identity function $f(\omega) = \omega$, $\omega \in \Omega$, mapping each element of the set of combinatorial objects into itself. Since the uniform distribution is assumed on Ω, the probability that the identity function f takes any fixed value ω is the same for all $\omega \in \Omega$. Hence the notion of a random combinatorial object of Ω, such as the identity function $f(\omega) = \omega$, agrees with the usual notion of a random element of a set as an element sampled from all elements of the set with equal probabilities.

Note that a random combinatorial object with the same distribution could also be defined on larger probability spaces. For our purposes, however, the natural construction presented here is sufficient for the most part. The exceptions are those few cases that involve several independent random combinatorial objects and in which it would be necessary to resort to a richer probability space, such as the direct product of the natural probability spaces.

Since we use probability spaces with uniform distributions, in spite of the probabilistic terminology, the problems considered are in essence enumeration problems of combinatorial analysis. The probabilistic approach furnishes a convenient form of representation and helps us effectively use the methods of asymptotic analysis that have been well developed in the theory of probability.

Thus, in the probabilistic approach, numerical characteristics of a random combinatorial object are random variables. The main characteristic of a random variable ξ is its distribution function $F(x)$ defined for any real x as the probability of the event $\{\xi \leq x\}$, that is,

$$F(x) = \mathbf{P}\{\xi \leq x\}.$$

The distribution function $F(x)$ defines a probability distribution on the real line called the distribution of the random variable ξ. With respect to this distribution, given a function $g(x)$, the Lebesgue–Stieltjes integral

$$\int_{-\infty}^{\infty} g(x)\, dF(x)$$

can be defined. The probabilistic approach has advantages in the asymptotic investigations of combinatorial problems. As a rule, we have a sequence of random variables $\xi_n, n = 1, 2, \ldots$, each of which describes a characteristic of the random combinatorial object under consideration, and we are interested in the asymptotic behavior of the distribution functions $F_n(x) = \mathbf{P}\{\xi_n \leq x\}$ as $n \to \infty$.

A sequence of distributions with distribution functions $F_n(x)$ converges weakly to a distribution with the distribution function $F(x)$ if, for any bounded continuous function $g(x)$,

$$\int_{-\infty}^{\infty} g(x)\, dF_n(x) \to \int_{-\infty}^{\infty} g(x)\, dF(x)$$

as $n \to \infty$.

The weak convergence of distributions is directly connected with the pointwise convergence of the distribution functions as follows.

Theorem 1.1.1. *A sequence of distribution functions $F_n(x)$ converges to a distribution function $F(x)$ at all continuity points if and only if the corresponding sequence of distributions converges weakly to the distribution with distribution function $F(x)$.*

In a sense, the distribution, or the distribution function $F(x)$, characterizes the random variable ξ. The moments of ξ are simple characteristics. If

$$\int_{-\infty}^{\infty} |x|\, dF(x)$$

exists, then

$$\mathbf{E}\xi = \int_{-\infty}^{\infty} x\, dF(x)$$

is called the mathematical expectation, or mean, of the random variable ξ. Further,

$$m_r = \mathbf{E}\xi^r = \int_{-\infty}^{\infty} x^r\, dF(x)$$

is called the rth moment, or the moment of rth order (if the integral of $|x|^r$ exists).

In probabilistic combinatorics, one usually considers nonnegative integer-valued random variables. For such a random variable, the factorial moments are natural characteristics. We denote the rth factorial moment by

$$m_{(r)} = \mathbf{E}\xi(\xi - 1) \cdots (\xi - r + 1).$$

If a distribution function $F(x)$ can be represented in the form

$$F(x) = \int_{-\infty}^{x} p(u)\, du,$$

where $p(u) \geq 0$, then we say that the distribution has a density $p(u)$. In addition to the distribution function, it is convenient to represent the distribution of an integer-valued random variable ξ by the probabilities of its individual values. For ξ, we will use the notation

$$p_k = \mathbf{P}\{\xi = k\}, \quad k = 0, 1, \ldots,$$

and for integer-valued nonnegative random variables ξ_n,

$$p_k^{(n)} = \mathbf{P}\{\xi_n = k\}, \quad k = 0, 1, \ldots.$$

It is clear that

$$\mathbf{E}\xi = \sum_{n=0}^{\infty} k p_k,$$

if this series converges.

It is not difficult to see that the following assertion is true.

Theorem 1.1.2. *A sequence of distributions $\{p_k^{(n)}\}$, $n = 1, 2, \ldots$, converges weakly to a distribution $\{p_k\}$ if and only if for every fixed $k = 1, 2, \ldots$,*

$$p_k^{(n)} \to p_k$$

as $n \to \infty$.

If an estimate of the probability $\mathbf{P}\{\xi > 0\}$ is needed for a nonnegative integer-valued random variable ξ, then the simple inequality

$$\mathbf{P}\{\xi > 0\} = \sum_{k=1}^{\infty} \mathbf{P}\{\xi = k\} \leq \sum_{k=1}^{\infty} k p_k = \mathbf{E}\xi \tag{1.1.1}$$

can be useful. In particular, for a sequence ξ_n, $n = 1, 2, \ldots$, of such random variables with $\mathbf{E}\xi_n \to 0$ as $n \to \infty$, it follows that

$$\mathbf{P}\{\xi_n > 0\} \to 0.$$

Since it is generally easier to calculate the moments of a random variable than the whole distribution, one wants a criterion for the convergence of a sequence of distributions based on the corresponding moments. But, first, it should be noted that even if a random variable ξ has moments of all orders, its distribution cannot, in general, be reconstructed on the basis of these moments, since there exist distinct distributions that have the same sequences of moments. For example, it is not difficult to confirm that for any $n = 1, 2, \ldots$,

$$\int_0^{\infty} x^n e^{-1/4} \sin x^{1/4} \, dx = 0.$$

Hence, for $-1 \leq \alpha \leq 1$, the function

$$p_\alpha(x) = \tfrac{1}{24} e^{-1/4} \left(1 + \alpha \sin x^{1/4}\right)$$

is the density of a distribution on $[0, \infty)$ whose moments do not depend on α.

Thus the distribution functions with moments of all orders are divided into two classes: The first class contains the functions that may be uniquely reconstructed from their moments, and the second class contains the functions that cannot be reconstructed from their moments. There are several sufficient conditions for the moment problem to have a unique solution. Let

$$M_n = \int_{-\infty}^{\infty} |x|^n \, dF(x).$$

A distribution function $F(x)$ is uniquely reconstructed by the sequence m_r, $r = 1, 2, \ldots$, of its moments if there exists λ such that

$$\frac{1}{n} M_n^{1/n} \leq \lambda. \tag{1.1.2}$$

The following theorem describing the so-called method of moments is applicable only to the first class of distribution functions.

Theorem 1.1.3.   *If distribution functions $F_n(x)$, $n = 1, 2, \ldots$, have the moments of all orders and for any fixed $r = 1, 2, \ldots$,*

$$m_r^{(n)} = \int_{-\infty}^{\infty} x^r \, d F_n(x) \to m_r, \quad |m_r| < \infty,$$

as $n \to \infty$, then there exists a distribution function $F(x)$ such that for any fixed $r = 1, 2, \ldots$,

$$m_r = \int_{-\infty}^{\infty} x^r \, d F(x),$$

and from the sequence $F_n(x)$, $n = 1, 2, \ldots$, it is possible to select a subsequence $F_{n_k}(x)$, $k = 1, 2, \ldots$, that converges to $F(x)$ as $n \to \infty$ at every continuity point of $F(x)$.

   If the sequence m_r, $r = 1, 2, \ldots$, uniquely determines the distribution function $F(x)$, then $F_n(x) \to F(x)$ as $n \to \infty$ at every continuity point of $F(x)$.

   Note that the normal (Gaussian) and Poisson distributions are uniquely reconstructible by their moments.

   To use the method of moments, it is necessary to calculate moments of random variables. One useful method of calculating moments of integer-valued random variables is to represent them as sums of random variables that take only the values 0 and 1.

Theorem 1.1.4.   *If*

$$S_n = \xi_1 + \cdots + \xi_n,$$

and the random variables $\xi_1, \ldots, \xi_n$ take only the values 0 and 1, then for any $m = 1, 2, \ldots, n$,

$$S_n(S_n - 1) \cdots (S_n - m + 1) = \sum_{\{i_1, \ldots, i_m\}} \xi_{i_1} \cdots \xi_{i_m},$$

where the summation is taken over all different ordered sets of different indices $\{i_1, \ldots, i_m\}$, the number of which is equal to $\binom{n}{m} m!$.

   Generating functions also provide a useful tool for solving many problems related to distributions of nonnegative integer-valued random variables. The complex-valued function

$$\phi(z) = \phi_\xi(z) = \sum_{k=0}^{\infty} p_k z^k = \mathbf{E} z^\xi \tag{1.1.3}$$

is called the generating function of the distribution of the random variable ξ. It is defined at least for $|z| \leq 1$. For example, for the Poisson distribution with parameter λ, which is defined by the probabilities

$$p_k = \frac{\lambda^k}{k!} e^{-\lambda}, \quad k = 0, 1, \ldots,$$

the generating function is $e^{\lambda(z-1)}$.

Relation (1.1.3) determines a one-to-one correspondence between the generating functions and the distributions of nonnegative integer-valued random variables, since the distribution can be reconstructed by using the formula

$$p_k = \frac{1}{k!} \phi^{(k)}(0), \quad k = 0, 1, \ldots. \tag{1.1.4}$$

Generating functions are especially convenient for the investigation of sums of independent random variables. If $\xi_1, \ldots, \xi_n$ are independent nonnegative integer-valued random variables and $S_n = \xi_1 + \cdots + \xi_n$, then

$$\phi_{S_n}(z) = \phi_{\xi_1}(z) \cdots \phi_{\xi_n}(z).$$

The correspondence between the generating functions and the distributions is continuous in the following sense.

Theorem 1.1.5. *Let $\{p_k^{(n)}\}$, $n = 1, 2, \ldots$, be a sequence of distributions. If for any $k = 0, 1, \ldots$,*

$$p_k^{(n)} \to p_k$$

as $n \to \infty$, then the sequence of corresponding generating functions $\phi_n(z)$, $n = 1, 2, \ldots$, converges to the generating function of the sequence $\{p_k\}$ uniformly in any circle $|z| \leq r < 1$.

In particular, if $\{p_k\}$ is a distribution, then the sequence of corresponding generating functions converges to the generating function $\phi(z)$ of the distribution $\{p_k\}$ uniformly in any circle $|z| \leq r < 1$.

Theorem 1.1.6. *If the sequence of generating functions $\phi_n(z)$, $n = 1, 2, \ldots$, of the distributions $\{p_k^{(n)}\}$ converges to a generating function $\phi(z)$ of a distribution $\{p_k\}$ on a set M that has a limit point inside of the circle $|z| \leq 1$, then the distributions $\{p_k^{(n)}\}$ converge weakly to the distribution $\{p_k\}$.*

Since a generating function $\phi(z) = \sum_{k=0}^{\infty} p_k z^k$ is analytic, its coefficients can be represented by the Cauchy formula

$$p_n = \frac{1}{n!} \phi^{(n)}(0) = \frac{1}{2\pi i} \int_C \frac{\phi(z)\, dz}{z^{n+1}}, \quad n = 0, 1, \ldots,$$

where the integral is over a contour C that lies inside the domain of analyticity of $\phi(z)$ and contains the point $z = 0$.

Thus, if we are interested in the behavior of p_n as $n \to \infty$, then we have to be able to estimate contour integrals of the form

$$G(\lambda) = \frac{1}{2\pi i} \int_C g(z) e^{\lambda f(z)} \, dz,$$

where $g(z)$ and $f(z)$ are analytic in the neighborhood of the curve of integration C and λ is a real parameter tending to infinity.

The saddle-point method is used to estimate such integrals. The contour of integration C may be chosen in different ways. The saddle-point method requires choosing the contour C in such a way that it passes through the point z_0, which is a root of the equation $f'(z) = 0$. Such a point is called the saddle point, since the function $\Re f(z)$ has a graph similar to a saddle or mountain pass. The saddle-point method requires choosing the contour of integration such that it crosses the saddle point z_0 in the direction of the steepest descent. However, finding such a contour and applying it are complicated problems, so for the sake of simplicity one usually does not choose the best contour, hence losing some accuracy in the remainder term when estimating the integral.

A parametric representation of the contour transforms the contour integral to an integral with a real variable of integration. Therefore the following theorem on estimating integrals with increasing parameters, based on Laplace's method, sometimes provides an answer to the initial question on estimating integrals.

Theorem 1.1.7. *If the integral*

$$G(\lambda) = \int_{-\infty}^{\infty} g(t) e^{\lambda f(t)} \, dt$$

converges absolutely for some $\lambda = \lambda_0$, *that is,*

$$\int_{-\infty}^{\infty} |g(t)| e^{\lambda_0 f(t)} \, dt \leq M;$$

if the function $f(t)$ *attains its maximum at a point* t_0 *and in a neighborhood of this point*

$$f(t) = f(t_0) + a_2(t - t_0)^2 + a_3(t - t_0)^3 + \cdots$$

with $a_2 < 0$;

if for an arbitrary small $\delta > 0$, *there exists* $h = h(\delta) > 0$ *such that*

$$f(t_0) - f(t) \geq h,$$

for $|t - t_0| > \delta$;

and if, as $t \to t_0$,

$$g(t) = c(t - t_0)^{2m} (1 + O(|t - t_0|)),$$

where c is a nonzero constant and m is a nonnegative integer, then, as $\lambda \to \infty$,

$$G(\lambda) = e^{\lambda f(t_0)} \lambda^{-m-1/2} cc_1^{2m+1} \Gamma(m + 1/2)\big(1 + O\big(1/\sqrt{\lambda}\big)\big),$$

where $\Gamma(x)$ is the Euler gamma function and

$$c_1 = \frac{1}{\sqrt{-a_2}} = \frac{1}{\sqrt{-f''(t_0)/2}}.$$

In particular, if $m = 0$, then $c = g(t_0)$, and as $\lambda \to \infty$,

$$G(\lambda) = e^{\lambda f(t_0)} \frac{g(t_0)}{\sqrt{-f''(t_0)/2}} \sqrt{\pi/\lambda}\big(1 + O\big(1/\sqrt{\lambda}\big)\big). \tag{1.1.5}$$

To demonstrate that this rather complicated theorem can really be used, let us estimate the integral

$$\Gamma(\lambda + 1) = \int_0^\infty x^\lambda e^{-x} \, dx$$

as $\lambda \to \infty$, and obtain the Stirling formula. The change of variables $x = \lambda t$ leads to the equation

$$\Gamma(\lambda + 1) = \lambda^{\lambda+1} e^{-\lambda} \int_0^\infty e^{-\lambda(t-1-\log t)} \, dt.$$

Here $g(t) = 1$, and $f(t) = -(t - 1 - \log t)$, $f(1) = 0$, $f'(1) = 0$, $f''(1) = -1$. The conditions of the theorem are fulfilled; therefore, by (1.1.5),

$$G(\lambda) = \int_0^\infty e^{\lambda f(t)} \, dt = \sqrt{2\pi/\lambda}\big(1 + O\big(1/\sqrt{\lambda}\big)\big),$$

and for the Euler gamma function, we obtain the representation

$$\Gamma(\lambda + 1) = \lambda^{\lambda+1/2} e^{-\lambda} \sqrt{2\pi}\big(1 + O\big(1/\sqrt{\lambda}\big)\big)$$

as $\lambda \to \infty$, coinciding with the Stirling formula, except for the remainder term, which can be improved to $O(1/\lambda)$.

Generating functions are only suited for nonnegative integer-valued random variables. A more universal method of proving theorems on the convergence of sequences of random variables is provided by characteristic functions. The characteristic function of a random variable ξ or the characteristic function of its distribution is defined as

$$\varphi(t) = \varphi_\xi(t) = \mathbf{E}e^{it\xi} = \int_{-\infty}^\infty e^{itx} \, dF(x), \tag{1.1.6}$$

where $-\infty < t < \infty$ and $F(x)$ is the distribution function of ξ.

If the rth moment m_r exists, then the characteristic function $\varphi(t)$ is r times differentiable, and

$$\varphi^{(r)}(0) = i^r m_r.$$

Characteristic functions are convenient for investigating sums of independent random variables, since if $S_n = \xi_1 + \cdots + \xi_n$, where $\xi_1, \ldots, \xi_n$ are independent random variables, then

$$\varphi_{S_n}(t) = \varphi_{\xi_1}(t) \cdots \varphi_{\xi_n}(t).$$

The characteristic function of the normal distribution with parameters (m, σ^2) and density

$$p(x) = \frac{1}{\sqrt{2\pi}\sigma} e^{-(x-m)^2/(2\sigma^2)}$$

is $e^{imt - \sigma^2 t^2/2}$.

Relation (1.1.6) defines a one-to-one correspondence between characteristic functions and distributions. There are different inversion formulas that provide a formal possibility of reconstructing a distribution from its characteristic function, but they have limited practical applications. We state the simplest version of the inversion formulas.

Theorem 1.1.8.   *If a characteristic function $\varphi(t)$ is absolutely integrable, then the corresponding distribution has the bounded density*

$$p(x) = \frac{1}{2\pi} \int_{-\infty}^{\infty} e^{-itx} \varphi(t)\, dt.$$

The correspondence defined by (1.1.6) is continuous in the following sense.

Theorem 1.1.9.   *A sequence of distributions converges weakly to a limit distribution if and only if the corresponding sequence of characteristic functions $\varphi_n(t)$ converges to a continuous function $\varphi(t)$ as $n \to \infty$ at every fixed t, $-\infty < t < \infty$. In this case, $\varphi(t)$ is the characteristic function of the limit distribution, and the convergence $\varphi_n(t) \to \varphi(t)$ is uniform in any finite interval.*

For a sequence ξ_n of characteristics of random combinatorial objects, applying Theorem 1.1.9 gives the limit distribution function. But for integer-valued characteristics, one would rather have an indication of the local behavior, that is, the behavior of the probabilities of individual values. To this end the so-called local limit theorems of probability theory are used.

Let ξ be an integer-valued random variable and $p_n = \mathbf{P}\{\xi = n\}$. It is clear that $\mathbf{P}\{\xi \in \Gamma_1\} = 1$, where Γ_1 is the lattice of all integers. If there exists a lattice Γ_d with a span d such that $\mathbf{P}\{\xi \in \Gamma_d\} = 1$ and there is no lattice Γ with span greater than d such that $\mathbf{P}\{\xi \in \Gamma\}$, then d is called the maximal span of the distribution of ξ. The characteristic function $\varphi(t)$ of the random variable ξ is periodic with period $2\pi/d$ and $|\varphi(t)| < 1$ for $0 < t < 2\pi/d$.

For integer-valued random variables, the inversion formula has the following form:

$$p_n = \frac{1}{2\pi} \int_{-\pi}^{\pi} e^{-itn} \varphi(t) \, dt.$$

Consider the sum $S_N = \xi_1 + \cdots + \xi_N$ of independent identically distributed integer-valued random variables $\xi_1, \ldots, \xi_N$. When the distributions of the summands are identical and do not depend on N, the problem of estimating the probabilities $P\{S_N = n\}$, as $N \to \infty$, has been completely solved. If there exist sequences of centering and normalizing numbers A_N and B_N such that the distributions of the random variables $(S_N - A_N)/B_N$ converge weakly to some distribution, then the limit distribution has a density. Moreover, a local limit theorem holds on the lattice with a span equal to the maximal span of the distribution of the random variable ξ_1. If the maximal span of the distribution of ξ_1 is 1, then the local theorem holds on the lattice of integers.

Theorem 1.1.10.   *Let $\xi_1, \xi_2, \ldots$ be a sequence of independent identically distributed integer-valued random variables and let there exist A_N and B_N such that, as $N \to \infty$ for any fixed x,*

$$P\left\{ \frac{S_N - A_N}{B_N} \le x \right\} \to \int_{-\infty}^{x} p(u) \, du.$$

Then, if the maximal span of the distribution of ξ_1 is 1,

$$B_N P\{S_N = n\} - p\big((n - A_N)/B_N\big) \to 0$$

uniformly in n.

Local limit theorems are of primary importance in what follows. Therefore, let us prove a local theorem on convergence to the normal distribution as a model for proofs of local limit theorems in more complex cases, which will be discussed later in the book.

Theorem 1.1.11.   *Let the independent identically distributed integer-valued random variables $\xi_1, \xi_2, \ldots$ have a mathematical expectation a and a positive variance σ^2. Then, if the maximal span of the distribution of ξ_1 is 1,*

$$\sigma\sqrt{N} P\{\xi_1 + \cdots + \xi_N = n\} - \frac{1}{\sqrt{2\pi}} \exp\left\{ -\frac{(n - aN)^2}{2\sigma^2 N} \right\} \to 0$$

uniformly in n as $N \to \infty$.

Proof. Let

$$z = \frac{n - aN}{\sigma\sqrt{N}} \quad \text{and} \quad P_N(n) = P\{\xi_1 + \cdots + \xi_N = n\}.$$

If $\varphi(t)$ is the characteristic function of the random variable ξ_1, then the characteristic function of the sum $S_N = \xi_1 + \cdots + \xi_N$ is equal to $\varphi^N(t)$, and

$$\varphi^N(t) = \sum_{n=-\infty}^{\infty} P_N(n) e^{itn}.$$

By the inversion formula,

$$P_N(n) = \frac{1}{2\pi} \int_{-\pi}^{\pi} e^{-itn} \varphi^N(t) \, dt. \tag{1.1.7}$$

Let $\varphi^*(t)$ denote the characteristic function of the centered random variable $\xi_1 - a$, which equals $\varphi(t) \exp\{-ita\}$. Since $n = aN + \sigma z\sqrt{N}$, it follows from (1.1.7) that

$$P_N(n) = \frac{1}{2\pi} \int_{-\pi}^{\pi} e^{it\sigma z\sqrt{N}} (\varphi^*(t))^N \, dt.$$

After the substitution $x = t\sigma\sqrt{N}$, this equality takes the form

$$\sigma\sqrt{N} P_N(n) = \frac{1}{2\pi} \int_{-\pi\sigma\sqrt{N}}^{\pi\sigma\sqrt{N}} e^{ixz} \left(\varphi^*\left(x/(\sigma\sqrt{N})\right)\right)^N \, dx. \tag{1.1.8}$$

By the inversion formula,

$$\frac{1}{\sqrt{2\pi}} e^{-z^2/2} = \frac{1}{2\pi} \int_{-\infty}^{\infty} e^{-ixz - x^2/2} \, dx. \tag{1.1.9}$$

It follows from (1.1.8) and (1.1.9) that the difference

$$R_N = 2\pi \left(\sigma\sqrt{N} P_N(n) - \frac{1}{\sqrt{2\pi}} e^{-z^2/2}\right) \tag{1.1.10}$$

can be written as the sum of the following four integrals:

$$I_1 = \int_{-A}^{A} e^{-ixz} \left(\left(\varphi^*\left(x/(\sigma\sqrt{N})\right)\right)^N - e^{-x^2/2}\right) dx,$$

$$I_2 = -\int_{A \le |x|} e^{-ixz - x^2/2} \, dx,$$

$$I_3 = \int_{A \le |x| \le \varepsilon\sigma\sqrt{N}} e^{-ixz} \left(\varphi^*\left(x/(\sigma\sqrt{N})\right)\right)^N \, dx,$$

$$I_4 = \int_{\varepsilon\sigma\sqrt{N} \le |x| \le \pi\sigma\sqrt{N}} e^{-ixz} \left(\varphi^*\left(x/(\sigma\sqrt{N})\right)\right)^N \, dx,$$

where the constants A and ε will be chosen later.

To see that $R_N \to 0$ as $N \to \infty$, we take an arbitrary $\delta > 0$ and show that R_N can be made less than δ for sufficiently large N.

For I_2, we have

$$|I_2| \le \int_{A \le |x|} e^{-x^2/2}\, dx,$$

and $|I_2|$ can be made arbitrarily small by the choice of sufficiently large A.

Since $\mathbf{E}\xi_1 = a$ and $\mathbf{D}\xi_1 = \sigma^2$, for the characteristic function $\varphi^*(t)$ as $t \to 0$, we have

$$\varphi^*(t) = 1 - \frac{\sigma^2 t^2}{2} + o(t^2). \qquad (1.1.11)$$

Let $\varphi_N(t)$ denote the characteristic function of $(S_N - aN)/(\sigma\sqrt{N})$, which equals $(\varphi^*(x/(\sigma\sqrt{N})))^N$. For any fixed x and $N \to \infty$, we obtain from (1.1.11) the relation

$$\log \varphi_N(x) = N \log \varphi^*\left(x/(\sigma\sqrt{N})\right)$$

$$= N \log \left(1 - \frac{x^2}{2N} + o(1/N)\right)$$

$$= -\frac{x^2}{2} + o(1),$$

implying that for any fixed x as $N \to \infty$,

$$\varphi_N(x) \to e^{-x^2/2}. \qquad (1.1.12)$$

Moreover, as seen from (1.1.11), there exists $\varepsilon > 0$ such that, for $|t| \le \varepsilon$,

$$|\varphi^*(t)| \le 1 - \frac{\sigma^2 t^2}{4} \le e^{-\sigma^2 t^2/4}. \qquad (1.1.13)$$

Using this inequality to estimate I_3, we find that

$$I_3 \le \int_{A \le |x| \le \varepsilon\sigma\sqrt{N}} \left|\varphi^*(x/(\sigma\sqrt{N}))\right|^N dx \le \int_{A \le |x| \le \varepsilon\sigma\sqrt{N}} e^{-x^2/4}\, dx,$$

and by the choice of sufficiently large A, $|I_3|$ can be made arbitrarily small.

Let ε be such that (1.1.13) is satisfied and let A be large enough so that $|I_2| \le \delta/4$ and $|I_3| \le \delta/4$. Let us now estimate the integrals I_1 and I_4 for fixed ε and A. Relation (1.1.12) implies that the distribution of $(S_N - aN)/(\sigma\sqrt{N})$ converges weakly, as $N \to \infty$, to the normal distribution with parameters $(0, 1)$. The convergence of the characteristic functions $\varphi_N(x)$ to the characteristic function of the normal law is uniform in any finite interval, and the integral I_1 tends to zero as $N \to \infty$.

For I_4, we have

$$I_4 = \int_{\varepsilon\sigma\sqrt{N} \le |x| \le \pi\sigma\sqrt{N}} \left|\varphi^*(x/(\sigma\sqrt{N}))\right|^N dx = \sigma\sqrt{N} \int_{\varepsilon \le |t| \le \pi} |\varphi(t)|^N\, dt.$$

Since the maximal span of the distribution of ξ_1 is 1,

$$\max_{\varepsilon \le |t| \le \pi} |\varphi(t)| = q < 1.$$

Hence,

$$|I_4| \leq \sigma\sqrt{N}2\pi q^N,$$

and $I_4 \to 0$ as $N \to \infty$.

The estimates of I_1 and I_4 show that there exists N_0 such that $|I_1| \leq \delta/4$ and $|I_4| \leq \delta/4$ for $N \geq N_0$.

Thus the difference R_N tends to zero as $N \to \infty$ uniformly for all integers n.

∎

In most applications of local theorems in this text, the distribution of the summands of the sum $S_N = \xi_1 + \cdots + \xi_N$ depends on the number of summands N. In such cases, there is no complete answer to the question of when the local theorem holds for S_N. Even in the case of convergence to the normal law, the known sufficient conditions for the validity of a local theorem cannot be deemed fully satisfactory. Hence, for each specific distribution whose parameters depend on the number of summands in the sum, it is necessary to invoke the classical scheme given above as a model. In the hope of finding simple sufficient conditions for the validity of local theorems for integer-valued identically distributed summands, as in Theorems 1.1.10 and 1.1.11, we will often omit the particularly cumbersome calculations arising in estimating characteristic functions.

If $\xi_1, \ldots, \xi_N$ are independent identically distributed random variables such that

$$\mathbf{P}\{\xi_1 = 1\} = p \quad \text{and} \quad \mathbf{P}\{\xi_1 = 0\} = q = 1 - p \quad \text{for} \quad 0 < p < 1,$$

then $S_N = \xi_1 + \cdots + \xi_N$ has the binomial distribution with parameters (N, p), that is, for any $k = 0, 1, \ldots, N$,

$$\mathbf{P}\{S_N = k\} = \binom{N}{k}p^k q^{N-k}.$$

If $Npq \to \infty$, then the binomial distribution is approximated by the normal law. The following theorem, known as the local de Moivre–Laplace theorem, can be obtained by a direct analysis of the explicit formula.

Theorem 1.1.12. *If $N \to \infty$ and $(1 + u^6)/(Npq) \to 0$, where*

$$u = \frac{k - Np}{\sqrt{Npq}},$$

then

$$\binom{N}{k}p^k q^{N-k} = \frac{1}{\sqrt{2\pi Npq}}e^{-u^2/2}\left(1 + \frac{q - p}{6\sqrt{Npq}}(3u - u^3) + O\left(\frac{1 + u^6}{Npq}\right)\right).$$

Theorem 1.1.12 implies the well-known integral de Moivre–Laplace theorem.

Theorem 1.1.13. *If* $N \to \infty$ *and* $(1 + u^6)/(Npq) \to 0$, *where*

$$u = \frac{k - Np}{\sqrt{Npq}},$$

then

$$\mathbf{P}\{S_N \le k\} = \frac{1}{\sqrt{2\pi}} \int_{-\infty}^{u} e^{-x^2/2} \, dx (1 + o(1)).$$

If $p \to 0$, then the binomial distribution is approximated by the Poisson law. It is well known that if $N \to \infty$ and $Np \to \lambda$, $0 < \lambda < \infty$, then

$$\binom{N}{k} p^k q^{N-k} \to \frac{\lambda^k}{k!} e^{-\lambda}$$

for any fixed $k = 0, 1, \ldots$. The Poisson approximation is also valid if Np tends to infinity not too quickly.

Theorem 1.1.14. *If* $N \to \infty$, $Np \to \infty$, $(1 + u^2)p \to 0$, *where*

$$u = \frac{k - Np}{\sqrt{Np}},$$

then

$$\binom{N}{k} p^k q^{N-k} = \frac{(Np)^k}{k!} e^{-Np}(1 + o(1)).$$

The Poisson distribution converges to the normal law as its parameter tends to infinity.

Theorem 1.1.15. *If* $(1 + u^6)/\lambda \to 0$, *where* $u = (k - \lambda)/\sqrt{\lambda}$, *then*

$$\frac{\lambda^k e^{-\lambda}}{k!} = \frac{1}{\sqrt{2\pi\lambda}} e^{-u^2/2} \left(1 + \frac{u^3 - 3u}{6\sqrt{\lambda}} + O\left(\frac{1 + u^6}{\lambda}\right)\right).$$

Sometimes it is necessary to estimate the tails of the binomial distribution in the form of an inequality with an explicit constant.

Theorem 1.1.16. *For any* $x > 0$,

$$\mathbf{P}\{S_N - \mathbf{E}S_N \ge Nx\} \le e^{-2Nx^2}.$$

1.2. The generalized scheme of allocation

In the past three decades, the so-called generalized scheme of allocation of particles has been applied to many probabilistic problems of combinatorics, and many of the results in this text were obtained by reducing combinatorial problems to such a generalized scheme.

Consider n independent trials, each having N equiprobable outcomes, $1, 2,$ $\ldots, N$. Let η_i denote the number of occurrences of the ith outcome in this sequence of trials, $i = 1, 2, \ldots, N$. The random variables $\eta_1, \ldots, \eta_N$ have the multinomial distribution: If the nonnegative integers $k_1, \ldots, k_N$ are such that $k_1 + \cdots + k_N = n$, then

$$\mathbf{P}\{\eta_1 = k_1, \ldots, \eta_N = k_N\} = \frac{n!}{k_1! \cdots k_N! \, N^n}. \tag{1.2.1}$$

The situation in which the multinomial distribution arises can be described in terms of an equiprobable scheme of allocating particles to cells. If n particles are independently distributed with equal probabilities into N cells labeled $1, 2, \ldots, N$, then the contents of cells $\eta_1, \ldots, \eta_N$ have the multinomial distribution (1.2.1).

In the scheme of allocating particles to cells yielding the multinomial distribution, the contents of cells can be obtained by independent sequential allocation of particles. If one does not require that the contents of cells can be obtained by some sequential allocation of particles, with a simple probability law governing the sequential trials, then any set of integer-valued nonnegative random variables $\eta_1, \ldots, \eta_N$, such that $\eta_1 + \cdots + \eta_N = n$, can be viewed as a scheme of allocating n particles to N cells, and one can interpret η_i as the number of particles in the cell with index i, $i = 1, 2, \ldots, N$.

Some probabilistic problems of combinatorics can be treated by using generalized schemes of allocation in which the joint distribution of the contents of cells $\eta_1, \ldots, \eta_N$ can be represented in the form

$$\mathbf{P}\{\eta_1 = k_1, \ldots, \eta_N = k_N\} = \mathbf{P}\{\xi_1 = k_1, \ldots, \xi_N = k_N \mid \xi_1 + \cdots + \xi_N = n\},$$

$$\tag{1.2.2}$$

where $\xi_1, \ldots, \xi_N$ are independent identically distributed integer-valued random variables.

The generalized scheme of allocating particles to cells is given by the parameters n and N and the distribution of the random variables $\xi_1, \ldots, \xi_N$, which by relation (1.2.2) determines the joint distribution of the contents of the cells $\eta_1, \ldots, \eta_N$. Set

$$p_k = \mathbf{P}\{\xi_1 = k\}, \quad k = 0, 1, \ldots. \tag{1.2.3}$$

For the random variables $\eta_1, \ldots, \eta_N$ with the multinomial distribution (1.2.1), relation (1.2.2) is satisfied if ξ_1 has the Poisson distribution with arbitrary parameter λ:

$$p_k = \mathbf{P}\{\xi_1 = k\} = \frac{\lambda^k e^{-\lambda}}{k!}, \quad k = 0, 1, \ldots. \tag{1.2.4}$$

Therefore the distribution of $\eta_1, \ldots, \eta_N$ satisfying relation (1.2.2) for some distribution (1.2.3) can be viewed as a generalization of the multinomial distribution.

The term "classical scheme of allocation" has become common for the equiprobable scheme of allocating particles to cells leading to the multinomial distribution (1.2.1). The terminology of the classical scheme of allocating particles to cells proved to be convenient for describing a number of combinatorial problems where the multinomial distribution appears. Many results pertaining to the classical scheme of allocation can be obtained by applying relation (1.2.2) between the multinomial distribution and the Poisson distribution (1.2.4). Introducing generalized schemes of allocating particles not only broadens the scope of convenient language for describing combinatorial objects, but also offers the possibility of applying methods based on relation (1.2.2) that have been developed to analyze the classical scheme.

Let $\mu_r(n, N)$ denote the number of cells containing exactly r particles in the generalized scheme of allocation with distributions (1.2.2) and (1.2.3). We show that the representation (1.2.2) can be used to study this random variable.

Let $\xi_1^{(r)}, \ldots, \xi_N^{(r)}$ be independent identically distributed random variables whose distribution is linked with the distribution of $\xi_1, \ldots, \xi_N$ as follows:

$$\mathbf{P}\{\xi_1^{(r)} = k\} = \mathbf{P}\{\xi_1 = k \mid \xi_1 \neq r\}, \quad k = 0, 1, \ldots.$$

Also let

$$S_n = \xi_1 + \cdots + \xi_N, \qquad S_N^{(r)} = \xi_1^{(r)} + \cdots + \xi_N^{(r)}.$$

The following lemma expresses the distribution of $\mu_r(n, N)$ in terms of the probabilities of sums of independent identically distributed random variables.

Lemma 1.2.1.

$$\mathbf{P}\{\mu_r(n, N) = k\} = \binom{N}{k} p_r^k (1 - p_r)^{N-k} \frac{\mathbf{P}\{S_{N-k}^{(r)} = n - kr\}}{\mathbf{P}\{S_N = n\}}. \qquad (1.2.5)$$

Proof. Let $A_k^{(r)}$ be the event that exactly k of the random variables $\xi_1, \ldots, \xi_N$ take the value r. By equality (1.2.2),

$$\mathbf{P}\{\mu_r(n, N) = k\} = \mathbf{P}\{A_k^{(r)} \mid S_N = n\} = \frac{\mathbf{P}\{A_k^{(r)}, S_N = n\}}{\mathbf{P}\{S_N = n\}}.$$

The lemma is derived by obvious manipulations of the numerator: The events $A_k^{(r)}$ can occur for $\binom{N}{k}$ distinct choices of random variables taking the value r; therefore

$$\mathbf{P}\{A_k^{(r)}, S_N = n\}$$

$$= \binom{N}{k} p_r^k (1 - p_r)^{N-k}$$

$$\times \mathbf{P}\{S_N = n \mid \xi_1 \neq r, \ldots, \xi_{N-k} \neq r, \xi_{N-k+1} = r, \ldots, \xi_N = r\}$$

$$= \binom{N}{k} p_r^k (1 - p_r)^{N-k} \mathbf{P}\{S_{N-k}^{(r)} = n - kr\}. \qquad \blacksquare$$

In the generalized scheme of allocating particles, there is a rather simple approach to study the order statistics $\eta_{(1)} \leq \eta_{(2)} \leq \cdots \leq \eta_{(N)}$ constructed for the random variables $\eta_1, \ldots, \eta_N$ arranged in nondecreasing order.

Let $\xi_1^{(A)}, \ldots, \xi_N^{(A)}$ be independent identically distributed random variables such that

$$\mathbf{P}\{\xi_1^{(A)} = k\} = \mathbf{P}\{\xi_1 = k \mid \xi_1 \notin A\}, \quad k = 0, 1, \ldots,$$

where A is a subset of the set of natural numbers with $\mathbf{P}\{\xi_1 \notin A\} > 0$. In particular, if A consists of one value r, then $\xi_1^{(A)} = \xi_1^{(r)}$, where $\xi_1^{(r)}$ is the random variable defined preceding Lemma 1.2.1. Set

$$S_N^{(A)} = \xi_1^{(A)} + \cdots + \xi_N^{(A)}.$$

The following lemma reduces the study of distributions of order statistics to that of probabilities related to sums of independent random variables.

Lemma 1.2.2. *For any positive integer m,*

$$\mathbf{P}\{\eta_{(m)} \leq r\} = 1 - \sum_{l=0}^{m-1} \binom{N}{l}(1 - P_r)^l P_r^{N-l} \frac{\mathbf{P}\{S_l^{(\bar{A}_r)} + S_{N-l}^{(A_r)} = n\}}{\mathbf{P}\{S_N = n\}}, \quad (1.2.6)$$

$$\mathbf{P}\{\eta_{(N-m+1)} \leq r\} = \sum_{l=0}^{m-1} \binom{N}{l} P_r^l (1 - P_r)^{N-l} \frac{\mathbf{P}\{S_l^{(A_r)} + S_{N-l}^{(\bar{A}_r)} = n\}}{\mathbf{P}\{S_N = n\}}, \quad (1.2.7)$$

where A_r is the set of all nonnegative integers not exceeding r, $\bar{A}_r$ is its complement in the set of all nonnegative integers, and $P_r = \mathbf{P}\{\xi_1 > r\}$.

Proof. Let us prove (1.2.7) for $m = 1$. For the maximal order statistic $\eta_{(N)} = \max(\eta_1, \ldots, \eta_N)$, by (1.2.2) and the independence of $\xi_1, \ldots, \xi_N$, we have

$$\mathbf{P}\{\eta_{(N)} \leq r\} = \mathbf{P}\{\eta_1 \leq r, \ldots, \eta_N \leq r\}$$

$$= \mathbf{P}\{\xi_1 \leq r, \ldots, \xi_N \leq r \mid S_N = n\}$$

$$= \frac{(\mathbf{P}\{\xi_1 \leq r\})^N \mathbf{P}\{S_N = n \mid \xi_1 \leq r, \ldots, \xi_N \leq r\}}{\mathbf{P}\{S_N = n\}}.$$

By using the random variables $\xi_1^{(A_r)}, \ldots, \xi_N^{(A_r)}$, we finally obtain

$$\mathbf{P}\{\eta_{(N)} \leq r\} = \frac{(1 - P_r)^N \mathbf{P}\{S_N^{(A_r)} = n\}}{\mathbf{P}\{S_N = n\}}. \quad (1.2.8)$$

Relations (1.2.6) and (1.2.7) for other values of m are similarly proved. ∎

For the joint distribution of the random variables $\mu_{r_1}(n, N), \ldots, \mu_{r_s}(n, N)$, we can prove the following lemma as we did in Lemma 1.2.1.

Lemma 1.2.3.

$$P\{\mu_{r_1}(n, N) = k_1, \ldots, \mu_{r_s}(n, N) = k_s\}$$

$$= \frac{N! \, p_{r_1}^{k_1} \cdots p_{r_s}^{k_s} \left(1 - p_{r_1}^{k_1} - \cdots - p_{r_s}^{k_s}\right)^{N-k_1-\cdots-k_s}}{k_1! \cdots k_s! \, (N - k_1 - \cdots - k_s)!}$$

$$\times \frac{P\{S_{N-k_1-\cdots-k_s}^{(r_1,\ldots,r_s)} = n - k_1 r_1 - \cdots - k_s r_s\}}{P\{S_N = n\}},$$

where $s - 1, k_1, \ldots, k_s, r_1, \ldots, r_s$ *are nonnegative integers and* $r_1, \ldots, r_s$ *are distinct.*

Lemmas 1.2.1, 1.2.2, and 1.2.3 express the distributions of the random variables $\mu_r(n, N)$ and the order statistics $\eta_{(1)}, \eta_{(2)}, \ldots, \eta_{(N)}$ in the generalized scheme of allocating particles in terms of probabilities related to sums of independent random variables. Obtaining limit distributions for the random variables $\mu_r(n, N)$ and $\eta_{(1)}, \eta_{(2)}, \ldots, \eta_{(N)}$ is reduced to applying local limit theorems for sums of independent identically distributed integer-valued random variables.

We now give some examples of how combinatorial problems can be reduced to the generalized scheme of allocating particles to cells.

Example 1.2.1. Consider single-valued mappings of the set $X_n = \{1, 2, \ldots, n\}$ into itself. A single-valued mapping s of the set X_n into itself can be represented as

$$s = \begin{pmatrix} 1, & 2, & \ldots, & n \\ s_1, & s_2, & \ldots, & s_n \end{pmatrix},$$

where s_k denotes the image of k, $k = 1, 2, \ldots, n$, under the mapping s. The mapping s may be thought of as an oriented graph $\Gamma_n^{(s)} = \Gamma(X_n, W_n)$ with vertex set X_n and arcs $W_n = \{(k, s_k), k = 1, 2, \ldots, n\}$, where the arc (k, s_k) is directed from k to $s_k, k = 1, 2, \ldots, n$. The number of arcs entering the vertex k in the graph $\Gamma_n^{(s)}$, which is the number of pre-images of the element k under the mapping s, is called the multiplicity of the vertex k.

Let Σ_n denote the set of all single-valued mappings of X_n into itself, and Γ_n the set of all graphs of these mappings. The number of elements of Σ_n is obviously equal to n^n. If the uniform distribution is defined on the set Σ_n, then we obtain a probability space whose set of elementary events Ω is the set Σ_n; and the probability for any subset of Σ_n is the number of elements in the subset divided by n^n. The random mapping σ is any of the n^n possible mappings with probability $P\{\sigma = s\} = n^{-n}, s \in \Sigma_n$. If

$$\sigma = \begin{pmatrix} 1, & 2, & \ldots, & n \\ \sigma_1, & \sigma_2, & \ldots, & \sigma_n \end{pmatrix},$$

where the random variable σ_i is the random image of the element i, $i = 1, 2, \ldots, n$, then, for any s,

$$P\{\sigma = s\} = P\{\sigma_1 = s_1, \ldots, \sigma_n = s_n\} = n^{-n}.$$

Thus the random variables $\sigma_1, \ldots, \sigma_n$ are independent and take the values 1, 2, $\ldots, n$ with equal probabilities.

Let η_r denote the multiplicity of the vertex r in the random mapping σ, $r = 1, 2, \ldots, n$. The quantity η_r is equal to the number of random variables $\sigma_1, \ldots, \sigma_n$ taking the value r; thus, for nonnegative integers $k_1, \ldots, k_n$ with $k_1 + \cdots + k_n = n$, the probability $P\{\eta_1 = k_1, \ldots, \eta_n = k_n\}$ is equal to the sum of probabilities $P\{\sigma_1 = s_1, \ldots, \sigma_n = s_n\} = n^{-n}$, where among $s_1, \ldots, s_n$ there are exactly k_r values equal to r, $r = 1, 2, \ldots, n$. The number of summands in this sum is obviously $n!/(k_1! \cdots k_n!)$; therefore

$$P\{\eta_1 = k_1, \ldots, \eta_n = k_n\} = \frac{n!}{k_1! \cdots k_n! \, n^n}.$$

Thus the joint distribution of the multiplicities of the vertices $\eta_1, \ldots, \eta_n$ of a random mapping is the multinomial distribution. Taking the vertices as cells and the arcs going into these vertices as particles, we obtain the classical scheme of allocating n particles to n cells with multinomial distribution of the contents of the cells $\eta_1, \ldots, \eta_n$. For the random variables $\eta_1, \ldots, \eta_n$, relation (1.2.2) holds:

$$P\{\eta_1 = k_1, \ldots, \eta_n = k_n\} = P\{\xi_1 = k_1, \ldots, \xi_n = k_n \mid \xi_1 + \cdots + \xi_n = n\},$$

in which $\xi_1, \ldots, \xi_n$ are independent and identically Poisson-distributed.

The number of vertices $\mu_r(n)$ in a random mapping with multiplicity r corresponds to the number of cells containing exactly r particles in the classical scheme of allocating n particles to n cells; to study these variables, as well as the order statistics made up of the multiplicities of the vertices, one can invoke Lemmas 1.2.1, 1.2.2, and 1.2.3.

Example 1.2.2. Consider all distinct partitions of n into N summands not less than $r \geq 0$. The number of such partitions is $\binom{n-(r-1)N-1}{N-1}$. Let us define the uniform distribution on the set of these partitions by assigning the probability $\binom{n-(r-1)N-1}{N-1}^{-1}$ to each partition $n = n_1 + \cdots + n_N$, $n_1, \ldots, n_N \geq r$. Then n can be written in the form

$$n = \eta_1 + \cdots + \eta_N,$$

where the summands $\eta_1, \ldots, \eta_N$ are random variables. If $n_1, \ldots, n_N \geq r$ and $n = n_1 + \cdots + n_N$, then

$$P\{\eta_1 = n_1, \ldots, \eta_N = n_N\} = \binom{n - (r-1)N - 1}{N - 1}^{-1}.$$

The general scheme of allocation corresponding to this combinatorial problem is obtained if we use the geometric distribution for the distribution of the random variables $\xi_1, \ldots, \xi_N$:

$$\mathbf{P}\{\xi_1 = k\} = p^{k-r}(1-p), \quad k = r, r+1, \ldots, \quad 0 < p < 1.$$

Indeed, as is easily verified,

$$\mathbf{P}\{\xi_1 = n_1, \ldots, \xi_N = n_N \mid \xi_1 + \cdots + \xi_N = n\} = \binom{n - (r-1)N - 1}{N-1}^{-1},$$

since, for geometrically distributed summands,

$$\mathbf{P}\{\xi_1 + \cdots + \xi_N = n\} = \binom{n - (r-1)N - 1}{N-1} p^{n-Nr}(1-p)^N.$$

Example 1.2.3.   Note that it is not necessary for the random variables $\xi, \ldots, \xi_N$ in a generalized scheme to be identically distributed. Consider the following example. Draw n balls at random without replacement from an urn containing m_i balls of the ith color, $i = 1, \ldots, N$. Let η_i denote the number of balls drawn of the ith color, $i = 1, \ldots, N$. It is easily seen that for nonnegative integers $n_1, \ldots, n_N$ such that $n_1 + \cdots + n_N = n$,

$$\mathbf{P}\{\eta_1 = n_1, \ldots, \eta_N = n_N\} = \frac{\binom{m_1}{n_1} \cdots \binom{m_N}{n_N}}{\binom{m}{n}},$$

where $m = m_1 + \cdots + m_N$.

If in the generalized scheme of allocation the random variables $\xi_1, \ldots, \xi_N$ have the binomial distributions

$$\mathbf{P}\{\xi_i = k\} = \binom{m_i}{k} p^k (1-p)^{m_i - k},$$

where $0 < p < 1, k = 1, 2, \ldots, m_i, i = 1, \ldots, N$, then

$$\mathbf{P}\{\xi_1 = n_1, \ldots, \xi_N = n_N \mid \xi_1 + \cdots + \xi_N = n\} = \frac{\binom{m_1}{n_1} \cdots \binom{m_N}{n_N}}{\binom{m}{n}},$$

and the distribution of the random variables $\eta_1, \ldots, \eta_N$ coincides with the conditional distribution of the independent random variables $\xi_1, \ldots, \xi_N$ under the condition $\xi_1 + \cdots + \xi_N = n$. Thus $\eta_1, \ldots, \eta_N$ may be viewed as contents of cells in the generalized scheme of allocation, in which the random variables $\xi_1, \ldots, \xi_N$ have different binomial distributions.

Example 1.2.4. In a sense, the graph Γ_n of a random mapping consists of trees. Indeed, the graph can be naturally decomposed into connected components. Clearly, each connected component of the graph Γ_n contains exactly one cycle. Vertices in the cycle are called cyclic. If we remove the arcs joining the cyclic vertices, then the graph turns into a forest, that is, a graph consisting of rooted trees.

Recall that a rooted tree with $n + 1$ vertices is a connected undirected graph without cycles, with one special vertex called the root, and with n nonroot labeled vertices. A rooted tree with $n + 1$ vertices has n edges. In what follows, we view all edges of trees as directed away from the root, and the multiplicity of a vertex of a tree is defined as the number of edges emanating from it.

Let T_n denote the set of all rooted trees with $n + 1$ vertices whose roots are labeled zero, and the n nonroot vertices are labeled $1, 2, \ldots, n$. The number of elements of the set T_n is equal to $(n + 1)^{n-1}$.

A forest with N roots and n nonroot vertices is a graph, all of whose components are trees. The roots of these trees are labeled with $1, \ldots, N$ and the nonroot vertices with $1, \ldots, n$. We denote the set of all such forests by $T_{n,N}$. The number of elements in the set $T_{n,N}$ is $N(n+N)^{n-1}$. The number of forests in which the kth tree contains n_k nonroot vertices, $k = 1, 2, \ldots, n$, is

$$\frac{n!}{n_1! \cdots n_N!}(n_1 + 1)^{n_1-1} \cdots (n_N + 1)^{n_N-1},$$

where the factor $n!/(n_1! \cdots n_N!)$ is the number of partitions of n vertices into N ordered groups, and $(n_k + 1)^{n_k-1}$ is the number of trees that can be constructed from the kth group of vertices of each partition. Then

$$n! \sum_{n_1 + \cdots + n_N = n} \frac{(n_1 + 1)^{n_1-1} \cdots (n_N + 1)^{n_N-1}}{n_1! \cdots n_N!} = N(n + N)^{n-1}, \qquad (1.2.9)$$

where the summation is taken over nonnegative integers $n_1, \ldots, n_N$ such that $n_1 + \cdots + n_N = n$.

Next, we define the uniform distribution on $T_{n,N}$. Let η_k denote the number of nonroot vertices in the kth tree of a random forest in $T_{n,N}$, $k = 1, \ldots, N$. For the random variables $\eta_1, \ldots, \eta_N$, we have

$$\mathbf{P}\{\eta_1 = n_1, \ldots, \eta_N = n_N\} = \frac{n! \, (n_1 + 1)^{n_1} \cdots (n_N + 1)^{n_N}}{N(n + N)^{n-1}(n_1 + 1)! \cdots (n_N + 1)!}, \qquad (1.2.10)$$

where $n_1, \ldots, n_N$ are nonnegative integers and $n_1 + \cdots + n_N = n$.

Let us consider independent identically distributed random variables $\xi_1, \ldots, \xi_N$ for which

$$\mathbf{P}\{\xi_1 = k\} = \frac{(k + 1)^k}{(k + 1)!} x^k e^{-\theta(x)}, \qquad k = 0, 1, \ldots, \qquad (1.2.11)$$

where the parameter x lies in the interval $0 < x \le e^{-1}$ and the function $\theta(x)$ is defined as

$$\theta(x) = \sum_{k=1}^{\infty} \frac{k^{k-1}}{k!} x^k.$$

By using (1.2.9), we easily obtain

$$\mathbf{P}\{\xi_1 + \cdots + \xi_N = n\} = \sum_{n_1 + \cdots + n_N = n} \frac{(n_1 + 1)^{n_1} \cdots (n_N + 1)^{n_N}}{(n_1 + 1)! \cdots (n_N + 1)!} x^n e^{-N\theta(x)}$$

$$= \frac{N(n + N)^{n-1}}{n!} x^n e^{-N\theta(x)};$$

hence, for any x, $0 < x \le e^{-1}$, and for nonnegative integers $n_1, \ldots, n_N$ such that $n_1 + \cdots + n_N = n$,

$$\mathbf{P}\{\xi_1 = n_1, \ldots, \xi_N = n_N \mid \xi_1 + \cdots + \xi_N = n\}$$
$$= \frac{n! (n_1 + 1)^{n_1} \cdots (n_N + 1)^{n_N}}{N(n + N)^{n-1}(n_1 + 1)! \cdots (n_N + 1)!}. \qquad (1.2.12)$$

The right-hand sides of (1.2.10) and (1.2.12) are identical, and the joint distribution of $\eta_1, \ldots, \eta_N$ coincides with the distribution of $\xi_1, \ldots, \xi_N$ under the condition that $\xi_1 + \cdots + \xi_N = n$. Thus, for the random variables $\eta_1, \ldots, \eta_N$ and $\xi_1, \ldots, \xi_N$, relation (1.2.2) holds, enabling us to study tree sizes in a random forest by using the generalized scheme of allocating particles into cells, with the random variables $\xi_1, \ldots, \xi_N$ having the distribution given by (1.2.11).

1.3. Connectivity of graphs and the generalized scheme

Not pretending to give an exhaustive solution, let us describe a rather general model of a random graph by using the generalized scheme of allocation. Consider the set of all graphs $\Gamma_n(R)$ with n labeled vertices possessing a property R. We assume that connectivity is defined for the graphs from this set and that each graph is represented as a union of its connected components. In the formal treatment that follows, it may be helpful to keep in mind the graphs of random mappings or of random permutations. The former graphs consist of components that are connected directed graphs with exactly one cycle, whereas the latter graphs consist only of cycles.

Let a_n denote the number of graphs in the set $\Gamma_n(R)$ and let b_n be the number of connected graphs in $\Gamma_n(R)$. We denote by $\Gamma_{n,N}(R)$ the subset of graphs in $\Gamma_n(R)$ with exactly N connected components. Note that the components of a graph in $\Gamma_{n,N}(R)$ are unordered, and hence we can consider only the symmetric characteristics that do not depend on the order of the components. To avoid this restriction, we, instead, consider the set $\bar{\Gamma}_{n,N}(R)$ of combinatorial objects constructed by means of all possible orderings of the components of each graph from

$\Gamma_{n,N}(R)$. The elements of this set are ordered collections of N components, each of which is a connected graph possessing the property R, and the total number of vertices in the components is equal to n. Since the vertices of a graph in $\Gamma_{n,N}(R)$ are labeled, all the connected components of the graph are distinct; therefore the number of elements in $\bar{\Gamma}_{n,N}(R)$ is equal to $N!\,a_{n,N}$, where $a_{n,N}$ is the number of elements of the set $\Gamma_{n,N}(R)$ consisting of the unordered collection of components.

Now let us impose a restriction on the property R of graphs. Let a graph possess the property R if and only if the property holds for each connected component: The property R is then called decomposable.

Set $a_0 = 1$, $b_0 = 0$ and introduce the generating functions

$$A(x) = \sum_{n=1}^{\infty} \frac{a_n x^n}{n!}, \qquad B(x) = \sum_{n=0}^{\infty} \frac{b_n x^n}{n!}.$$

Lemma 1.3.1. *If the property R is decomposable, then*

$$a_{n,N} = \frac{n!}{N!} \sum_{n_1+\cdots+n_N=n} \frac{b_{n_1}\cdots b_{n_N}}{n_1!\cdots n_N!}, \qquad (1.3.1)$$

where the summation is taken over nonnegative integers $n_1, \ldots, n_N$ such that $n_1 + \cdots + n_N = n$.

Proof. With $n_1 + \cdots + n_N = n$ and $n_1, \ldots, n_N \geq 1$, let $\bar{a}_n(n_1, \ldots, n_N)$ denote the number of graphs in $\bar{\Gamma}_{n,N}(R)$ with ordered components of sizes $n_1, \ldots, n_N$. We construct all $\bar{a}_n(n_1, \ldots, n_N)$ such graphs and decompose the n labeled vertices into N groups so that there are n_i vertices in the ith group, $i = 1, \ldots, N$. This can be done in $n!/(n_1!\cdots n_N!)$ ways. From n_i vertices, we construct a connected graph possessing the property R; this can be done in b_{n_i} ways. Thus the number of ordered sets of connected components of sizes $n_1, \ldots, n_N$ is

$$\bar{a}_n(n_1, \ldots, n_N) = \frac{n!\,b_{n_1}\cdots b_{n_N}}{n_1!\cdots n_N!}.$$

Since N components can be ordered in $N!$ ways, the number $a_n(n_1, \ldots, n_N)$ of unordered sets, or the number of graphs in $\Gamma_{n,N}(R)$ having exactly N components of sizes $n_1, \ldots, n_N$, is

$$a_n(n_1, \ldots, n_N) = \frac{1}{N!}\bar{a}_n(n_1, \ldots, n_N) = \frac{n!\,b_{n_1}\cdots b_{n_N}}{N!\,n_1!\cdots n_N!}. \qquad (1.3.2)$$

■

Lemma 1.3.2. *If the property R is decomposable, then*

$$A(x) = e^{B(x)}.$$

Proof. As follows from (1.3.1), the number a_n of all graphs in $\Gamma_n(R)$ is

$$a_n = \sum_{N=1}^{n} \frac{n!}{N!} \sum_{n_1+\cdots+n_N=n} \frac{b_{n_1}\cdots b_{n_N}}{n_1!\cdots n_N!}. \qquad (1.3.3)$$

By dividing both sides of this equality by $n!$, multiplying by x^n, and summing over n, we get the chain of equalities

$$A(x) - 1 = \sum_{n=1}^{\infty} \frac{a_n x^n}{n!}$$

$$= \sum_{n=1}^{\infty} \sum_{N=1}^{n} \frac{1}{N!} \sum_{n_1+\cdots+n_N=n} \frac{b_{n_1} x^{n_1} \cdots b_{n_N} x^{n_N}}{n_1!\cdots n_N!}$$

$$= \sum_{N=1}^{n} \frac{1}{N!} \left(\sum_{n=1}^{\infty} \frac{b_n x^n}{N!} \right)^N$$

$$= e^{B(x)} - 1,$$

which proves the lemma.   ∎

Let us define the uniform distribution on the set $\Gamma_n(R)$ and consider the random variables α_m equal to the number of components of size m in a random graph from $\Gamma_n(R)$. The total number of components ν_n of a random graph from $\Gamma_n(R)$ is related to these variables by $\nu_n = \alpha_1 + \cdots + \alpha_n$. Arrange the components in order of nondecreasing sizes and denote by β_m the size of the mth components in the ordered series; if $m > \nu_n$, set $\beta_m = 0$.

We will also consider the random variables defined on the set $\bar{\Gamma}_{n,N}(R)$ of ordered sets of N components. The ordered components labeled with the numbers from 1 to N play the role of cells in the generalized scheme of allocating particles. Define the uniform distribution on $\bar{\Gamma}_{n,N}(R)$ and denote by $\eta_1, \ldots, \eta_N$ the sizes of the ordered connected components of a random element in $\bar{\Gamma}_{n,N}(R)$. It is then clear that

$$\mathbf{P}\{\eta_1 = n_1, \ldots, \eta_N = n_N\} = \frac{N! a_n(n_1, \ldots, n_N)}{N! a_{n,N}} = \frac{a_n(n_1, \ldots, n_N)}{a_{n,N}}. \qquad (1.3.4)$$

Theorem 1.3.1.   *If the series*

$$B(x) = \sum_{n=0}^{\infty} \frac{b_n x^n}{n!} \qquad (1.3.5)$$

has a nonzero radius of convergence, then the random variables $\eta_1, \ldots, \eta_N$ are the contents of cells in the generalized scheme of allocation in which the independent *identically distributed random variables $\xi_1, \ldots, \xi_N$ have the distribution*

$$\mathbf{P}\{\xi_1 = k\} = \frac{b_k x^k}{k!\, B(x)}, \tag{1.3.6}$$

where the positive value x from the domain of convergence of (1.3.5) *may be taken arbitrarily.*

Proof. Let us find the conditional joint distribution of the random variables $\xi_1, \ldots, \xi_N$ with distribution (1.3.6) under the condition $\xi_1 + \cdots + \xi_N = n$. For such random variables,

$$\mathbf{P}\{\xi_1 + \cdots + \xi_N = n\} = \frac{x^n}{(B(x))^N} \sum_{n_1 + \cdots + n_N = n} \frac{b_{n_1} \cdots b_{n_N}}{n_1! \cdots n_N!}, \tag{1.3.7}$$

and by virtue of (1.3.1),

$$\mathbf{P}\{\xi_1 + \cdots + \xi_N = n\} = \frac{x^n N!}{(B(x))^N n!} a_{n,N}. \tag{1.3.8}$$

Hence, if $n_1, \ldots, n_N \geq 1$ and $n_1 + \cdots + n_N = n$, then

$$\mathbf{P}\{\xi_1 = n_1, \ldots, \xi_N = n_N \mid \xi_1 + \cdots + \xi_N = n\}$$

$$= \frac{b_{n_1} \cdots b_{n_N} x^n}{n_1! \cdots n_N!\, (B(x))^N \mathbf{P}\{\xi_1 + \cdots + \xi_N = n\}}$$

$$= \frac{b_{n_1} \cdots b_{n_N} n!}{n_1! \cdots n_N!\, N!\, a_{n,N}},$$

and according to (1.3.2),

$$\mathbf{P}\{\xi_1 = n_1, \ldots, \xi_N = n_N \mid \xi_1 + \cdots + \xi_N = n\} = \frac{a_n(n_1, \ldots, n_N)}{a_{n,N}}. \tag{1.3.9}$$

From (1.3.4) and (1.3.9), we obtain the relation (1.2.2) between the random variables $\eta_1, \ldots, \eta_N$ and $\xi_1, \ldots, \xi_N$ in the generalized scheme of allocating particles to cells. ∎

In the generalized scheme of allocating particles, we usually study the random variables $\mu_r(n, N)$ equal to the number of cells containing exactly r particles and the order statistics $\eta_{(1)}, \eta_{(2)}, \ldots, \eta_{(N)}$ obtained by arranging the contents of cells in nondecreasing order. In this case, $\mu_r(n, N)$ is the number of components of size r, and $\eta_{(1)}, \eta_{(2)}, \ldots, \eta_{(N)}$ are the sizes of the components in a random element from $\bar{\Gamma}_{n,N}(R)$ arranged in nondecreasing order. The random variables help in studying distributions of the random variables $\alpha_1, \ldots, \alpha_n$ and the associated variables defined on the set $\Gamma_n(R)$ of all graphs possessing the property R.

Lemma 1.3.3.   *For any positive x from the domain of convergence of* (1.3.5),

$$\mathbf{P}\{\nu_n = N\} = \frac{n!\,(B(x))^N}{N!\,a_n x^n}\mathbf{P}\{\xi_1 + \cdots + \xi_N = n\}. \qquad (1.3.10)$$

Proof. Relation (1.3.10) follows from (1.3.8) because $\mathbf{P}\{\nu_n = N\} = a_{n,N}/a_n$ by definition. ∎

It is clear by virtue of (1.3.3) that the number a_n can also be expressed in terms of probabilities related to $\xi_1, \ldots, \xi_N$:

$$a_n = \sum_{N=1}^{\infty} \frac{n!\,(B(x))^N}{N!\,x^n}\mathbf{P}\{\xi_1 + \cdots + \xi_N = n\}. \qquad (1.3.11)$$

Lemma 1.3.4.   *For any nonnegative integers $N, m_1, \ldots, m_n$,*

$$\mathbf{P}\{\alpha_1 = m_1, \ldots, \alpha_n = m_n \mid \nu_n = N\}$$
$$= \mathbf{P}\{\mu_1(n, N) = m_1, \ldots, \mu_n(n, N) = m_n\}.$$

Proof. The conditional distribution on $\Gamma_n(R)$ under the condition $\nu_n = N$ is concentrated on the set $\Gamma_{n,N}(R)$ of graphs having exactly N connected components and is uniform on this set. Hence,

$$\mathbf{P}\{\alpha_1 = m_1, \ldots, \alpha_n = m_n \mid \nu_n = N\} = \frac{c_N(m_1, \ldots, m_n)}{a_{n,N}}, \qquad (1.3.12)$$

where $a_{n,N}$ is the number of elements in $\Gamma_{n,N}(R)$ and $c_N(m_1, \ldots, m_n)$ is the number of graphs in $\Gamma_{n,N}(R)$ such that the number of components of size r is m_r, $r = 1, 2, \ldots, n$.

Consider the above set $\bar{\Gamma}_{n,N}(R)$ composed of ordered sets of N components. Let $\bar{c}_N(m_1, \ldots, m_n)$ denote the number of elements in $\bar{\Gamma}_{n,N}(R)$ such that the number of components of size r is $m_r, r = 1, 2, \ldots, n$. It is clear that

$$\mathbf{P}\{\mu_1(n, N) = m_1, \ldots, \mu_n(n, N) = m_n\} = \frac{\bar{c}_N(m_1, \ldots, m_n)}{\bar{a}_{n,N}}, \qquad (1.3.13)$$

where $\bar{a}_{n,N}$ is the number of elements in $\bar{\Gamma}_{n,N}(R)$. The assertion of the lemma follows from (1.3.12) and (1.3.13) because $\bar{a}_{n,N} = N!\,a_{n,N}$ and $\bar{c}_N(m_1, \ldots, m_n) = N!\,c_N(m_1, \ldots, m_n)$. ∎

Thus, if the series (1.3.5) has a nonzero radius of convergence, then all of the random variables expressed by $\alpha_1, \ldots, \alpha_n$ can be studied by using the generalized scheme of allocating particles in which the random variables $\xi_1, \ldots, \xi_N$ have the distribution (1.3.6). Roughly speaking, under the condition that the number ν_n of connected components of the graph $\Gamma_n(R)$ is N, the sizes of these components (under a random ordering) have the same joint distribution as the random variables

$\eta_1, \ldots, \eta_N$ in the generalized scheme of allocating particles that are defined by the independent random variables $\xi_1, \ldots, \xi_N$ with distribution (1.3.6). Thus, for $v_n = N$ the random variables $\beta_1, \ldots, \beta_N$ are expressed in terms of $\alpha_1, \ldots, \alpha_n$ in exactly the same way as the order statistics $\eta_{(1)}, \ldots, \eta_{(N)}$ in the generalized scheme of allocating particles are expressed in terms of $\mu_1(n, N), \ldots, \mu_n(n, N)$. Hence, Lemma 1.3.4 implies the following assertion.

Lemma 1.3.5. *For any nonnegative integers* $N, k_1, \ldots, k_N$,

$$\mathbf{P}\{\beta_1 = k_1, \ldots, \beta_N = k_N \mid v_n = N\} = \mathbf{P}\{\eta_{(1)} = k_1, \ldots, \eta_{(N)} = k_N\}.$$

$$(1.3.14)$$

We now consider the joint distribution of $\mu_1(n, N), \ldots, \mu_n(n, N)$.

Lemma 1.3.6. *For nonnegative integers* $m_1, \ldots, m_n$ *such that* $m_1 + \cdots + m_n = N$ *and* $m_1 + 2m_2 + \cdots + nm_n = n$,

$$\mathbf{P}\{\mu_1(n, N) = m_1, \ldots, \mu_n(n, N) = m_n\}$$

$$= \frac{n! \, b_1^{m_1} \ldots b_n^{m_n}}{m_1! \cdots m_n! \, (1!)^{m_1} \cdots (n!)^{m_n} a_n \, \mathbf{P}\{v_n = N\}}. \qquad (1.3.15)$$

Proof. To obtain (1.3.15), it suffices to calculate $\bar{c}_N(m_1, \ldots, m_n)$ in (1.3.13). It is clear that

$$\bar{c}_N(m_1, \ldots, m_n) = \sum \bar{a}_n(n_1, \ldots, n_N),$$

where the summation is taken over all sets $(n_1, \ldots, n_N)$ containing the element r exactly m_r times, $r = 1, \ldots, n$. The number of such sets is $N!/(m_1! \cdots m_n!)$, and for each of them, by (1.3.2),

$$\bar{a}_n(n_1, \ldots, n_N) = \frac{n! \, b_1^{m_1} \cdots b_n^{m_n}}{(1!)^{m_1} \cdots (n!)^{m_n}}.$$

Hence,

$$\bar{c}_N(m_1, \ldots, m_n) = \frac{N! \, n! \, b_1^{m_1} \cdots b_n^{m_n}}{m_1! \cdots m_n! \, (1!)^{m_1} \cdots (n!)^{m_n}}.$$

To obtain formula (1.3.15), it remains to note that

$$\mathbf{P}\{v_n = N\} = \frac{a_{n,N}}{a_n} = \frac{\bar{a}_{n,N}}{N! \, a_n}.$$

∎

Lemmas 1.3.4 and 1.3.6 enable us to express the joint distribution of the random variables $\alpha_1, \ldots, \alpha_n$ in a random graph from $\Gamma_n(R)$.

Lemma 1.3.7. *If $m_1, \ldots, m_n$ are nonnegative integers, then*

$$
\mathbf{P}\{\alpha_1 = m_1, \ldots, \alpha_n = m_n\} =
\begin{cases}
\dfrac{n!}{a_n} \displaystyle\prod_{r=1}^{n} \dfrac{b_r^{m_r}}{m_r!\,(r!)^{m_r}} & \text{if } \sum_{r=1}^{n} r m_r = n, \\[4mm]
0 & \text{otherwise.}
\end{cases}
$$

Proof. By the total probability formula,

$$
\mathbf{P}\{\alpha_1 = m_1, \ldots, \alpha_n = m_n\}
$$

$$
= \sum_{k=1}^{N} \mathbf{P}\{\nu_n = k\}\mathbf{P}\{\alpha_1 = m_1, \ldots, \alpha_n = m_n \mid \nu_n = k\}
$$

$$
= \mathbf{P}\{\nu_n = N\}\mathbf{P}\{\alpha_1 = m_1, \ldots, \alpha_n = m_n \mid \nu_n = N\},
$$

where $N = m_1 + \cdots + m_n$. By using Lemma 1.3.4, we find that

$$
\mathbf{P}\{\alpha_1 = m_1, \ldots, \alpha_n = m_n\}
$$

$$
= \mathbf{P}\{\nu_n = N\}\mathbf{P}\{\mu_1(n, N) = m_1, \ldots, \mu_n(n, N) = m_n\}. \qquad (1.3.16)
$$

It remains to note that $\mathbf{P}\{\mu_1(n, N) = m_1, \ldots, \mu_n(n, N) = m_n\} = 0$ if $m_1 + 2m_2 + \cdots + nm_n \neq n$ and that equality (1.3.15) from Lemma 1.3.6 holds for the probability $\mathbf{P}\{\mu_1(n, N) = m_1, \ldots, \mu_n(n, N) = m_n\}$ if $m_1 + \cdots + m_n = N$ and $m_1 + 2m_2 + \cdots + nm_n = n$. The substitution of (1.3.15) into (1.3.16) proves Lemma 1.3.7. ∎

We now turn to some examples.

Example 1.3.1. The set S_n of one-to-one mappings corresponds to the set $\Gamma_n(R)$ of graphs with n vertices for which we have the property R: Graphs are directed with exactly one arc entering each vertex and exactly one arc emanating from each vertex. This property is decomposable. The connected components of such a graph are (directed) cycles. In this case, $a_n = n!$, $b_n = (n-1)!$, and the generating functions

$$
A(x) = \frac{1}{1-x}, \qquad B(x) = -\log(1-x)
$$

satisfy the relations of Lemma 1.3.2:

$$
A(x) = e^{B(x)}. \qquad (1.3.17)
$$

To study the lengths of cycles of a random permutation and the associated variables, one can use the generalized scheme of allocating particles in which the random variables $\xi_1, \ldots, \xi_N$ have the distribution

$$
\mathbf{P}\{\xi_1 = k\} = -\frac{x^k}{k \log(1-x)}, \qquad k = 1, 2, \ldots, \qquad 0 < x < 1.
$$

Example 1.3.2. The set Σ_n of all single-valued mappings corresponds to the set $\Gamma_n(R)$ of graphs with n vertices with property R: The graphs are directed with exactly one arc emanating from each vertex. This property is decomposable. Since the number of elements of Σ_n is n^n, from relation (1.3.17) for the generating functions we find that

$$B(x) = \log A(x) = \log \sum_{n=0}^{\infty} \frac{n^n x^n}{n!},$$

yielding

$$b_n = (n-1)! \sum_{k=0}^{n-1} \frac{n^k}{k!}.$$

The radius of convergence of $A(x)$ and $B(x)$ is e^{-1}, and at the point $x = e^{-1}$, they diverge.

To study the characteristics of a random mapping, we can use the generalized scheme of allocating particles in which the random variables $\xi_1, \ldots, \xi_N$ have the distribution

$$\mathbf{P}\{\xi_1 = k\} = \frac{b_k x^k}{k! \, B(x)}, \quad k = 1, 2, \ldots, \quad 0 < x < e^{-1}.$$

Example 1.3.3. Consider the set of all unordered partitions of the set $X_n = \{1, 2, \ldots, n\}$ into disjoint subsets, the union of which is X_n. The partition of X_n into unordered subsets $Y_1, \ldots, Y_N$ corresponds to the hypergraph of $\Gamma_{n,N}(R)$ with n vertices and N hyperedges $Y_1, \ldots, Y_N$. Since all of the $N!$ orderings of the hyperedges $Y_1, \ldots, Y_N$ are distinct, each hypergraph of $\Gamma_{n,N}(R)$ gives us $N!$ distinct objects of $\bar{\Gamma}_{n,N}(R)$ that are hypergraphs with n vertices and N ordered hyperedges $A_1, \ldots, A_N$, with the sets of hyperedges being permutations of $Y_1, \ldots, Y_N$. The property R determining this class of graphs requires that a graph be a hypergraph whose distinct hyperedges have no common vertices. Each connected component of such a graph is a hyperedge. Clearly, the number of connected graphs possessing the property R with n vertices is 1, that is, $b_n = 1$, so

$$B(x) = \sum_{n=1}^{\infty} \frac{x^n}{n!} = e^x - 1.$$

Since R is decomposable,

$$A(x) = e^{e^x - 1}.$$

This equality, or (1.3.3), yields

$$a_n = \sum_{N=1}^{n} \frac{n!}{N!} \sum_{n_1 + \cdots + n_N = n} \frac{1}{n_1! \cdots n_N!},$$

where the second summation is over positive integers $n_1, \ldots, n_N$.

Thus, to study random partitions, we can use the generalized scheme of allocation in which the random variables $\xi_1, \ldots, \xi_N$ have the truncated Poisson distribution

$$P\{\xi_1 = k\} = \frac{x^k}{k!(e^x - 1)}, \quad k = 1, 2, \ldots, \quad 0 < x < \infty.$$

Example 1.3.4. A tree is a connected graph without cycles. As the set $\Gamma_{n,N}(R)$, let us consider the set $\mathcal{F}_{n,N}$ of all forests consisting of N trees with the total number n of labeled vertices. The trees in a forest are not ordered. The property R determining this class of graphs requires that a graph be undirected without cycles. The property R is decomposable. The number b_n of connected graphs possessing the property R is the number of nonrooted trees with n vertices and $b_n = n^{n-2}$, so the generating function is

$$B(x) = \sum_{n=1}^{\infty} \frac{n^{n-2} x^n}{n!}, \quad 0 < x \le e^{-1}.$$

Thus, to study a random forest from $\mathcal{F}_{n,N}$, we can use the generalized scheme of allocation in which the random variables $\xi_1, \ldots, \xi_N$ have the distribution

$$P\{\xi_1 = k\} = \frac{k^{k-2} x^k}{k! B(x)}, \quad k = 1, 2, \ldots, \quad 0 < x \le e^{-1}.$$

1.4. Forests of nonrooted trees

The graphs consisting of nonrooted trees and unicyclic components play the same role in investigating graphs as the forests of rooted trees do for graphs of mappings. Hence, the following sections concentrate on these objects, using the generalized scheme of allocation.

As in Example 1.3.4, let $\mathcal{F}_{n,N}$ be the set of all forests of N nonrooted trees with n vertices. It is known that the number of forests of N ordered rooted trees with total number n of nonroot vertices is $N(N + n)^{n-1}$. In contrast to the forests of rooted trees, there is no simple formula for the number $F_{n,N} = |\mathcal{F}_{n,N}|$ of forests of nonrooted trees. Therefore the first step is to study the asymptotic behavior of $F_{n,N}$.

Denote by T the number of edges in a forest belonging to $\mathcal{F}_{n,N}$. It is easy to see that $T = n - N$. Following the general algorithm for applying the generalized scheme of allocation, let us consider the set $\bar{\mathcal{F}}_{n,N}$, which consists of N ordered nonrooted trees, and define the uniform distribution on this set. Denote by $\eta_1, \ldots, \eta_N$ the sizes of ordered trees in a random graph from $\bar{\mathcal{F}}_{n,N}$. By Cayley's formula for counting trees, the number b_n of nonrooted trees with n vertices is n^{n-2}. Denote by $\bar{a}_n(n_1, \ldots, n_N)$ the number of elements in $\bar{\mathcal{F}}_{n,N}$ for which $\{\eta_1 = n_1, \ldots, \eta_N = n_N\}$. It is easy to see that for positive integers $n_1, \ldots, n_N$

with $n_1 + \cdots + n_N = n$,

$$\bar{a}_n(n_1, \ldots, n_N) = \frac{n!}{n_1! \cdots n_N!} b_{n_1} \cdots b_{n_N}, \qquad (1.4.1)$$

and the number of elements in $\bar{\mathcal{F}}_{n,N}$ is

$$\bar{F}_{n,N} = \sum_{n_1 + \cdots + n_N = n} \bar{a}_n(n_1, \ldots, n_N) = \sum_{n_1 + \cdots + n_N = n} \frac{n! \, b_{n_1} \cdots b_{n_N}}{n_1! \cdots n_N!}.$$

Thus, for the number of forests $F_{n,N}$, we obtain the formula

$$F_{n,N} = \frac{n!}{N!} \sum_{n_1 + \cdots + n_N = n} \frac{n_1^{n_1 - 2} \cdots n_N^{n_N - 2}}{n_1! \cdots n_N!}, \qquad (1.4.2)$$

where the summation is over positive integers $n_1, \ldots, n_N$ such that $n_1 + \cdots + n_N = n$.

Introduce independent identically distributed random variables $\xi_1, \ldots, \xi_N$ for which

$$\mathbf{P}\{\xi_1 = k\} = \frac{b_k x^k}{k! \, B(x)} = \frac{k^{k-2} x^k}{k! \, B(x)}, \qquad k = 1, 2, \ldots, \qquad (1.4.3)$$

where

$$B(x) = \sum_{k=1}^{\infty} \frac{b_k x^k}{k!} = \sum_{k=1}^{\infty} \frac{k^{k-2} x^k}{k!}, \qquad 0 < x \le e^{-1}. \qquad (1.4.4)$$

In accordance with the results of the previous section and Example 1.3.4, the generalized scheme of allocation can be applied to investigating random forests of nonrooted trees, that is, relation (1.2.2) is valid: For any integers $n_1, \ldots, n_N$,

$$\mathbf{P}\{\eta_1 = n_1, \ldots, \eta_N = n_N\} = \mathbf{P}\{\xi_1 = n_1, \ldots, \xi_N = n_N \mid \xi_1 + \cdots + \xi_N = n\}.$$

For the number of forests $F_{n,N}$, formula (1.3.8) is valid, which, of course, can be obtained directly from (1.4.2) and (1.4.3):

$$F_{n,N} = \frac{n!(B(x))^N}{N! x^n} \mathbf{P}\{\xi_1 + \cdots + \xi_N = n\}, \qquad (1.4.5)$$

where $B(x)$ is defined by (1.4.4), and the value of the parameter x in the distribution (1.4.3) of the random variables $\xi_1, \ldots, \xi_N$ can be chosen arbitrarily from the domain of convergence of the series $B(x)$.

Thus, to obtain the asymptotics of $F_{n,N}$, it is sufficient to choose an appropriate value of x, $0 < x \le e^{-1}$, and analyze the asymptotic behavior of the probability $\mathbf{P}\{\xi_1 + \cdots + \xi_N = n\}$ for the sum of the random variables $\xi_1, \ldots, \xi_N$ that have the distribution (1.4.3) with the chosen value of the parameter x.

The first two moments of the random variable ξ_1 have the following expressions:

$$\mathbf{E}\xi_1 = \frac{1}{B(x)} \sum_{k=1}^{\infty} \frac{k^{k-1}x^k}{k!},$$

$$\mathbf{E}\xi_1^2 = \frac{1}{B(x)} \sum_{k=1}^{\infty} \frac{k^k x^k}{k!}.$$

Therefore, along with $B(x)$, we consider two functions

$$a(x) = \sum_{k=1}^{\infty} \frac{k^k x^k}{k!}, \qquad \theta(x) = \sum_{k=1}^{\infty} \frac{k^{k-1}x^k}{k!}.$$

The function $\theta(x)$ is the solution of the equation

$$\theta e^{-\theta} = x \tag{1.4.6}$$

if we choose the solution that is less than 1.

The functions $a(x)$ and $B(x)$ can be represented in terms of this function. Differentiating (1.4.6) gives

$$\theta'(x)e^{-\theta(x)} - \theta(x)\theta'(x)e^{-\theta(x)} = 1;$$

hence,

$$\theta'(x) = \frac{\theta(x)}{x(1 - \theta(x))}. \tag{1.4.7}$$

On the other hand,

$$x\theta'(x) = \sum_{k=1}^{\infty} \frac{k^k x^k}{k!} = a(x).$$

Thus

$$a(x) = \frac{\theta(x)}{1 - \theta(x)}. \tag{1.4.8}$$

Slightly more complicated calculations are needed to obtain the relation

$$B(x) = \tfrac{1}{2}\big(1 - (1 - \theta(x))^2\big). \tag{1.4.9}$$

Consider the function

$$h(x) = (1 - \theta(x))^2.$$

By using (1.4.7), we obtain

$$h'(x) = -2(1 - \theta(x))\theta'(x) = -\frac{2\theta(x)}{x} = -2\sum_{k=1}^{\infty} \frac{k^{k-1}x^{k-1}}{k!}.$$

When we integrate both sides of this equality, we obtain

$$\int_0^x h'(t)\,dt = h(x) - 1 = -2\sum_{k=1}^\infty \frac{k^{k-1}}{k!}\int_0^x t^{k-1}\,dt$$

$$= -2\sum_{k=1}^\infty \frac{k^{k-2}x^k}{k!} = -2B(x),$$

which implies equality (1.4.9).

Relations (1.4.8) and (1.4.9) allow us to calculate the mean $\mathbf{E}\xi_1$ and the variance $\mathbf{D}\xi_1$. For $0 < \theta < 1$, we set

$$x = \theta e^{-\theta}.$$

For such a choice of the parameter x,

$$\theta(x) = \theta, \quad a(x) = \frac{\theta}{1-\theta}, \quad B(x) = \frac{\theta(2-\theta)}{2};$$

therefore

$$m = \mathbf{E}\xi_1 = \frac{\theta(x)}{B(x)} = \frac{2}{2-\theta},$$

$$\sigma^2 = \mathbf{D}\xi_1 = \frac{a(x)}{B(x)} - \left(\frac{\theta(x)}{B(x)}\right)^2 = \frac{2\theta}{(1-\theta)(2-\theta)^2}.$$

If the parameter θ is fixed, then Theorem 1.1.11 may be applied to the sum

$$\zeta_N = \xi_1 + \cdots + \xi_N.$$

In fact, the theorem on local convergence to the normal law is valid in a wider region.

Theorem 1.4.1. *If $N \to \infty$ and $\theta = \theta(N)$ varies such that $\theta N \to \infty$ and $(1-\theta)^3 N \to \infty$, then*

$$\mathbf{P}\{\zeta_N = k\} = \frac{1}{\sigma\sqrt{2\pi N}} e^{-u^2/2}(1 + o(1))$$

uniformly in the integers k such that $u = (k - Nm)/(\sigma\sqrt{N})$ lies in any fixed finite interval.

Proof. First we prove that, under the conditions of the theorem, the distribution of $(\zeta_N - mN)/(\sigma\sqrt{N})$ converges weakly to the normal distribution with parameters $(0, 1)$. According to Theorem 1.1.9, it is sufficient to demonstrate convergence of the corresponding characteristic function $\varphi_N(t)$ to the characteristic function $e^{-t^2/2}$ of the standard normal distribution.

The characteristic function of ξ_1 equals

$$\varphi(t) = \frac{1}{B(x)} \sum_{k=1}^{\infty} \frac{k^{k-2} x^k e^{itk}}{k!} = \frac{B(xe^{it})}{B(x)}.$$

By virtue of (1.4.7), (1.4.8), and (1.4.9),

$$B(x) = \tfrac{1}{2}\left(1 - (1 - \theta(x))^2\right),$$

$$x B'(x) = \theta(x),$$

$$x^2 B''(x) = \theta^2(x)(1 - \theta(x))^{-1},$$

$$x^3 B'''(x) = \theta^3(x)(1 - 2\theta(x))(1 - \theta(x))^{-3}.$$

Therefore

$$\varphi'(t) = \frac{i\theta(xe^{it})}{B(x)},$$

$$\varphi''(t) = \frac{\theta(xe^{it})}{B(x)(1 - \theta(xe^{it}))}, \qquad (1.4.10)$$

$$\varphi'''(t) = -\frac{i\theta(xe^{it})}{B(x)(1 - \theta(xe^{it}))^3}.$$

For $x = \theta e^{-\theta}$,

$$\theta(x) = \theta, \qquad B(x) = \frac{\theta(2 - \theta)}{2}.$$

Denote by $\psi(t)$ the characteristic function of the centered random variable $\xi_1 - \theta(x)/B(x)$. Then

$$\psi'(0) = 0, \qquad \psi''(0) = -\sigma^2 = \frac{2\theta}{(1 - \theta)(2 - \theta)^2}. \qquad (1.4.11)$$

Let

$$g(t) = \log \psi(t).$$

It is not difficult to check that

$$g'''(t) = \frac{2i\theta(xe^{it})(2\theta^2(xe^{it}) - \theta(xe^{it}) - 2)}{(1 - \theta(xe^{it}))^3(2 - \theta(xe^{it}))^2}.$$

Therefore, if $x = \theta e^{-\theta}$, then there exists a constant c such that

$$|g'''(t)| \le \frac{c\theta}{(1 - \theta)^3},$$

and

$$\psi(t) = e^{g(t)} = \exp\left\{-\frac{\sigma^2 t^2}{2} + O\left(\frac{\theta|t|^3}{(1-\theta)^3}\right)\right\}. \tag{1.4.12}$$

The characteristic function $\varphi_N(t)$ of the random variable $(\zeta_N - mN)/(\sigma\sqrt{N})$ satisfies the equality $\varphi_N(t) = \psi^N(t/(\sigma\sqrt{N}))$; hence, for any fixed t, as $N \to \infty$,

$$\varphi_N(t) = \exp\left\{-\frac{t^2}{2} + O\left(\frac{1}{\sqrt{\theta(1-\theta)^3 N}}\right)\right\}. \tag{1.4.13}$$

The conditions of the theorem specify that $N\theta \to \infty$, $N(1-\theta)^3 \to \infty$; hence, for any fixed t, as $N \to \infty$,

$$\varphi_N(t) \to e^{-t^2/2},$$

and the distribution of $(\zeta_N - mN)/(\sigma\sqrt{N})$ converges weakly to the standard normal law.

To prove the local convergence of these distributions, we need additional estimates of the characteristic function $\varphi(t)$. It is reasonable to assume that the local theorem is valid in the same regions as the integral theorem proved above, but the necessary estimates are complicated to find, and therefore we restrict ourselves to a proof of the local theorem only in the case where $\theta \le \theta_0 < 1$ and $\theta N \to \infty$.

From (1.4.12), it follows that there exists $\varepsilon > 0$ such that for $|t| \le \varepsilon$ and $\theta \le \theta_0 < 1$,

$$|\psi(t)| \le e^{-c\sigma^2 t^2}. \tag{1.4.14}$$

We now show that for any ε, $0 < \varepsilon < \pi$, there exists a positive constant c such that for $\varepsilon \le |t| \le \pi$ and $0 < \theta \le 1$,

$$|\varphi(t)| \le e^{-c\theta}. \tag{1.4.15}$$

If $\theta \to 0$, then

$$x = \theta e^{-\theta} = \theta - \theta^2 + O(\theta^3),$$

$$\varphi(t) = \frac{B(xe^{it})}{B(x)} = \frac{xe^{it} + x^2 e^{2it}/2 + O(\theta^3)}{\theta(1 - \theta/2)}$$

$$= e^{it} + (e^{2it} - e^{it})\theta/2 + O(\theta^2).$$

Now

$$\left|e^{it} + (e^{2it} - e^{it})\theta/2\right|^2 = 1 - 2\theta\sin^2(t/2) + O(\theta^2),$$

as $\theta \to 0$; therefore

$$|\varphi(t)| = 1 - \theta\sin^2(t/2) + O(\theta^2),$$

uniformly in t, and for $\varepsilon \leq |t| \leq \pi$ there exists $\delta > 0$ and $c_1 > 0$ such that

$$|\varphi(t)| \leq e^{-c_1 \theta} \qquad (1.4.16)$$

for $\theta < \delta$.

For any θ, $0 < \theta \leq 1$, the distribution of ξ_1 has maximal span 1 and $\varphi(t)$ is continuous in t and θ in the region

$$B = \{(t, \theta) : \varepsilon \leq |t| \leq \pi, \ 0 < \delta \leq \theta \leq 1\}.$$

Therefore

$$q = \sup_B |\varphi(t)| < 1,$$

and there exists $c_2 > 0$ such that

$$|\varphi(t)| \leq e^{-c_2 \theta} \qquad (1.4.17)$$

for $(t, \theta) \in B$.

This estimate and (1.4.16) imply (1.4.15).

Proving the local theorem, we follow the proof of Theorem 1.1.11 as a model for similar proofs. We set

$$u = \frac{k - mN}{\sigma \sqrt{N}}, \qquad P_N(k) = \mathbf{P}\{\zeta_N = k\}$$

and represent the difference

$$R_N = 2\pi \left(\sigma \sqrt{N} P_N(k) - \frac{1}{\sqrt{2\pi}} e^{-u^2/2} \right)$$

as a sum of the following four integrals:

$$I_1 = \int_{-A}^{A} e^{-itu} \left(\left(\psi \left(t / (\sigma \sqrt{N}) \right) \right)^N - e^{-t^2/2} \right) dt,$$

$$I_2 = - \int_{A \leq |t|} e^{-itu - t^2/2} \, dt,$$

$$I_3 = \int_{A \leq |t| \leq \varepsilon \sigma \sqrt{N}} e^{-itu} \left(\psi \left(t / (\sigma \sqrt{N}) \right) \right)^N dt,$$

$$I_4 = \int_{\varepsilon \sigma \sqrt{N} \leq |t| \leq \pi \sigma \sqrt{N}} e^{-itu} \left(\psi \left(t / (\sigma \sqrt{N}) \right) \right)^N dt,$$

where the constants A and ε will be chosen later.

To see that $R_N \to 0$ as $N \to \infty$, we show that R_N can be made arbitrarily small by choosing of ε, A, and N. It is clear that

$$|I_2| \leq \int_{A \leq |t|} e^{-t^2/2} \, dt,$$

and $|I_2|$ can be made arbitrarily small by choosing a sufficiently large A.

Choose $\varepsilon > 0$ such that estimate (1.4.14) is fulfilled. Then, for $\theta \leq \theta_0 < 1$, $|t| \leq \varepsilon$,

$$\left| \psi\left(t/(\sigma\sqrt{N}) \right) \right| \leq e^{-ct^2},$$

so that

$$|I_3| \leq \int_{A \leq |t| \leq \varepsilon\sigma\sqrt{N}} \left| \psi\left(t/(\sigma\sqrt{N}) \right) \right|^N dt \leq \int_{A \leq |t|} e^{-ct^2} dx,$$

and $|I_3|$ can be made arbitrarily small by the choice of sufficiently large A.

For fixed A, the integral I_1 tends to zero because $\varphi(t) \to e^{-t^2/2}$ uniformly with respect to t in any finite interval.

Finally, with the help of estimate (1.4.17), we obtain that as $N \to \infty$,

$$|I_4| \leq \int_{\varepsilon\sigma\sqrt{N} \leq |t| \leq \pi\sigma\sqrt{N}} \left| \psi\left(\frac{t}{\sigma\sqrt{N}} \right) \right|^N dt$$

$$\leq \sigma\sqrt{N} \int_{\varepsilon \leq |t| \leq \pi} |\varphi(t)|^N dt$$

$$\leq \sigma\sqrt{N} e^{-c\theta N} \to 0.$$

$\blacksquare$

Denote by $p(u; \alpha, \beta)$ the density of the stable law with parameters α and β in Zolotarev's parameterization (see [60]). If $\alpha \neq 1$, the characteristic function $f(t)$ of this distribution can be represented in the form

$$f(t) = \exp\left\{ -|t|^\alpha \exp\left\{ -\frac{i\pi}{2} K(\alpha)\beta \frac{t}{|t|} \right\} \right\},$$

where $K(\alpha) = 1 - |1 - \alpha|$. By the inversion formula,

$$p(u; \alpha, \beta) = \frac{1}{2\pi} \int_{-\infty}^{\infty} e^{-itu} \exp\left\{ -|t|^\alpha \exp\left\{ -\frac{i\pi}{2} K(\alpha)\beta \frac{t}{|t|} \right\} \right\} dt. \quad (1.4.18)$$

If $N \to \infty$ and $\theta = 1$, then the distribution of $(\zeta_N - 2N)/(bN^{3/2})$, where $b = 2(2/3)^{2/3}$, is approximated by the stable distribution with parameters $\alpha = 3/2$, $\beta = -1$.

Theorem 1.4.2. *If $N \to \infty$, $\theta = 1$, $b = 2(2/3)^{2/3}$, then*

$$bN^{2/3}\mathbf{P}\{\zeta_N = n\} = p(u; 3/2, -1)(1 + o(1))$$

uniformly in the integers n such that $u = (n - 2N)/(bN^{2/3})$ lies in any fixed finite interval.

Proof. The terms of the sum $\zeta_N = \xi_1 + \cdots + \xi_N$ are independent identically distributed random variables, and for $\theta = 1$,

$$\mathbf{P}\{\xi_1 = k\} = \frac{2k^{k-2}e^{-1}}{k!}, \quad k = 1, 2, \ldots, \tag{1.4.19}$$

and $\mathbf{E}\xi_1 = 2$, since $\theta(e^{-1}) = 1$ and $B(e^{-1}) = 1/2$. The maximal span of the distribution is 1; therefore, by Theorem 1.1.10, it suffices to prove that the distribution of $(\theta_N - 2N)/(bN^{2/3})$ converges weakly to the stable law given in the theorem.

In addition to $\theta(x)$, $a(x)$, and $B(x)$ defined above, we consider the function

$$C(z) = \sum_{k=1}^{\infty} \frac{k^{k-3}z^k}{k!}, \quad |z| \le 1.$$

This can be expressed in terms of $\theta(z)$: Let

$$g(z) = (1 - \theta(z))^3.$$

By using the equalities

$$z\theta'(z) = \frac{\theta(z)}{1 - \theta(z)}$$

and

$$B(z) = \tfrac{1}{2}\bigl(1 - (1 - \theta(z))^2\bigr),$$

we easily obtain

$$zg'(z) = -3\theta(z) + 3\theta^2(z) = 3\theta(z) - 6B(z).$$

Integration then gives

$$\int_0^z g'(u)du = g(z) - 1 = 3\int_0^z \frac{\theta(u)du}{u} - 6\int_0^z \frac{B(u)du}{u} = 3B(z) - 6C(z).$$

Expressing $B(z)$ in terms of $\theta(z)$ demonstrates that, for $|z| \le 1$,

$$C(z) = \tfrac{5}{12} - \tfrac{1}{4}(1 - \theta(z))^2 - \tfrac{1}{6}(1 - \theta(z))^3.$$

Since $\theta(e^{-1}) = 1$, we find that $C(e^{-1}) = 5/12$.

Set

$$u(z) = 1 - \theta(z), \quad v(z) = C(z) - C(e^{-1}).$$

We have shown that

$$v(z) = -\tfrac{1}{4}u^2(z) - \tfrac{1}{6}u^3(z).$$

If we invert this expression, we obtain two formal solutions

$$u(z) = \pm 2i\sqrt{v(z)} + \tfrac{4}{3}v(z) + O\bigl(|v(z)|^{3/2}\bigr);$$

since $u(x) > 0$ and $v(x) < 0$ for $0 < x \leq e^{-1}$, we choose the solution

$$u(z) = -2i\sqrt{v(z)} + \tfrac{4}{3}v(z) + O\big(|v(z)|^{3/2}\big). \tag{1.4.20}$$

Hence,

$$(1 - \theta(z))^2 = u^2(z) = -4v(z) - \frac{16i}{3}(v(z))^{3/2} + O\big(|v(z)|^2\big). \tag{1.4.21}$$

The first two derivatives of $C(z)$ are

$$C'(z) = \sum_{k=1}^{\infty} \frac{k^{k-2}z^{k-1}}{k!} = \frac{B(z)}{z},$$

$$C''(z) = \sum_{k=1}^{\infty} \frac{k^{k-1}z^{k-2}}{k!} = \frac{\theta(z)}{z^2}.$$

Therefore, for real t,

$$C\big(e^{-1+it}\big) - C\big(e^{-1}\big) = it/2 + O(t^2). \tag{1.4.22}$$

Now we find an expression of the characteristic function $\varphi(t)$ of the random variable ξ_1 with distribution (1.4.19). It is clear that

$$\varphi(t) = B\big(e^{-1+it}\big)/B\big(e^{-1}\big).$$

From (1.4.20), (1.4.21), and (1.4.22), we find that for $z = e^{it-1}$,

$$\varphi(t) = 1 - (1 - \theta(z))^2$$

$$= 1 + 4v(z) + \frac{16i}{3}(v(z))^{3/2} + O\big(|v(z)|^2\big)$$

$$= 1 + 2it + \frac{2i}{3}2\sqrt{2}|t|^{3/2}\left(\frac{it}{|t|}\right)^{3/2} + O(t^2)$$

$$= 1 + 2it + |bt|^{3/2}i\left(\frac{it}{|t|}\right)^{3/2} + O(t^2),$$

where $b = 2(2/3)^{2/3}$. By virtue of the equality

$$i\left(\frac{it}{|t|}\right)^{3/2} = -\exp\left\{\frac{i\pi t}{4|t|}\right\},$$

we can rewrite the last relation as

$$\varphi(t) = \frac{B\big(e^{-1+it}\big)}{B\big(e^{-1}\big)} = 1 + 2it - |bt|^{3/2}\exp\left\{\frac{i\pi t}{4|t|}\right\} + O(t^2).$$

Since

$$e^{-2it} = 1 - 2it + O(t^2)$$

as $t \to 0$, we find that

$$\psi(t) = e^{-2it}\varphi(t) = 1 - |bt|^{3/2}\exp\left\{\frac{i\pi t}{4|t|}\right\} + O(t^2).$$

The characteristic function of the random variable $(\zeta_N - 2N)/(bN^{2/3})$ is

$$\psi^N\left(t/(bN^{2/3})\right) = \left(1 - \frac{|t|^{3/2}}{N}\exp\left\{\frac{i\pi t}{4|t|}\right\} + O(N^{-4/3})\right)^N$$

and converges to

$$f(t) = \exp\left\{-|t|^{3/2}\exp\left\{\frac{i\pi t}{4|t|}\right\}\right\}$$

at any fixed t. The function $f(t)$ is the characteristic function of the stable law $p(u; \alpha, \beta)$ with parameters $\alpha = 3/2$, $\beta = -1$. Therefore, according to Theorem 1.1.10, as $N \to \infty$,

$$bN^{2/3}\mathbf{P}\{\zeta_N = n\} - p(u; 3/2, -1) \to 0$$

uniformly in k, where $u = (k - 2N)/(bN^{2/3})$. The function $p(x; 3/2, -1)$ is positive for any x; hence,

$$bN^{2/3}\mathbf{P}\{\zeta_N = k\} = p(u; 3/2, -1)(1 + o(1))$$

uniformly in k such that $u = (k - 2N)/(bN^{2/3})$ lies in any fixed finite interval.

∎

We now turn to the estimate of the number of forests $F_{n,N}$ with n vertices, N trees, and $T = n - N$ edges. Theorems 1.4.1 and 1.4.2 allow us to estimate the number of forests.

Theorem 1.4.3. *If* $n \to \infty$ *and* $\theta = 2T/n$ *varies such that* $\theta N \to \infty$ *and* $N(1 - \theta)^3 \to \infty$, *then*

$$F_{n,N} = \frac{n^{2T}\sqrt{1 - \theta}}{2^T T!}(1 + o(1)). \tag{1.4.23}$$

Proof. Put

$$\theta = 2T/n, \qquad x = \theta e^{-\theta}. \tag{1.4.24}$$

By virtue of (1.4.5),

$$F_{n,N} = \frac{n!(B(x))^N}{N!x^n}\mathbf{P}\{\zeta_N = n\}, \tag{1.4.25}$$

where the parameters are chosen so that

$$B(x) = \frac{1}{2}\left(1 - (1 - \theta)^2\right) = \frac{\theta(2 - \theta)}{2} = \frac{2TN}{n^2}. \tag{1.4.26}$$

Since $m = \mathbf{E}\xi_1 = 2/(2 - \theta) = n/N$, by Theorem 1.4.1,

$$\mathbf{P}\{\zeta_N = n\} = \frac{1}{\sigma\sqrt{2\pi N}}(1 + o(1)), \qquad (1.4.27)$$

where

$$\sigma^2 = \mathbf{D}\xi_1 = \frac{2\theta}{(1 - \theta)(2 - \theta)^2} = \frac{nT}{(1 - \theta)N^2}.$$

If we substitute (1.4.24), (1.4.25), and (1.4.27) into (1.4.25), we can conclude that under the conditions of the theorem,

$$F_{n,N} = \frac{n!(B(x))^N\sqrt{N(1 - \theta)}}{N!x^n\sqrt{2\pi nT}}(1 + o(1)) = \frac{n^{2T}\sqrt{1 - \theta}}{2^T T^T e^{-T}\sqrt{2\pi T}}(1 + o(1)).$$

■

Theorem 1.4.4. *If $n \to \infty$ and $2T/n \to 1$ so that*

$$(1 - 2T/n)N^{1/3} \to b^{2/3}v/2, \quad -\infty < v < \infty,$$

then

$$F_{n,N} = \frac{n^n\sqrt{\pi}}{N!2^N N^{1/6}(2/3)^{2/3}}p(-v; 3/2, -1)(1 + o(1)). \qquad (1.4.28)$$

Proof. Under the conditions of the theorem,

$$u = \frac{n - 2N}{(bN)^{2/3}} = -(1 - 2T/n)N^{1/3}\frac{n}{b^{2/3}N} \to -v;$$

thus, by Theorem 1.4.2, continuity, and positivity of the density $p(u; 3/2, -1)$,

$$bN^{2/3}\mathbf{P}\{\zeta_N = n\} = p(-v; 3/2, -1)(1 + o(1)). \qquad (1.4.29)$$

We chose $\theta = 1$ in Theorem 1.4.2; hence, $x = e^{-1}$ and $B(x) = B(e^{-1}) = 1/2$. Having substituted these values and (1.4.29) into (1.4.25), we conclude that, under the conditions of Theorem 1.4.4,

$$F_{n,N} = \frac{n!}{N!2^N e^{-n} bN^{2/3}}p(-v; 3/2, -1)(1 + o(1))$$

$$= \frac{n^n\sqrt{\pi}}{N!2^N N^{1/6}(2/3)^{2/3}}p(-v; 3/2, -1)(1 + o(1)).$$

■

Although the density $p(x; 3/2, -1)$ cannot be represented in terms of simple functions, we can use the relation $p(x; \alpha, \beta) = p(-x; \alpha, -\beta)$ and the following series expansion for $x > 0$ and $1 < \alpha < 2$ for our calculations:

$$p(x; \alpha, \beta) = \frac{1}{\pi}\sum_{n=0}^{\infty}(-1)^n\frac{\Gamma((n + 1)/\alpha)}{\alpha n!}x^n \cos\frac{\pi n}{2}\left(1 + \left(1 + \frac{1}{n}\right)\frac{2 - \alpha}{\alpha}\beta\right).$$

1.5. Trees of given sizes in a random forest

Let $\mu_r = \mu_r(n, N)$ be the number of trees with r vertices in a random forest with n labeled vertices and N nonrooted trees, $r = 1, 2 \ldots$. Recall that such a forest has $T = n - N$ edges. In this section, we consider the asymptotic behavior of the random variables $\mu_r(n, N)$. Following the approach established in the previous section, we use the generalized scheme of allocation of n particles to N cells determined by identically distributed random variables $\xi_1, \ldots, \xi_N$ such that

$$\mathbf{P}\{\xi_1 = k\} = p_k = p_k(\theta) = \frac{2k^{k-2}\theta^{k-1}e^{-k\theta}}{k!(2 - \theta)}, \quad k = 1, 2, \ldots, \quad 0 < \theta < 2.$$

As we have calculated,

$$\mu = \mu(\theta) = \mathbf{E}\xi_1 = \frac{2}{2 - \theta},$$

and for $0 < \theta < 1$,

$$\sigma^2 = \sigma(\theta) = \mathbf{D}\xi_1 = \frac{2\theta}{(1 - \theta)(2 - \theta)^2}.$$

We will also use the notation

$$s_r^2 = s_r^2(\theta) = p_r\left(1 - p_r - \frac{(\mu - r)^2}{\sigma^2}p_r\right), \quad r = 1, 2 \ldots.$$

The random variables μ_r behave much like the corresponding variables for a random forest of rooted trees. We highlight some of these results; see [30] for a complete description. As before, let $\theta = 2T/n$. Again the value $\theta = 1$ is of particular interest, so we introduce the following notations: For $r = 1, 2, \ldots,$

$$\pi_r = \pi_r(\theta) = \begin{cases} p_r(\theta), & 0 < \theta < 1, \\ p_r(1), & 1 \le \theta < 2, \end{cases}$$

$$\sigma_{rr}^2 = \sigma_{rr}^2(\theta) = \begin{cases} s_r(\theta), & 0 < \theta < 1, \\ p_r(1)(1 - p_r(1)), & 1 \le \theta < 2. \end{cases}$$

The truncated values $\pi_r(\theta)$ and $\sigma_{rr}^2(\theta)$ allow us to summarize the rather complicated behavior of $\mu_r, r \ge 3$, in the following two theorems.

Theorem 1.5.1. *If $n, N \to \infty$ and $r = r(n) \ge 3$ varies such that $N\pi_r(\theta) \to \infty$, then*

$$\mathbf{P}\{\mu_r = k\} = \frac{1}{\sigma_{rr}(\theta)\sqrt{2\pi N}}e^{-u_r^2/2}(1 + o(1))$$

uniformly in the integers k such that

$$u_r = \frac{k - N\pi_r(\theta)}{\sigma_{rr}(\theta)N^{1/2}}$$

lies in any fixed finite interval.

Theorem 1.5.2. *If $n, N \to \infty$ and $r = r(n) \geq 3$ varies such that $N\pi_r(\theta) \to \lambda$ for some λ, $0 < \lambda < \infty$, then for any fixed $k = 0, 1, \ldots$,*

$$\mathbf{P}\{\mu_r = k\} = \frac{\lambda^k e^{-\lambda}}{k!}(1 + o(1)).$$

The random variables μ_1 and μ_2, like their analogs for forests of rooted trees, have some special properties, but we will not discuss them.

When edges are added sequentially to a forest, then by Theorems 1.5.1 and 1.5.2, the asymptotic behavior of μ_r does not depend on θ if $\theta \geq 1$. If $Np_r(1) \to \infty$, then the limit distribution of μ_r, with similar centering and normalizing, is the standard normal distribution for all θ, $1 \leq \theta < 2$.

There are similar results for the case $\theta \geq 1$ and $Np_r(1) \to \lambda$ for some λ, $0 < \lambda < \infty$, with the limit distribution of the μ_r for all θ, $1 \leq \theta < 2$, being the Poisson distribution with parameter λ. Thus the point $\theta = 1$ can be interpreted as a critical point in the evolution of a random forest.

We now prove Theorems 1.5.1 and 1.5.2.

Proof of Theorems 1.5.1 and 1.5.2. According to Example 1.3.4 and Lemma 1.2.1,

$$\mathbf{P}\{\mu_r = k\} = \binom{N}{k} p_r^k (1 - p_r)^{N-k} \frac{\mathbf{P}\{\zeta_{N-k}^{(r)} = n - kr\}}{\mathbf{P}\{\zeta_N = n\}}, \qquad (1.5.1)$$

where $\zeta_N = \xi_1 + \cdots + \xi_N$, $\zeta_N^{(r)} = \xi_1^{(r)} + \cdots + \xi_N^{(r)}$, the random variables $\xi_1, \ldots, \xi_N; \xi_1^{(r)}, \ldots, \xi_N^{(r)}$ are independent and identically distributed,

$$p_r = \mathbf{P}\{\xi_1 = k\} = \frac{k^{k-2} x^k}{k! B(x)}, \qquad k = 1, 2, \ldots,$$

$$B(x) = \sum_{k=1}^{\infty} \frac{k^{k-2} x^k}{k!}, \qquad 0 < x \leq e^{-1},$$

$$\mathbf{P}\{\xi_1^{(r)} = k\} = \mathbf{P}\{\xi_1 = k \mid \xi_1 \neq r\}, \qquad (1.5.2)$$

and the parameter x of the distribution of $\xi_1, \ldots, \xi_N$ may be taken arbitrarily from the domain of convergence of the series $B(x)$.

We set $\theta = 2T/n$. It is convenient to choose $x = \theta e^{-\theta}$ for $0 < \theta < 1$ and $x = e^{-1}$ for $1 \le \theta < 2$. With these choices, (1.5.1) gives

$$\mathbf{P}\{\mu_r = k\} = \binom{N}{k}(\pi_r(\theta))^k(1 - \pi_r(\theta))^{N-k}\frac{\mathbf{P}\{\zeta^{(r)}_{N-k} = n - kr\}}{\mathbf{P}\{\zeta_N = n\}}, \qquad (1.5.3)$$

where

$$\mathbf{P}\{\xi_1 = k\} = \pi_k(\theta), \quad k = 1, 2, \ldots,$$

and the distribution of $\xi_1^{(r)}$ is defined by (1.5.2).

Reasoning by contradiction, we see that it is sufficient to prove Theorems 1.5.1 and 1.5.2 under the assumption that θ lies in any of the following three domains: first, where $N\theta \to \infty$ and $(1 - \theta)^3 N \to \infty$; second, where $(1 - \theta)^3 N$ is bounded by an arbitrary constant; and, third, where $(1 - \theta)^3 N \to -\infty$. Negating either theorem implies the existence of a subsequence of the parameters n, N such that θ lies in one of these three domains for which the other conditions are satisfied but for which the conclusion is false. Therefore we assume that n, $N \to \infty$ in such a way that θ lies in one of the domains and prove the assertions of Theorems 1.5.1 and 1.5.2 in the corresponding three cases.

Consider first Theorem 1.5.1 in the first domain of θ. By the de Moivre–Laplace theorem, the binomial distribution is approximated by a normal or Poisson distribution. More precisely, if $N\pi_r(\theta) \to \infty$, then

$$\binom{N}{k}(\pi_r(\theta))^k(1 - \pi_r(\theta))^{N-k} = \frac{1 + o(1)}{\sqrt{2\pi n\pi_r(\theta)(1 - \pi_r(\theta))}}e^{-z^2/2} \qquad (1.5.4)$$

uniformly in k such that

$$z = \frac{(k - N\pi_r(\theta))^2}{2N\pi_r(\theta)(1 - \pi_r(\theta))}$$

lies in any fixed finite interval.

The probability $\mathbf{P}\{\zeta_N = n\}$ from the denominator of (1.5.3) has been estimated in the previous section. Applying Theorem 1.4.1, we have for θ in the first domain,

$$\mathbf{P}\{\zeta_N = n\} = \frac{1}{\sigma\sqrt{2\pi N}}(1 + o(1)), \qquad (1.5.5)$$

where

$$\sigma^2 = \sigma^2(\theta) = \mathbf{D}\xi_1 = \frac{2\theta}{(1 - \theta)(2 - \theta)}.$$

To find the asymptotics of the numerator of (1.5.3), we begin by calculating the first and second moments:

$$m_r = m_r(\theta) = \mathbf{E}\xi_1^{(r)} = \frac{\mu - r\pi_r}{1 - \pi_r},$$

$$\sigma_r^2 = \sigma_r^2(\theta) = \mathbf{D}\xi_1^{(r)} = \frac{\sigma^2}{1 - \pi_r}\left(1 - \frac{\pi_r(\mu - r)^2}{(1 - \pi_r)\sigma^2}\right),$$

where $\mu = \mathbf{E}\xi_1 = 2/(2 - \theta)$.

A proof similar to that of Theorem 1.4.1 shows that a normal approximation is valid for the sum $\zeta_N^{(r)} = \xi_1^{(r)} + \cdots + \xi_N^{(r)}$. More precisely, if n, $N \to \infty$ such that $\theta N \to \infty$ and $(1 - \theta)^3 N \to \infty$, then

$$\mathbf{P}\{\zeta_N^{(r)} = s\} = \frac{1}{\sigma_r\sqrt{2\pi N}}e^{-(s - Nm_r)^2/(2\sigma_r^2 N)}(1 + o(1)) \qquad (1.5.6)$$

uniformly in $r \geq 3$ and s such that $(s - Nm_r)/(\sigma\sqrt{N})$ lies in any fixed finite interval.

We now use (1.5.6) with $s = n - kr$ and $N - k$ summands to obtain an asymptotic expression for $\mathbf{P}\{\zeta_{N-k}^{(r)} = n - kr\}$. Since

$$k = N\pi_r + u_r\sigma_{rr}\sqrt{N},$$

where

$$\sigma_{rr}^2 = \sigma_{rr}^2(\theta) = p_r(1 - p_r - (\mu - r)^2 p_r/\sigma^2),$$

we have

$$N - k = N(1 - p_r) - u_r\sigma_{rr}\sqrt{N} = N(1 - p_r)\left(1 - \frac{u_r\sigma_{rr}}{(1 - p_r)\sqrt{N}}\right). \qquad (1.5.7)$$

It is easy to see that $\sigma_{rr}/(1 - p_r)$ is bounded, and for u_r lying in any finite interval,

$$N - k = N(1 - p_r)(1 + O(N^{-1/2})). \qquad (1.5.8)$$

The exponent in (1.5.6) may now be written as

$$u^2 = \frac{(n - kr - Nm_r)^2}{2\sigma_r^2(N - k)}.$$

Taking into account (1.5.7), (1.5.8), and the equalities

$$n = N\mu, \qquad m_r - \mu = \frac{p_r(\mu - r)}{1 - p_r}, \qquad m_r - r = \frac{\mu - r}{1 - p_r},$$

which hold for θ in the first domain, we obtain

$$u = \frac{k(m_r - r) - N(m_r - \mu)}{\sigma_r(N - k)^{1/2}} = \frac{p_r^{1/2}(\mu - r)(k - Np_r)}{\sigma\sigma_{rr}(N - k)^{1/2}}$$

$$= \frac{p_r^{1/2}(\mu - r)(k - Np_r)}{\sigma(1 - p_r)^{1/2}\sigma_{rr}\sqrt{N}}(1 + o(1)) = \frac{p_r^{1/2}(\mu - r)}{\sigma(1 - p_r)^{1/2}}u_r(1 + o(1)).$$

Applying (1.5.7) gives

$$\mathbf{P}\{\zeta_{N-k}^{(r)} = n - kr\} = \frac{1}{\sigma^r \sqrt{2\pi N(1 - p_r)}} e^{-p_r(\mu-r)^2 u_r^2/(2\sigma^2(1-p_r))}(1 + o(1)).$$

(1.5.9)

When we substitute (1.5.4), (1.5.5), and (1.5.9) into (1.5.3), we see that under the conditions of Theorem 1.5.1 with θ in the first domain, this expression transforms into the product of an exponent and a coefficient. The coefficient of the exponent is

$$\frac{\sigma\sqrt{2\pi N}}{\sqrt{2\pi N p_r(1 - p_r)}\sigma_r\sqrt{2\pi N(1 - p_r)}}$$

$$= \frac{1}{\sqrt{2\pi N p_r(1 - p_r)\left(1 - p_r(\mu - r)^2/((1 - p_r)\sigma^2)\right)}} = \frac{1}{\sigma_{rr}\sqrt{2\pi N}}.$$

Combining the exponents from (1.5.4) and (1.5.9) yields the resulting exponent

$$-\frac{(k - Np_r)^2}{2Np_r(1 - p_r)} - \frac{p_r(\mu - r)^2(k - Np_r)^2}{2\sigma^2(1 - p_r)\sigma_{rr}^2 N} + o(1) = -\frac{(k - Np_r)^2}{2\sigma_{rr}N} + o(1).$$

Thus Theorem 1.5.1 is proved for θ varying in the first domain.

Under the conditions of Theorem 1.5.2, k is fixed, and when we apply (1.5.5) and (1.5.6) with the corresponding parameters, we obtain the ratio

$$\frac{\mathbf{P}\{\zeta_{N-k}^{(r)} = n - kr\}}{\mathbf{P}\{\zeta = n\}} \to 1.$$

Therefore the assertion of Theorem 1.5.2 follows from the Poisson approximation of the first factor in (1.5.3).

In the second domain, we choose the parameter of the distribution of $\xi_1, \ldots, \xi_N$ to be 1. If $Np_r(1) \to \infty$, then

$$\binom{N}{k}(p_r(1))^k(1 - p_r(1))^{N-k} = \frac{1}{\sqrt{2\pi Np_r(1)(1 - p_r(1))}} e^{-z^2/2}(1 + o(1))$$

(1.5.10)

uniformly in k such that

$$z = \frac{k - Np_r(1)}{\sqrt{Np_r(1)(1 - p_r(1))}}$$

lies in any fixed finite interval.

Applying Theorem 1.4.2 gives

$$bN^{2/3}\mathbf{P}\{\zeta_N = n\} = p(u; 3/2, -1)(1 + o(1))$$

(1.5.11)

uniformly in n such that $u = (n - 2N)/(bN^{2/3})$ lies in any fixed finite interval.

Restricting the random variables $\xi_1^{(r)}, \ldots, \xi_N^{(r)}$ does not affect their maximum span and convergence to the stable law with density $p(u; 3/2, -1)$. The only difference is that now the mean of a summand is $\mathbf{E}\xi_1^{(r)} = m_r(1) = 2/(1 - p_r(1))$. Therefore, as $j \to \infty$,

$$\frac{bj^{2/3}}{(1 - p_r(1))^{2/3}} \mathbf{P}\{\zeta_j^{(r)} = l\} = \frac{bj^{2/3}}{(1 - p_r(1))^{2/3}}$$

$$\times \mathbf{P}\left\{\frac{(\zeta_j^{(r)} - jm_r(1))(1 - p_r(1))^{2/3}}{bj^{2/3}} = \frac{l - jm_r(1)(1 - r(1))^{2/3}}{bj^{2/3}}\right\}$$

$$= p(v; 3/2, -1)(1 + o(1))$$

uniformly in l such that

$$v = \frac{(l - jm_r(1))(1 - p_r(1))^{2/3}}{bj^{2/3}}$$

lies in any fixed finite interval.

By substituting $N - k$ for j and $n - kr$ for l and recalling that

$$k = Np_r(1) + z\sqrt{Np_r(1)(1 - p_r(1))},$$

where z is bounded, we have

$$bN^{2/3}\mathbf{P}\{\zeta_{N-k}^{(r)} = n - kr\} = p(u; 3/2, -1)(1 + o(1)) \qquad (1.5.12)$$

uniformly in $r \geq 3$, where, as in (1.5.11), $u = (n - 2N)/(bN^{2/3})$, since

$$v = \frac{(l - jm_r(1))(1 - p_r(1))^{2/3}}{bj^{2/3}} = \frac{n - 2N}{bN^{2/3}} + o(1).$$

Thus the asymptotics of $\mathbf{P}\{\zeta_N = n\}$ and $\mathbf{P}\{\zeta_{n-k}^{(r)} = n - kr\}$ is the same and their ratio in (1.5.3) tends to 1. Therefore the asymptotics of $\mathbf{P}\{\mu_r = k\}$ is determined by the first factor and coincides with the asymptotics of the corresponding binomial probability. Theorems 1.5.1 and 1.5.2 have now been proved in the second domain.

It remains to prove the theorems for the third domain, where $(1 - 2T/n)^3 N \to -\infty$. We choose $\theta = 1$ in the distribution of the random variables $\xi_1, \ldots, \xi_N$ and prove that in (1.5.3) the ratio

$$\mathbf{P}\{\zeta_{N-k}^{(r)} = n - kr\}/\mathbf{P}\{\zeta_N = n\} \to 1 \qquad (1.5.13)$$

uniformly in r, and $k = Np_r(1) + z\sqrt{Np_r(1)(1 - p_r(1))}$, where z lies in any fixed finite interval.

In this case, $(1 - 2T/n)^3 N \to -\infty$, so the values n for the sum ζ_N and the values $n - kr$ for the sum $\zeta_{N-k}^{(r)}$ lie in what is called the region of large deviations. Therefore we need to apply the theorem on large deviations. We will not give the proof, but the main idea is simple: If the distribution of a sum of independent identically distributed integer-valued random variables with zero mean converges to a stable law with parameter α, $1 < \alpha < 2$, then the major contribution to a large deviation of the sum is made by only one of the summands (see [137]). Applying this theorem to the sum ζ_N gives the following result for θ in the third domain.

If $n, N \to \infty$ such that $N(1 - 2T/n)^3 \to -\infty$, then

$$
\begin{aligned}
\mathbf{P}\{\zeta_N = n\} &= \mathbf{P}\{\zeta_N - 2N = n - 2N\} \\
&= N\mathbf{P}\{\xi_1 - 2 = n - 2N\}(1 + o(1)) \\
&= \left(\frac{2}{\pi}\right)^{1/2} \frac{N}{(n - 2N)^{5/2}}(1 + o(1)).
\end{aligned}
$$

The theorem given in [137] cannot be applied to the sum $\zeta_N^{(r)}$, since its summands become noninteger after centering by the expectation m_r. Britikov, using the method given in [137], along with ideas from [58] and [113], proved in [30] that the probability $\mathbf{P}\{\zeta_{N-k}^{(r)} = n - kr\}$ has the same asymptotics as $\mathbf{P}\{\zeta_N = n\}$. More precisely, if $n, N \to \infty$ such that $N(1 - 2T/n)^3 \to -\infty$, then

$$
\begin{aligned}
\mathbf{P}\{\zeta_{N-k}^{(r)} = n - kr\} &= \mathbf{P}\{\zeta_{N-k}^{(r)} - (N - k)m_r = n - kr - (N - k)m_r\} \\
&= (N - k)\mathbf{P}\{\xi_1^{(r)} - m_r = n - kr - (N - k)m_r\} \\
&= (N - k)\mathbf{P}\{\xi_1^{(r)} = n - 2N + O(\sqrt{N})\} \\
&= \left(\frac{2}{\pi}\right)^{1/2} \frac{N}{(n - 2N)^{5/2}}(1 + o(1))
\end{aligned}
$$

uniformly in $r \geq 1$ and k such that

$$
(k - Np_r(1))/(Np_r(1)(1 - p_r(1)))^{1/2}
$$

lies in any fixed finite interval.

Thus the ratio in (1.5.3) tends to 1, and the asymptotics of $\mathbf{P}\{\mu_r = k\}$ is determined by the first factor and coincides with the asymptotics of the corresponding binomial probability. This proves Theorems 1.5.1 and 1.5.2 in the third domain.

The proof of Theorems 1.5.1 and 1.5.2 is now complete. ∎

1.6. Maximum size of trees in a random forest

The results of the previous section give some information on the behavior of the maximum size $\eta_{(N)}$ of trees in a random forest from $\mathcal{F}_{n,N}$ with $T = n - N$ edges. Indeed, if $\theta = 2T/n \to 0$ and there exists $r = r(n, N)$ such that $Np_r(\theta) \to \infty$ and $Np_{r+1}(\theta) \to \lambda$, $0 < \lambda < \infty$, then the distribution of the number μ_r of trees of size r approaches a normal distribution, and the distribution of μ_{r+1} approaches the Poisson distribution with parameter λ. This implies that the limit distribution of the random variable $\eta_{(N)}$ is concentrated on the points r and $r + 1$.

If $\theta = 2T/n \to \gamma, \gamma > 0$, then there are infinitely many $r = r(n, N)$ such that the distribution of μ_r approaches a Poisson distribution; hence, the distribution of $\eta_{(N)}$ is scattered more and more as γ increases. If $0 < \gamma < 1$, then the limit distribution is concentrated on a countable set of integers, whereas if $\gamma \geq 1$, then $\eta_{(N)}$ must be normalized to have a limit distribution, and the normalizing values tend to infinity at different rates, depending on the region of θ.

Thus, it should be possible to prove the limit theorems for $\eta_{(N)}$ when $T/n \to 0$ by using results on μ_r from the previous section. But if $2T/n \to \gamma$ for $\gamma > 0$, this approach may not work, and even if it did, the proofs would not be simple.

Therefore we choose instead to use the approach based on the generalized scheme of allocation. Let $\xi_1, \ldots, \xi_N$ be random variables with distribution

$$p_r(\theta) = \mathbf{P}\{\xi_1 = k\} = \frac{2r^{r-2}\theta^{r-1}e^{-r\theta}}{r!(2-\theta)}, \quad 0 < \theta < 2, \tag{1.6.1}$$

where $k = 1, 2, \ldots$. We choose $\theta = 2T/n$. Then, according to Lemma 1.2.2,

$$\mathbf{P}\{\eta_{(N)} \leq r\} = (1 - P_r)^N \frac{\mathbf{P}\{\bar{\zeta}_N^{(R)} = n\}}{\mathbf{P}\{\zeta_N = n\}}, \tag{1.6.2}$$

where

$$\zeta_N = \xi_1 + \cdots + \xi_N, \qquad \bar{\zeta}_N^{(r)} = \bar{\xi}_1^{(r)} + \cdots + \bar{\xi}_N^{(r)},$$

with $\bar{\xi}_1^{(r)}, \ldots, \bar{\xi}_N^{(r)}$ being independent identically distributed random variables such that

$$\mathbf{P}\{\bar{\xi}_1^{(r)} = k\} = \mathbf{P}\{\xi_1 = k \mid \xi_1 \leq r\}, \quad k = 1, \ldots, r, \tag{1.6.3}$$

and

$$P_r = P_r(\theta) = \mathbf{P}\{\xi_1 \leq r\} = \sum_{k=1}^{r} p_k(\theta). \tag{1.6.4}$$

We now state the theorems that completely describe the behavior of $\eta_{(N)}$, deferring their proofs. Our procedure follows Britikov [28].

Theorem 1.6.1. *If $n, N \to \infty$, $\theta = 2T/n \to 0$, and the integers*

$$r = r(n, N) > 1$$

vary such that $Np_r(\theta) \to \infty$ and $Np_{r+1}(\theta) \to \lambda$ for $0 \leq \lambda < \infty$, then

$$\mathbf{P}\{\eta_{(N)} = r\} = e^{-\lambda} + o(1),$$
$$\mathbf{P}\{\eta_{(N)} = r + 1\} = 1 - e^{-\lambda} + o(1).$$

Note that if $\lambda \neq 0$ in the conditions of the theorem, then $N p_r(\theta) \to \infty$ without any additional requirements. In particular, the conditions of the theorem are fulfilled if

$$T/n^{(r-1)/r} \to \rho, \quad 0 < \rho < \infty.$$

Under this condition, Theorem 1.6.1 was proved by Erdős and Renyi [37], whose well-known paper provided the only results on the behavior of $\eta_{(N)}$ until Britikov's work seventeen years later [28].

Theorem 1.6.2. *If $n, N \to \infty$, $\theta = 2T/n \to \gamma$, $0 < \gamma < 1$, then for any fixed integer k,*

$$\mathbf{P}\{\eta_{(N)} - [a] \leq k\} = \exp\left\{-\frac{(\gamma - 1 - \log \gamma)^{5/2}}{(e^{\gamma-1} - \gamma)\sqrt{2\pi}} e^{(k+\{a\})(\gamma-1-\log \gamma)}\right\} (1 + o(1)),$$

where

$$a = \frac{\log n - \frac{5}{2} \log \log n}{\theta - 1 - \log \theta},$$

and $[a]$ and $\{a\}$ denote, respectively, the integer and fractional parts of a.

Theorem 1.6.3. *If $n, N \to \infty$, $\theta = 2T/n \to 1$, and $N(1 - \theta)^3 \to \infty$, then for any fixed z,*

$$\mathbf{P}\{\beta\eta_{(N)} - u \leq z\} \to e^{-e^{-z}},$$

where $\beta = -\log(\theta e^{1-\theta})$ and u is the root of the equation

$$\left(\frac{2}{\pi}\right)^{1/2} N\beta^{3/2} = u^{5/2} e^u.$$

Theorem 1.6.4. *If $n, N \to \infty$ such that $N^{1/3}(1 - 2T/n) \to v$, $-\infty < v < \infty$, then for any fixed positive z,*

$$\mathbf{P}\left\{\frac{\eta_{(N)}}{bN^{2/3}} \leq z\right\} \to 1 + \frac{1}{p(v; 3/2, -1)} \sum_{s=1}^{\infty} \frac{1}{s!} \left(-\frac{3}{4\sqrt{\pi}}\right)^s I_s(z, v),$$

where $b = 2(2/3)^{2/3}$,

$$I_s(w, y) = \int_{\mathcal{A}} \frac{p(y - x_1 - \cdots - x_s; 3/2, -1)}{(x_1 \cdots x_s)^{5/2}} dx_1 \cdots dx_s,$$

$$\mathcal{A} = \{(x_1, \ldots, x_s): x_j \geq w, \ j = 1, \ldots, s\},$$

and $p(y; 3/2, -1)$ is the density of the stable law with parameters $\alpha = 3/2$, $\beta = -1$.

Theorem 1.6.5. If $n, N \to \infty$, $N(1 - 2T/n)^3 \to -\infty$, then for any fixed z,

$$\mathbf{P}\left\{\frac{n - 2N - \eta_{(N)}}{bN^{2/3}} \le z\right\} \to \int_{-\infty}^{z} p(y; 3/2, -1)\, dy.$$

We will prove Theorems 1.6.1–1.6.5 with the help of relation (1.6.2). Under the conditions of Theorems 1.6.1–1.6.3,

$$\mathbf{P}\{\bar\zeta_N^{(r)} = n\}/\mathbf{P}\{\zeta_N = n\} \to 1, \tag{1.6.5}$$

and the limit distribution of $\eta_{(N)}$ is the same as the limit distribution of the maximum of the random variables $\xi_1, \ldots, \xi_N$. Therefore we first obtain some auxiliary results on the asymptotic behavior of

$$P_r = P_r(\theta) = \sum_{k=r+1}^{\infty} p_r(\theta).$$

Lemma 1.6.1. If $n, N \to \infty$, $\theta = 2T/n \to 0$, and the integers $r = r(n, N) > 1$ vary such that $Np_r(\theta) \to \infty$, $Np_{r+1}(\theta) \to \lambda$, $0 \le \lambda < \infty$, then

$$NP_{r-1} \to \infty, \qquad NP_r \to \lambda, \qquad NP_{r+1} \to 0.$$

Proof. Under the conditions of the lemma, $x = \theta e^{-\theta} \to 0$. It follows from (1.6.3) that

$$P_r = \sum_{s=1}^{\infty} p_{r+s}(\theta) = p_{r+1}(\theta)\left(1 + \sum_{s=2}^{\infty} \frac{p_{r+s}(\theta)}{p_{r+1}(\theta)}\right), \tag{1.6.6}$$

$$P_{r+1} = p_{r+1}(\theta) \sum_{s=2}^{\infty} \frac{p_{r+s}(\theta)}{p_{r+1}(\theta)}. \tag{1.6.7}$$

Taking into account the bounds for factorials

$$r^r \sqrt{r} e^{-r} < r! < \sqrt{2\pi} r^r \sqrt{r} e^{-r} e^{1/(12r)},$$

we find from (1.6.1) that

$$\frac{p_{r+s}(\theta)}{p_{r+1}(\theta)} \le c_1 (xe)^{s-1},$$

where c_1 is a constant. Hence,

$$\sum_{s=2}^{\infty} \frac{p_{r+s}(\theta)}{p_{r+1}(\theta)} \le \frac{c_1 xe}{1 - xe} = o(1)$$

as $\theta \to 0$. Now by virtue of (1.6.6) and (1.6.7),

$$N P_r = N p_{r+1}(\theta) = \lambda + o(1),$$

$$N P_{r-1} > N p_r(\theta) \to \infty, \qquad N P_{r+1} \to 0.$$

∎

Note that if $\lambda \neq 0$, then $N p_r(\theta) \to \infty$ without any additional conditions, so this requirement may be excluded from the conditions of the lemma if $\lambda \neq 0$. Indeed,

$$N p_r(\theta) = N p_{r+1}(\theta) \frac{p_r(\theta)}{p_{r+1}(\theta)}.$$

Since $x \to 0$ and

$$\frac{p_r(\theta)}{p_{r+1}(\theta)} = \left(\frac{r}{r+1}\right)^{r-2} \frac{1}{x} = \left(1 - \frac{1}{r+1}\right)^{r-2} \frac{1}{x},$$

there exists a constant c_2 such that $N p_r(\theta) > c_2 N p_{r+1}(\theta)/x$ and $N p_r(\theta) \to \infty$.

Lemma 1.6.2. *If $n, N \to \infty$, $\theta = 2T/n \to \gamma$, $0 < \gamma < 1$, and $r = r(n, N) \to \infty$, then*

$$N P_r = N p_r(\theta) c(1 - c)^{-1}(1 + o(1)),$$

where $c = \gamma e^{1-\gamma}$.

Proof. It is clear that

$$N P_r = N p_r(\theta) \sum_{s=1}^{\infty} \frac{p_{r+s}(\theta)}{p_r(\theta)},$$

and

$$\frac{p_{r+s}(\theta)}{p_r(\theta)} = \left(\frac{r}{r+s}\right)^{5/2} (xe)^s (1 + O(1/r)).$$

Moreover, there exist constants $c_3 > 0$ and $q < 1$ such that

$$p_{r+s}(\theta)/p_r(\theta) \le c_3 (xe)^s \le c_3 q^s.$$

Therefore the series $\sum_{s=1}^{\infty} p_{r+s}(\theta)/p_r(\theta)$ converges uniformly and we can pass to the limit under the sum so that

$$\sum_{s=1}^{\infty} p_{r+s}(\theta)/p_r(\theta) \to \sum_{s=1}^{\infty} c^s = \frac{c}{1 - c}.$$

∎

Lemma 1.6.3. *If $n, N \to \infty$, $\theta = 2T/n \to 1$, and $N(1 - \theta)^3 \to \infty$, then for any fixed z,*

$$NP_r \to e^{-z},$$

where r is an integer such that $\beta r = u + z + o(1)$, $\beta = -\log(\theta e^{1-\theta})$, and u is the root of the equation

$$\left(\frac{2}{\pi}\right)^{1/2} N\beta^{3/2} = u^{5/2} e^u.$$

Proof. It is clear under the conditions of the lemma that $\beta = -\log(\theta e^{1-\theta}) \to 0$ and $u \to \infty$, since $N\beta^{3/2} \to \infty$ by virtue of the condition $N(1 - \theta)^3 \to \infty$. We apply Stirling's formula and obtain

$$NP_r = N \sum_{k=r+1}^{\infty} \frac{k^{k-2}x^k}{k!B(x)} = \left(\frac{2}{\pi}\right)^{1/2} N\beta^{3/2} \sum_{k>r} \frac{e^{-\beta k}\beta}{(\beta k)^{5/2}} \left(1 + O\left(\frac{1}{r}\right)\right).$$

The sum

$$\sum_{k>r}(\beta k)^{-5/2}e^{-\beta k}\beta$$

is an integral sum of the function $f(y) = y^{-5/2}e^{-y}$ with step β and is approximated by the corresponding integral:

$$\sum_{k>r}(\beta k)^{-5/2}e^{-\beta k}\beta = \int_{r\beta}^{\infty} y^{-5/2}e^{-y}\,dy(1 + o(1))$$

$$= (r\beta)^{-5/2}e^{-r\beta}(1 + o(1)).$$

Therefore

$$NP_r = \left(\frac{2}{\pi}\right)^{1/2} N\beta^{3/2}(r\beta)^{-5/2}e^{-r\beta}(1 + o(1)). \tag{1.6.8}$$

By definition, $r\beta = u + z + o(1)$ and

$$\left(\frac{2}{\pi}\right)^{1/2} N\beta^{3/2} = u^{5/2}e^u.$$

Substituting these expressions into (1.6.8) yields

$$NP_r = e^{-z}(1 + o(1)).$$

$\blacksquare$

Now we are ready to prove the theorems of this section.

Proof of Theorems 1.6.1–1.6.3. By applying Lemma 1.6.1, we find that under the conditions of Theorem 1.6.1,

$$(1 - P_{r-1})^N \to 0, \qquad (1 - P_r)^N \to e^{-\lambda}, \qquad (1 - P_{r+1})^N \to 1$$

as $N \to \infty$. These relations, together with (1.6.5), whose proof is pending, imply the assertion of Theorem 1.6.1.

Let

$$a = \frac{\log n - \frac{5}{2} \log \log n}{\theta - 1 - \log \theta},$$

and choose $r = [a] + k$, where k is a fixed integer. Under the conditions of Theorem 1.6.2, $r = [a] + k \to \infty$ and according to Lemma 1.6.2,

$$N P_r = N p_r(\theta) c(1 - c)^{-1}(1 + o(1)),$$

where $c = \gamma e^{1-\gamma}$. It is easy to see that

$$
\begin{aligned}
N p_r(\theta) &= \frac{2Nr^{r-2}\theta^{r-1}e^{r\theta}}{r!(2-\theta)} = \frac{ne^{r(1-\theta+\log\theta)}}{r^{5/2}\theta\sqrt{2\pi}}(1 + o(1)) \\
&= \frac{(\gamma - 1 - \log \gamma)^{5/2}}{\gamma\sqrt{2\pi}} e^{-(k-\{a\})(\gamma-1-\log\gamma)}(1 + o(1)).
\end{aligned}
$$

Thus

$$N P_r = \frac{(\gamma - 1 - \log \gamma)^{5/2} c}{\gamma(1 - c)\sqrt{2\pi}} e^{-(k-\{a\})(\gamma-1-\log\gamma)}(1 + o(1)),$$

and consequently,

$$(1 - P_r)^N = \exp\left\{ -\frac{(\gamma - 1 - \log \gamma)^{5/2}}{(e^{\gamma-1} - \gamma)\sqrt{2\pi}} e^{-(k-\{a\})(\gamma-1-\log\gamma)} \right\}(1 + o(1)).$$

Under the conditions of Theorem 1.6.3, Lemma 1.6.3 shows that $N P_r \to e^{-z}$ and

$$(1 - P_r)^N \to e^{-e^{-z}}.$$

Thus, to complete the proof of Theorems 1.6.1–1.6.3, it remains to verify (1.6.5) under each set of conditions.

Since $\theta N \to \infty$ and $N(1 - \theta)^3 \to \infty$, by Theorem 1.4.1 the random sum ζ_N is asymptotically normal, and

$$\mathbf{P}\{\zeta_N = n\} = \frac{1}{\sigma(\theta)\sqrt{2\pi N}}(1 + o(1)),$$

where

$$\sigma^2(\theta) = \mathbf{D}\xi_1 = \frac{2\theta}{(1-\theta)(2-\theta)}.$$

While estimating the asymptotic behavior of $(1 - P_r)^N$ in Lemmas 1.6.1–1.6.3, we determined the choice of r. We now prove the central limit theorem for the sum $\bar{\zeta}_N^{(r)}$ for these choices of r. Set $B_N = \sigma(\theta)N^{1/2}$.

The characteristic function of the random variable $\bar{\xi}_1^{(r)} - m(\theta)$, where $m(\theta) = \mathbf{E}\xi_1$, is

$$\frac{e^{-itm(\theta)}}{1-P_r} \sum_{k=1}^r p_k(\theta)e^{itk} = \frac{e^{-itm(\theta)}}{1-P_r}\left(\varphi(t) - \sum_{k>r} p_k(\theta)e^{itk}\right),$$

where $\varphi(t)$ is the characteristic function of the random variable ξ_1. Hence, the characteristic function $\varphi_r(t, \theta)$ of the random variable $(\bar{\zeta}_N^{(r)} - Nm(\theta))/B_N$ can be written

$$\varphi_r(t, \theta) = \frac{e^{-itNm(\theta)/B_N}}{(1-P_r)^N}\varphi^N\left(\frac{t}{B_N}\right)\left(1 - \sum_{k>r} p_k(\theta)e^{itk/B_N}(1+o(1))\right)^N.$$

According to Theorem 1.4.1, the distribution of $(\zeta_N - Nm(\theta))/B_N$ converges to the standard normal law, and consequently,

$$e^{-itNm(\theta)/B_N}\varphi^N(t/B_N) \to e^{-t^2/2}. \tag{1.6.9}$$

It is clear that

$$\sum_{k>r} p_r(\theta)e^{itk/B_N} = P_r + \sum_{k>r} p_r(\theta)(e^{itk/B_N} - 1) = P_r + O\left(\frac{1}{B_N}\sum_{k>r} kp_r(\theta)\right),$$

and it is not difficult to prove in each of the three cases that

$$\frac{1}{B_N}\sum_{k>r} kp_k(\theta) = o(1/N). \tag{1.6.10}$$

Estimates (1.6.9) and (1.6.10) imply that for any fixed t,

$$\varphi_r(t, \theta) \to e^{-t^2/2},$$

and the distribution of $(\bar{\zeta}_N^{(r)} - Nm(\theta))/B_N$ converges to the standard normal distribution. The local convergence

$$\mathbf{P}\{\bar{\zeta}_N^{(r)} = n\} = \frac{1}{\sigma(\theta)\sqrt{2\pi N}}(1+o(1))$$

needed for the proof of (1.6.5) can be proved in the standard way.

Thus the ratio in (1.6.5) tends to 1, and this, together with the estimates of $(1 - P_r)^N$, completes the proof of Theorems 1.6.1–1.6.3. ∎

To prove Theorem 1.6.4, the following lemma is needed.

Lemma 1.6.4. *If $N \to \infty$, the parameter θ in the distribution (1.6.1) equals 1, $N^{1/3}(1 - 2T/n) \to v$, and $r = zN^{2/3}$, where z is a positive constant, then*

$$bN^{2/3}\mathbf{P}\{\bar{\zeta}_N^{(r)} = n\} = f(z, v) + o(1),$$

where

$$f(z, y) = \exp\left\{\frac{z^{-3/2}}{2\sqrt{\pi}}\right\}\left(p(y; 3/2, -1) + \sum_{s=1}^{\infty}\frac{1}{s!}\left(-\frac{3}{4\sqrt{\pi}}\right)^s I_s(z, y)\right),$$

and $I_s(z, y)$ is defined in Theorem 1.6.4.

Proof. As $N \to \infty$,

$$p_k = p_k(1) = \frac{2k^{r-2}e^{-k}}{k!} = \left(\frac{2}{\pi}\right)^{1/2}k^{-5/2}(1 + o(1)) \tag{1.6.11}$$

uniformly in $k > r$.

It is clear that

$$\sum_{k>r}\frac{1}{k^{5/2}}\exp\left\{\frac{itk}{bN^{2/3}}\right\} = \frac{1}{b^{3/2}N}\sum_{k>r}\left(\frac{k}{bN^{2/3}}\right)^{-5/2}\exp\left\{\frac{itk}{bN^{2/3}}\right\}\frac{1}{bN^{2/3}}.$$

The last sum is an integral sum of the function $y^{-5/2}e^{ity}$ with step $1/(bN^{2/3})$; hence,

$$\sum_{k>r}\frac{1}{k^{5/2}}\exp\left\{\frac{itk}{bN^{2/3}}\right\} = \frac{1}{b^{3/2}N}\left(\int_z^{\infty}y^{-5/2}e^{ity}\,dy + o(1)\right). \tag{1.6.12}$$

Set

$$H(t, z) = \frac{3}{4\sqrt{\pi}}\int_z^{\infty}y^{-5/2}e^{ity}\,dy.$$

Then

$$|H(t, z)| \le H(0, z) = \frac{3}{4\sqrt{\pi}}\int_z^{\infty}y^{-5/2}\,dy = \frac{z^{-3/2}}{2\sqrt{\pi}}. \tag{1.6.13}$$

Taking into account $b = 2(2/3)^{2/3}$, we obtain from (1.6.12) and (1.6.13) that

$$\sum_{k>r}p_k\exp\left\{\frac{itk}{bN^{2/3}}\right\} = \left(\frac{2}{\pi}\right)^{1/2}\sum_{k>r}\frac{1}{k^{5/2}}\exp\left\{\frac{itk}{bN^{2/3}}\right\} + o\left(\frac{1}{N}\right)$$

$$= \frac{H(t, z) + o(1)}{N}. \tag{1.6.14}$$

In particular,

$$NP_r = H(0, z)(1 + o(1)). \tag{1.6.15}$$

The characteristic function $\varphi_r(t, 1)$ of the random variable $(\bar{\zeta}_N^{(r)} - 2N)/(bN^{2/3})$ can be written

$$\varphi_r(t, 1) = \left(\varphi \left(\frac{t}{bN^{2/3}}, 1 \right) - \exp \left\{ -\frac{2it}{bN^{2/3}} \right\} \sum_{k>r} p_k \exp \left\{ \frac{itk}{bN^{2/3}} \right\} \right)^N$$
$$\times (1 - P_r)^{-N},$$

where $\varphi(t, 1)$ is the characteristic function of $\xi_1 - \mathbf{E}\xi_1$. Note that $\mathbf{E}\xi_1 = 2$ in this case. It follows from (1.6.13), (1.6.14), and Theorem 1.4.2 that

$$\varphi_r(t, 1) = \varphi^N \left(\frac{t}{bN^{2/3}}, 1 \right) \left(1 - \sum_{k>r} p_k \exp \left\{ \frac{itk}{bN^{2/3}} \right\} (1 + o(1)) \right)^N (1 - P_r)^{-N}$$

$$= \psi(t) \left(1 - \frac{H(t, z)}{N} + o \left(\frac{1}{N} \right) \right)^N$$

$$\times \left(1 - \frac{H(0, z)}{N} + o \left(\frac{1}{N} \right) \right)^{-N},$$

where $\psi(t)$ is the characteristic function of the stable distribution with density $p(y; 3/2, -1)$. Thus, for any fixed t, as $N \to \infty$,

$$\varphi_r(t, 1) \to g(t, z) = \psi(t) \exp\{-H(t, z) + H(0, z)\}.$$

The function $g(t, z)$ is continuous; therefore, by Theorem 1.1.9, it is a characteristic function. Since $|g(t, z)|$ is integrable, it corresponds to the density

$$f(z, y) = \frac{1}{2\pi} \int_{-\infty}^{\infty} e^{-ity} g(t, z) \, dt.$$

The span of the distribution of $\bar{\xi}_1^{(r)}$ is 1; therefore, by Theorem 1.1.10, the local convergence is valid.

Thus it remains to show that $f(z, y)$ has the form given in Theorem 1.6.4. Representing $e^{-H(t,z)}$ by its Taylor series gives

$$f(z, y) = e^{H(0,z)} \sum_{s=0}^{\infty} \frac{(-1)^s}{s!} f_s(z, y), \qquad (1.6.16)$$

where

$$f_s(z, y) = \frac{1}{2\pi} \int_{-\infty}^{\infty} e^{-ity} \psi(t) H^s(t, z) \, dt.$$

It is easy to see that the function $2\sqrt{\pi} z^{3/2} H(t, z)$ is the characteristic function of the distribution with density

$$p_z(y) = \tfrac{3}{2} z^{3/2} y^{-5/2}, \quad y \geq z. \qquad (1.6.17)$$

Therefore the function

$$\left(2\sqrt{\pi}z^{3/2}\right)^s \psi(t) H^s(t, z)$$

is the characteristic function of the sum $\beta + \beta_1 + \cdots + \beta_s$ of independent random variables, where β has the stable law with density $p(y; 3/2, -1)$ and $\beta_1, \ldots, \beta_s$ are identically distributed with density $p_z(y)$. The density of the sum $\beta + \beta_1 + \cdots + \beta_s$ is

$$h_s(y) = \left(\tfrac{3}{2}z^{3/2}\right)^s I_s(t, y),$$

where $I_s(t, y)$ is defined in Theorem 1.6.4. Thus

$$\frac{1}{2\pi} \int_{-\infty}^{\infty} e^{-ity} \psi(t) H^s(t, y) \, dt = \left(\frac{3}{4\sqrt{\pi}}\right)^s I_s(t, y).$$

When we substitute this expression into (1.6.16), we find that

$$f(z, y) = e^{H(0,z)} \sum_{s=0}^{\infty} \frac{1}{s!} \left(-\frac{3}{4\sqrt{\pi}}\right)^s I_s(t, y). \qquad (1.6.18)$$

∎

Taking into account (1.6.15), Theorem 1.4.2, and (1.6.18), we see that Theorem 1.6.4 follows from (1.6.2).

To prove Theorem 1.6.5 with the help of (1.6.2), we need to know the asymptotic behavior of large deviations of $P\{\bar{\zeta}_N^{(r)} = n\}$. We give that information without proof (see [28]).

Lemma 1.6.5. *If $n, N \to \infty$, the parameter θ in the distribution (1.6.1) equals 1, $N(1 - 2T/n)^3 \to -\infty$, and $r = n - 2N - bzN^{2/3}$, where z is a constant, then*

$$P\{\bar{\zeta}_N^{(r)} = n\} = \left(\frac{2}{\pi}\right)^{1/2} \frac{N}{(n - 2N)^{5/2}} \int_z^{\infty} p(y; 3/2, -1) \, dy \, (1 + o(1)).$$
$$(1.6.19)$$

The assertion of Theorem 1.6.5 follows from (1.6.19), Theorem 1.4.2, and the fact that $NP_r \to 0$ under the condition of Theorem 1.6.5.

1.7. Graphs with unicyclic components

A graph is called unicyclic if it is connected and contains only one cycle. The number of edges of a unicyclic graph coincides with the number of its vertices. Let $\mathcal{U}_n$ denote the set of all graphs with n vertices where every connected component is unicyclic. Any graph from $\mathcal{U}_n$ has n edges. In this section, we study the structure of a random graph from $\mathcal{U}_n$. We follow the general approach described in Section 1.2.

As usual, denote by u_n the number of graphs in $\mathcal{U}_n$; we will study u_n as $n \to \infty$. Let b_n be the number of unicyclic graphs with n vertices, and $b_n^{(r)}$ the number of unicyclic graphs with n vertices, where the cycle has size r. The cycle of a unicyclic graph is nondirected; in other aspects, a unicyclic graph is similar to the connected graph of a mapping of a finite set into itself. Let d_n be the number of connected graphs of mappings of a set with n labeled vertices into itself, and $d_n^{(r)}$ the number of such graphs with the cycle of size r. It is easy to see that

$$d_n^{(1)} = n^{n-1}, \qquad d_n^{(2)} = 2\binom{n}{2} n^{n-3} = n^{n-1} - n^{n-2},$$

$$b_n^{(1)} = d_n^{(1)}, \qquad b_n^{(2)} = d_n^{(2)}, \qquad b_n^{(r)} = d_n^{(r)}/2, \quad r \geq 3.$$

Introduce the generating functions

$$B(x) = \sum_{n=1}^{\infty} \frac{b_n x^n}{n!}, \qquad d(x) = \sum_{n=1}^{\infty} \frac{d_n x^n}{n!}, \qquad c(x) = \sum_{n=1}^{\infty} \frac{n^{n-2} x^n}{n!}.$$

These functions can be represented in terms of the function

$$\theta(x) = \sum_{n=1}^{\infty} \frac{n^{n-1} x^n}{n!},$$

which is the root of the equation $\theta e^{-\theta} = x$ in the interval $[0, 1]$. This function was used in Section 1.4. Taking into account the notation introduced here and using the results of Section 1.4, we see that

$$d(x) = -\log(1 - \theta(x)), \qquad c(x) = \tfrac{1}{2}\left(1 - (1 - \theta(x))^2\right).$$

Since $b_n = b_n^{(1)} + \cdots + b_n^{(n)}$, we have

$$B(x) = \sum_{n=1}^{\infty} \frac{b_n x^n}{n!} = \frac{1}{2}\sum_{n=1}^{\infty} \frac{d_n x^n}{n!} + \frac{1}{2}\sum_{n=1}^{\infty} \frac{d_n^{(1)} x^n}{n!} + \frac{1}{2}\sum_{n=1}^{\infty} \frac{d_n^{(2)} x^n}{n!}$$

$$= \frac{1}{2}d(x) + \theta(x) - \frac{1}{2}c(x)$$

$$= -\frac{1}{2}\log(1 - \theta(x)) + \theta(x) - \frac{1}{4}\left(1 - (1 - \theta(x))^2\right). \tag{1.7.1}$$

In accordance with the general model of Section 1.4, let us introduce independent identically distributed random variables $\xi_1, \ldots, \xi_N$ for which

$$\mathbf{P}\{\xi_1 = k\} = \frac{b_k x^k}{k!\, B(x)}, \quad k = 1, 2, \ldots. \tag{1.7.2}$$

The number of graphs in $\mathcal{U}_n$ with N components can be represented in the form

$$u_{n,N} = \frac{n!}{N!} \sum_{n_1 + \cdots + n_N = n} \frac{b_{n_1} \cdots b_{n_N}}{n_1! \cdots n_N!} = \frac{n!\,(B(x))^N}{N!\, x^n} \mathbf{P}\{\xi_1 + \cdots + \xi_N = n\}. \tag{1.7.3}$$

In what follows, we choose

$$x = \left(1 - 1/\sqrt{n}\right)e^{1-1/\sqrt{n}}.$$

Theorem 1.7.1. *As $n \to \infty$,*

$$u_n = \frac{\sqrt{2\pi}\,e^{3/4}}{2^{1/4}\Gamma(1/4)}n^{n-1/4}(1 + o(1)),$$

where

$$\Gamma(p) = \int_0^\infty x^{p-1}e^{-x}\,dx$$

is the Euler gamma function.

Before proving Theorem 1.7.1, we will prove some auxiliary results.

Lemma 1.7.1. *For $x = (1 - 1/\sqrt{n})e^{1-1/\sqrt{n}}$,*

$$\left(1 - \theta(xe^{it})\right)^2 = \frac{1}{n} - 2it + \varepsilon_1(t) + \varepsilon_2(t, n),$$

where $\varepsilon_1(t)/t \to 0$ as $t \to 0$ uniformly in n and $|\varepsilon(t, n)| \le 2|t|/\sqrt{n}$.

Proof. We found in Section 1.4 that

$$u(w) = (1 - \theta(w))^2 = 1 - 2\sum_{k=1}^{\infty}\frac{k^{k-2}w^k}{k!} = 1 - 2c(w), \quad |w| \le e^{-1}.$$

When we write $u(w)$ as

$$u(xe^{it}) = u(e^{-1+it}) + \frac{1}{n} + \varepsilon_2(t, n), \tag{1.7.4}$$

it is clear that $\theta(x) = 1 - 1/\sqrt{n}$ and $\theta(e^{-1}) = 1$; therefore $u(e^{-1}) - u(x) = -1/n$. With this equality and the observation that $x < e^{-1}$, we obtain the estimates

$$|\varepsilon_2(t, n)| = \left|u(xe^{it}) - u(e^{-1+it}) - 1/n\right|$$

$$= 2\left|\sum_{k=1}^{\infty}\frac{k^{k-2}(e^{-k} - x^k)(e^{itk} - 1)}{k!}\right|$$

$$\le 2\sum_{k=1}^{\infty}\frac{k^{k-1}(e^{-k} - x^k)|t|}{k!}$$

$$= 2|t|(\theta(e^{-1}) - \theta(x)) = a|t|/\sqrt{n}. \tag{1.7.5}$$

The function $u(e^{-1+it})$ has the first derivative $-2i$ at the point $t = 0$; thus, as $t \to 0$,

$$u\left(e^{-1+it}\right) = -2it + o(t). \tag{1.7.6}$$

The assertion of the lemma follows from (1.7.4), (1.7.5), and (1.7.6). ∎

Lemma 1.7.2. *If $n \to \infty$, $N = \alpha \log n + o(\log n)$, where α is a positive constant, then*

$$n\mathbf{P}\{\xi_1 + \cdots + \xi_N = k\} = \frac{1}{2^\alpha \Gamma(\alpha)} z^{\alpha-1} e^{-z/2} (1 + o(1))$$

uniformly in k such that $z = k/n$ lies in any fixed interval of the form $0 < z_0 \le z \le z_1 < \infty$.

Proof. The characteristic function of the sum $(\xi_1 + \cdots + \xi_N)/n$ is equal to $\varphi_N(t) = \varphi^N(t/n)$, where $\varphi(t)$ is the characteristic function of ξ_1. It is clear that

$$\varphi(t) = B\left(xe^{it}\right)/B(x).$$

Lemma 1.7.1 and equation (1.7.1) give

$$4B\left(xe^{it/n}\right) = -\log \frac{1-2it}{n} + 3 + o(1).$$

Therefore

$$\varphi\left(\frac{t}{n}\right) = \frac{B\left(xe^{it/n}\right)}{B(x)} = \frac{\log n - \log(1-2it) + 3 + o(1)}{\log n + 3 + o(1)}$$

$$= 1 - \frac{\log(1-2it) + o(1)}{\log n},$$

and if $N = \alpha \log n + o(1)$, then for any fixed t,

$$\varphi_N(t) = \varphi^N(t/n) = \left(1 - \frac{\log(1-2it) + o(1)}{\log n}\right)^N = \frac{1+o(1)}{(1-2it)^\alpha},$$

and the distribution of $(\xi_1 + \cdots + \xi_N)/n$ converges weakly to the distribution with density

$$\frac{1}{2^\alpha \Gamma(\alpha)} z^{\alpha-1} e^{-z/2},$$

that is, to the chi-square distribution with 2 degrees of freedom, which corresponds to the characteristic function $(1 - 2it)^{-\alpha}$.

The local convergence can be proved in the usual way by using Lemmas 1.12.3–1.12.7 from [78]. ∎

Let $u_{n,N}$ be the number of graphs in $\mathcal{U}_n$ with N components and

$$\lambda_n = \frac{1}{4}\log n, \qquad u = \frac{N - \lambda_n}{\sqrt{\lambda_n}}.$$

Lemma 1.7.3. *If $n \to \infty$, then*

$$u_{n,N} = \frac{\sqrt{2\pi}\,e^{3/4}}{2^{1/4}\Gamma(1/4)}\,n^{n-1/4}\frac{\lambda_n^N e^{-\lambda_n}}{N!}(1 + o(1))$$

uniformly in N such that $|u| \leq (\log n)^{1/4}$.

Proof. It is clear that

$$u_{n,N} = \frac{n!}{N!}\sum_{n_1+\cdots+n_N=n}\frac{b_{n_1}\cdots b_{n_N}}{n_1!\cdots n_N!}$$

$$= \frac{n!\,(B(x))^N}{N!x^n}\mathbf{P}\{\xi_1 + \cdots + \xi_N = n\}. \tag{1.7.7}$$

By putting $\alpha = 1/4$ in Lemma 1.7.2, we obtain

$$n\mathbf{P}\{\xi_1 + \cdots + \xi_N = n\} = \frac{1}{2^{1/4}\Gamma(1/4)}\,e^{-1/2}(1 + o(1)) \tag{1.7.8}$$

uniformly in N when $|u| \leq (\log n)^{1/4}$.

The assertion of the lemma follows from (1.7.7) and (1.7.8), since

$$B(x) = \tfrac{1}{4}\log n + \tfrac{3}{4} + o(1),$$

$$x^n = e^{-n-1/2}(1 + o(1)),$$

$$(B(x))^N = \lambda_n^N e^{3/4}(1 + o(1)). \tag{1.7.9}$$

■

The assertion of Theorem 1.7.1 can be obtained by summing $u_{n,N}$ over N. Lemma 1.7.3 estimates $u_{n,N}$ for N close to λ_n. The following lemmas give estimates of $u_{n,N}$ for the other values of N needed in the proof.

Lemma 1.7.4. *For any fixed $\alpha_0, \alpha_1, 0 < \alpha_0 \leq \alpha_1 < \infty$, there exists a constant c_1 such that for $\alpha_0 \log n \leq N \leq \alpha_1 \log n$,*

$$u_{n,N} \leq c_1 n^{n-1/4}\frac{\lambda_n^N e^{-\lambda_n}}{N!}.$$

Proof. It follows from Lemma 1.7.2 that there exists a constant A such that

$$n\mathbf{P}\{\xi_1 + \cdots + \xi_N = n\} \leq A \tag{1.7.10}$$

for $\alpha_0 \log n \leq N \leq \alpha_1 \log n$. Indeed, if (1.7.10) did not hold, then a sequence of the parameters $n \to \infty$, $N = \alpha \log n + o(\log n)$ would exist for which the assertion of Lemma 1.7.2 would not be true. Lemma 1.7.4 then follows from (1.7.7), (1.7.9), and (1.7.10). ∎

Lemma 1.7.5. *If $N \leq \log n$, then there exists a constant c_2 such that*

$$n\mathbf{P}\{\xi_1 + \cdots + \xi_N = n\} \leq c_2 \log n.$$

Proof. It is well known that

$$d_m = (m-1)! \sum_{k=0}^{m-1} \frac{m^k}{k!}.$$

Indeed, since the number of forests with n nonroot vertices and N rooted trees labeled $1, \ldots, N$ is $N(n+N)^{n-1}$, the number $d_m^{(r)}$ of connected graphs of mappings of an m-set into itself with the cycle of size r can be represented as

$$d_m^{(r)} = \binom{m}{r}(r-1)! \, r m^{m-r-1} = \frac{m! \, m^{m-r-1}}{(m-r)!}.$$

Here $\binom{m}{r}$ is the number of possible choices of r vertices that constitute the cycle; $(r-1)!$ is the number of cycles that can be constructed from r vertices; and rm^{m-r-1} is the number of forests with r cyclic vertices as the roots. Hence,

$$d_m = \sum_{r=1}^{m} d_m^{(r)} = \sum_{r=1}^{m} \frac{m! \, m^{m-r-1}}{(m-r)!} = (m-1)! \sum_{k=0}^{m-1} \frac{m^k}{k!}.$$

As $m \to \infty$,

$$d_m = \tfrac{1}{2}(m-1)! \, e^m (1 + o(1)),$$

and there exists a constant c_3 such that

$$b_m \leq d_m \leq c_3 (m-1)! \, e^m.$$

Moreover, $B(x) = \log n(1 + o(1))/4$ and $x^m \leq e^{-m}$ for all $m \geq 0$. Therefore there exists a constant c_2 such that

$$\mathbf{P}\{\xi_1 = m\} \leq \frac{c_2}{m \log n}. \tag{1.7.11}$$

It is clear that

$$\{\xi_1 + \cdots + \xi_N = n\} = \bigcup_{i=1}^{N} \bigcup_{k \geq \lceil n/N \rceil} \{\xi_i = k, \, \xi_1 + \cdots + \xi_N - \xi_i = n - k\}.$$

Since $\mathbf{P}\{\xi_1 = k\}$ decreases as k increases, we have

$$\mathbf{P}\{\xi_1 + \cdots + \xi_N = n\}$$

$$\leq N \sum_{k \geq [n/N]} \mathbf{P}\{\xi_1 = k\}\mathbf{P}\{\xi_2 + \cdots + \xi_N = n - k\}$$

$$\leq N\mathbf{P}\{\xi_1 = [n/N]\} \sum_{k \geq [n/N]} \mathbf{P}\{\xi_2 + \cdots + \xi_N = n - k\}$$

$$\leq N\mathbf{P}\{\xi_1 = [n/N]\}.$$

The lemma now follows from (1.7.11).   ∎

Lemma 1.7.6. *For $N \leq \log n$,*

$$u_{n,N} \leq c_4 n^{n-1/4} \log n \frac{\lambda_n^N e^{-\lambda_n}}{N!},$$

where c_4 is a constant.

This lemma follows from (1.7.7), (1.7.9), and Lemma 1.7.5.

Proof of Theorem 1.7.1. Roughly speaking, $u_{n,N} = c\lambda_n^N e^{-\lambda_n}/N!$, where c does not depend on N, and to obtain u_n, we sum the Poisson probabilities whose sum is 1. To do this rigorously, we divide the sum

$$u_n = \sum_{N=1}^{\infty} u_{n,N}$$

into four parts. Recall that $u = (N - \lambda_n)/\sqrt{\lambda_n}$. Let

$$S_1 = \sum_{A_1} u_{n,N}, \qquad S_2 = \sum_{A_2} u_{n,N}, \qquad S_3 = \sum_{A_3} u_{n,N}, \qquad S_4 = \sum_{A_4} u_{n,N},$$

where

$$A_1 = \{N: |u| \leq (\log n)^{1/4}\},$$

$$A_2 = \{N: |u| > (\log n)^{1/4}, \ \alpha_0 \log n \leq N \leq \alpha_1 \log n\},$$

$$A_3 = \{N: N < \alpha_0 \log n\},$$

$$A_4 = \{N: N > \alpha_1 \log n\}.$$

As $n \to \infty$,

$$\sum_{A_1} \frac{\lambda_n^N e^{-\lambda_n}}{N!} = 1 + o(1); \tag{1.7.12}$$

therefore it follows from Lemma 1.7.3 that

$$S_1 = \frac{\sqrt{2\pi} e^{3/4}}{2^{1/4}\Gamma(1/4)} n^{n-1/4}(1 + o(1)).$$

It remains to show that S_2, S_3, and S_4 are $o(n^{1-1/4})$. Lemma 1.7.4 implies that

$$S_2 \le c_1 n^{n-1/4} \left(1 - \sum_{A_1} \frac{\lambda_n^N e^{-\lambda_n}}{N!} \right),$$

and it follows from (1.7.12) that $S_2 = o(n^{n-1/4})$.

To obtain an estimate for S_3, we use the inequality

$$\sum_{1 \le N \le m} \frac{\lambda^N e^{-\lambda}}{N!} \le m \frac{\lambda^m e^{-\lambda}}{m!},$$

which is true for $m < \lambda$. Choose $\alpha_0 < 1/4$ such that

$$\alpha_0 - \alpha_0 \log \alpha_0 - \alpha_0 \log 4 < 1/8.$$

Then, for $m = \alpha_0 \log n$,

$$m \frac{\lambda_n^N e^{-\lambda_n}}{m!} \le \frac{c_5}{n^{1/8}},$$

where c_5 is a constant. By using the estimate from Lemma 1.7.6, we find that $S_3 \le c_4 c_5 n^{n-1/4-1/3} \log n$.

To obtain an estimate for S_4, we use the inequality

$$u_{n,N} \le \frac{n! (B(x))^N}{N! x^n} \le c_6 n^{n-1/4} n \frac{\lambda_n^N e^{-\lambda_n}}{N!}, \qquad (1.7.13)$$

where c_6 is a constant, which follows from (1.7.7) if $\mathbf{P}\{\xi_1 + \cdots + \xi_N = n\}$ is replaced by 1. For $m > \lambda$,

$$\sum_{N \ge m} \frac{\lambda^N e^{-\lambda}}{N!} \le \frac{\lambda^m}{m!}.$$

Choose $\alpha_1 > 1/4$ such that $\alpha_1 - \alpha_1 \log \alpha_1 - \alpha_1 \log 4 < -2$. Then for $m = \alpha_1 \log n$ and $\lambda_n = (\log n)/4$, we have the estimate $\lambda_n^m / m! \le n^{-2}$; thus (1.7.13) implies that $S_4 \le c_6 n^{n-5/4}$.

The assertion of the theorem follows from the estimates obtained for S_1, S_2, S_3, and S_4. ∎

We denote the number of components in a random graph of $\mathcal{U}_n$ by $\varkappa_n$. The following theorem is a direct corollary of Lemma 1.7.3 and Theorem 1.7.1.

Theorem 1.7.2. *As* $n \to \infty$,

$$\mathbf{P}\{\varkappa_n = N\} = \frac{2}{\sqrt{2\pi \log n}} e^{-u^2/2} (1 + o(1))$$

uniformly in N *for which* $u = (N - \frac{1}{4} \log n)/\sqrt{\frac{1}{4} \log n}$ *lies in any fixed finite interval.*

Indeed, Lemma 1.7.3 and Theorem 1.7.1 imply that

$$\mathbf{P}\{x_n = N\} = \frac{u_{n,N}}{u_n} = \frac{\lambda_n^N e^{-\lambda_n}}{N!}(1 + o(1))$$

uniformly in $|u| \leq (\log n)^{1/4}$, where $\lambda_n = \frac{1}{4}\log n$.

We now consider the maximum size β_n of the components of a random graph from $\mathcal{U}_n$.

Theorem 1.7.3. *If $n \to \infty$, then for any fixed γ, $0 < \gamma < 1$,*

$$\mathbf{P}\{\beta_n \leq \gamma n\} = \sum_{0 \leq s < 1/\gamma} \frac{(-1)^s}{4^s s!} W_s(1, \gamma) + o(1),$$

where $W_0(x, y) = 1$, and for $s = 1, 2, \ldots$,

$$W_s(z, \gamma) = \int_{\{x_i \geq \gamma, \; i=1,\ldots,s, \; x_1+\cdots+x_s \leq z\}} \frac{dx_1 \cdots dx_s}{x_1 \cdots x_s (z - x_1 - \cdots - x_s)^{3/4}}.$$

Proof. To study β_n, we use the general approach of Section 1.2. Let $\eta_1, \ldots, \eta_N$ be random variables with distribution

$$\mathbf{P}\{\eta_1 = n_1, \ldots, \eta_N = n_N\} = \mathbf{P}\{\xi_1 = n_1, \ldots, \xi_N = n_N \mid \xi_1 + \cdots + \xi_N = n\}.$$
$$(1.7.14)$$

It follows from (1.7.7) that these variables can be interpreted as the sizes of the ordered components of a random graph from $\mathcal{U}_n$ (see Section 1.2), in which x_n is N. Therefore

$$\mathbf{P}\{\beta_n \leq \gamma n\} = \sum_{N=1}^{\infty} \mathbf{P}\{x_n = N\} \mathbf{P}\{\eta_{(N)} \leq \gamma n\}, \qquad (1.7.15)$$

where $0 < \gamma < 1$ and $\eta_{(N)} = \max_{1 \leq i \leq N} \eta_i$. By Lemma 1.2.2,

$$\mathbf{P}\{\eta_{(N)} \leq \gamma n\} = (\mathbf{P}\{\xi_1 \leq \gamma n\})^N \frac{\mathbf{P}\{\bar{\xi}_1 + \cdots + \bar{\xi}_N = n\}}{\mathbf{P}\{\xi_1 + \cdots + \xi_N = n\}}, \qquad (1.7.16)$$

where $\bar{\xi}_1, \ldots, \bar{\xi}_N$ are independent identically distributed random variables for which

$$\mathbf{P}\{\bar{\xi}_1 = k\} = \mathbf{P}\{\xi_1 = k \mid \xi_1 \leq \gamma n\},$$

and the random variables $\xi_1, \ldots, \xi_N$ have distribution (1.7.2). We now estimate

$$H_{\gamma n}(t) = \sum_{k > \gamma n} \frac{b_k x^k}{k!} e^{itk/n}$$

for $x = (1 - 1/\sqrt{n})e^{-1+1/\sqrt{n}}$. By (1.7.1) for any fixed γ, $0 < \gamma < 1$, as $n \to \infty$,

$$H_{\gamma n}(t) = \frac{1}{2} \sum_{k>\gamma n} \frac{b_k x^k}{k!} e^{itk/n} + o(1).$$

Let us prove that

$$H_{\gamma n}(t) = H(\gamma, t) + o(1),$$

where

$$H(\gamma, t) = \frac{1}{4} \int_\gamma^\infty u^{-1} e^{-(1-2it)u/2} \, du.$$

It is easily seen that

$$\left(1 - 1/\sqrt{n}\right)^k e^{k/\sqrt{n}} = e^{-k/(2n)}\left(1 + o(1/\sqrt{n})\right),$$

$$\sum_{m=0}^{k-1} \frac{k^m e^{-k}}{m!} = \frac{1}{2} + o(1/\sqrt{n})$$

uniformly in $k \ge \gamma n$. Therefore, as $n \to \infty$,

$$H_{\gamma n}(t) = \sum_{k>\gamma n} \frac{1}{k}\left(1 - \frac{1}{\sqrt{n}}\right)^k \exp\left\{\frac{k}{\sqrt{n}} + \frac{itk}{n}\right\} \sum_{m=0}^{k-1} \frac{k^m e^{-k}}{m!}$$

$$= \frac{1}{2} \sum_{k>\gamma n} \frac{1}{k} e^{-(1-2it)k/(2n)}\left(1 + o(1/\sqrt{n})\right).$$

This sum is an integral sum of the function $u^{-1} e^{(1-2it)u/2}$ with step $1/n$. Hence,

$$H_{\gamma n}(t) = \frac{1}{2} \int_\gamma^\infty u^{-1} e^{-(1-2it)u/2} \, du + o(1) = H(\gamma, t) + o(1).$$

In particular, we obtain the following estimate for the tail of the distribution (1.7.2):

$$P\{\xi_1 > \gamma n\} = \frac{1}{B(x)} \sum_{k>\gamma n} \frac{b_k x^k}{k!}$$

$$= \frac{4H_{\gamma n}(0) + o(1)}{\log n} = \frac{4H(\gamma, 0) + o(1)}{\log n} \qquad (1.7.17)$$

as $n \to \infty$.

We now find the limit distribution of the sum $(\bar{\xi}_1 + \cdots + \bar{\xi}_N)/n$. The characteristic function of $\bar{\xi}_1/n$ is

$$\psi(t) = \frac{\varphi(t/n) - H_{\gamma n}(t)/B(x)}{1 - H_{\gamma n}(0)/B(x)}.$$

Using the estimates

$$\varphi(t/n) = 1 - \log(1 - 2it)/\log n + o(1/\log n), \qquad 4B(x) = \log n + O(1),$$

from (1.7.16) and (1.7.17), as $n \to \infty$, yields

$$\psi(t) = \left(1 - \frac{\log(1 - 2it) - 4H(\gamma, t) + o(1)}{\log n}\right)\left(1 - \frac{4H(\gamma, 0) + o(1)}{\log n}\right)^{-1},$$

and for any fixed t and $N = \frac{1}{4}\log n + o(\log n)$,

$$\psi^N(t) \to \varphi_\gamma(t) = (1 - 2it)^{-1/4}e^{-H(\gamma, t) + H(\gamma, 0)}.$$

When we expand $e^{-H(\gamma, t)}$ into its Taylor series, as we did in the proof of Lemma 1.6.4, we find that the characteristic function $\varphi_\gamma(t)$ corresponds to the density

$$f_\gamma(z) = \frac{e^{H(\gamma, 0) - z/2}}{2^{1/4}\Gamma(1/4)} \sum_{0 \le s < 1/\gamma} \frac{(-1)^s}{4^s s!} W_s(z, \gamma).$$

Thus, for any $\gamma, 0 < \gamma < 1$, the distribution of $(\bar{\xi}_1 + \cdots + \bar{\xi}_N)/n$ converges weakly to the distribution whose density is $f_\gamma(z)$ as $n \to \infty$ and $N = \frac{1}{4}\log n + o(\log n)$. We can show that local convergence of these distributions holds. If $n \to \infty$ and $N = \frac{1}{4}\log n + o(\log n)$ and $0 < \gamma < 1$, then

$$n\mathbf{P}\{\bar{\xi}_1 + \cdots + \bar{\xi}_N = k\} = f_\gamma(z) + o(1) \tag{1.7.18}$$

holds uniformly in k for which $z = k/n$ lies in any given interval of the form $0 < z_0 \le z \le z_1 < \infty$.

Using (1.7.17), we find that for $n \to \infty$ and $N = \frac{1}{4}\log n + o(\log n)$,

$$(\mathbf{P}\{\xi_1 \le \gamma n\})^N = \left(1 - \frac{4H(\gamma, 0) + o(1)}{\log n}\right)^N = e^{-H(\gamma, 0)} + o(1). \tag{1.7.19}$$

Substituting estimates (1.7.19), (1.7.18), and (1.7.8) into (1.7.16) gives

$$\mathbf{P}\{\eta_{(N)} \le \gamma n\} = \sum_{0 \le s < 1/\gamma} \frac{(-1)^s}{4^s s!} W_s(1, \gamma) + o(1). \tag{1.7.20}$$

To obtain the distribution of β_n, we need to average the distribution of $\eta_{(N)}$ with respect to the distribution of $\varkappa_n$. By Theorem 1.7.2, the number of components $\varkappa_n$ is asymptotically normal with parameters $(\frac{1}{4}\log n, \frac{1}{4}\log n)$, and for $N = \frac{1}{4}\log n + o(\log n)$, the probability $\mathbf{P}\{\eta_{(N)} \le \gamma n\}$ is asymptotically constant; therefore the assertion of the theorem follows from (1.7.15).   ∎

Denote by $\mathcal{U}_{n,2}$ and $\mathcal{U}_{n,3}$ the sets of all graphs with n labeled vertices consisting of unicyclic components where each cycle has more than one or more than two vertices, respectively. It is not difficult to see that we can treat $\mathcal{U}_{n,i}$, $i = 2, 3$, in the same way as $\mathcal{U}_n$ (which, following the above notation, we have to denote by

$\mathcal{U}_{n,1}$). The role of $B(x)$ for $\mathcal{U}_{n,i}$, $i = 2, 3$, is played by the generating functions

$$B_i(x) = \sum_{n=1}^{\infty} \frac{b_{n,i} x^n}{n!}, \quad i = 2, 3,$$

where $b_{n,i}$ is the number of unicyclic graphs with n vertices and cycle lengths not less than i.

It is clear that

$$b_{n,2} = \sum_{r=2}^{\infty} b_n^{(r)} = d_n^{(2)} + \frac{1}{2} \sum_{r=3}^{\infty} d_n^{(r)},$$

$$B_2(x) = -\frac{1}{2} c(x) + d(x) = -\frac{1}{4}\left(1 - (1 - \theta(x))^2\right) - \frac{1}{2} \log(1 - \theta(x)),$$

and for $x = (1 - 1/\sqrt{n}) e^{-1+1/\sqrt{n}}$,

$$B_2(x) = \frac{1}{4} \log n - \frac{1}{4} + o(1),$$

$$(B_2(x))^N = \lambda_n^N e^{-1/4}(1 + o(1)).$$

Similarly,

$$b_{n,3} = \sum_{r=3}^{\infty} b_n^{(r)} = \frac{1}{2} \sum_{r=3}^{\infty} d_n^{(r)},$$

$$B_3(x) = \frac{1}{2} d(x) - \theta(x) + c(x)$$

$$= -\frac{1}{2} \log(1 - \theta(x)) - \theta(x) - \frac{1}{2}\left(1 - (1 - \theta(x))^2\right), \quad (1.7.21)$$

and for $x = (1 - 1/\sqrt{n}) e^{-1+1/\sqrt{n}}$,

$$B_3(x) = \frac{1}{4} \log n - \frac{3}{4} + o(1),$$

$$(B_3(x))^N = \lambda_n^N e^{-3/4}(1 + o(1)).$$

Therefore, if $n \to \infty$, then for the numbers $u_n^{(i)}$ of the graphs in $\mathcal{U}_{n,i}$ and for the number $u_{n,N}^{(i)}$ of such graphs with N components, we have

$$u_n^{(i)} = A_i n^{n-1/4}(1 + o(1)),$$

$$u_{n,N}^{(i)} = A_i \frac{\lambda_n^N e^{-\lambda_n}}{N!} n^{n-1/4}(1 + o(1))$$

uniformly in the integers N such that $|N - \lambda_n|/\sqrt{\lambda_n}$ lies in any fixed finite interval,

where

$$A_1 = \frac{\sqrt{2\pi}\,e^{3/4}}{2^{1/4}\Gamma(1/4)}, \qquad A_2 = \frac{\sqrt{2\pi}\,e^{-1/4}}{2^{1/4}\Gamma(1/4)}, \qquad A_3 = \frac{\sqrt{2\pi}\,e^{-3/4}}{2^{1/4}\Gamma(1/4)}. \qquad (1.7.22)$$

Theorems 1.7.2 and 1.7.3 are valid for the random variables x_n and β_n in $\mathcal{U}_{n,2}$ and $\mathcal{U}_{n,2}$.

1.8. Graphs with components of two types

The generalized scheme of allocation can be used in the investigations of random graphs with nonhomogeneous structure. Consider the set $\mathcal{A}_{n,T}$ of all graphs with n vertices and T edges where each connected component contains no more than one cycle. As usual, we assign equal probabilities to the elements of $\mathcal{A}_{n,T}$ and consider a random graph with values from $\mathcal{A}_{n,T}$. Since any graph from the set $\mathcal{A}_{n,T}$ consists of trees and unicyclic components, we can use the results of the previous sections to study various characteristics of a random graph from $\mathcal{A}_{n,T}$. Consider first the number of elements in $\mathcal{A}_{n,T}$. As in the previous sections, we will denote by a_n the number of graphs under consideration with n vertices and by b_n the numbers of connected graphs under consideration with n vertices.

Instead of $\mathcal{A}_{n,T}$, we will use, where necessary, the notation $\mathcal{A}_{n,T}^{(1)}$ if cycles of lengths 1 and 2 are allowed; $\mathcal{A}_{n,T}^{(2)}$ if cycles of length 1 are forbidden; and $\mathcal{A}_{n,T}^{(3)}$ if cycles of lengths 1 and 2 are forbidden. Denote the number of graphs in $\mathcal{A}_{n,T}^{(i)}$ by $a_{n,T}^{(i)}$ and preserve the notation $a_{n,T}$ if the specialization is not needed. In accordance with the previous sections, the number of forests with n vertices, T edges, and $N = n - T$ trees is denoted by $F_{n,N}$. We use $u_n^{(i)}$ to denote the number of graphs with n vertices and unicyclic components if they are included in $\mathcal{A}_{n,T}^{(i)}$, $i = 1, 2, 3$, and preserve the notation u_n for the number of such graphs in $\mathcal{A}_{n,T}$ if the specialization is not important.

It is clear that

$$a_{n,T} = \sum_{m=0}^{n} \binom{n}{m} u_m F_{n-m,N}. \qquad (1.8.1)$$

Theorem 1.8.1. *If $n, T \to \infty$ such that $T/n \to 0$, then*

$$a_{n,T} = F_{n,N}(1 + o(1)) = \frac{n^{2T}}{2^T T!}(1 + o(1)).$$

Proof. It follows from Theorem 1.7.1 that there exists a constant c_1 such that

$$u_m \le c_1 m^{m-1/4}. \qquad (1.8.2)$$

Theorem 1.4.3 shows that under the conditions of Theorem 1.8.1,

$$F_{n,N} = \frac{n^{2T}}{2^T T!}(1 + o(1)).$$

(1.8.3)

The condition $T/n \to 0$ implies that $(T - m)/(n - m) \to 0$ uniformly in m, $0 \le m \le T$. Therefore, under the conditions of Theorem 1.8.1, there exists a constant c_2 such that

$$F_{n-m,N} \le \frac{c_2(n - m)^{2(T-m)}}{2^{T-m}(T - m)!}$$

(1.8.4)

for all m, $0 \le m \le T$.

We obtain from (1.8.1), (1.8.2), (1.8.3), and (1.8.4) that

$$a_{n,T} = F_{n,N} + \sum_{m=1}^{T} \binom{n}{m} u_m F_{n-m,N}$$

$$= F_{n,N}\left(1 + O\left(\sum_{m=1}^{T} \frac{m^m}{m!}\left(\frac{2eTn}{(n - T)^2}\right)^m\right)\right).$$

(1.8.5)

This completes the proof because $2Tn/(n - T)^2 \to 0$. ■

Let $\omega_{n,T}$ be the number of vertices contained in the unicyclic components of the random graph in $\mathcal{A}_{n,T}$. It is easily seen from Theorem 1.8.1 that if $n, T \to \infty$ and $T/n \to 0$, then

$$P\{\omega_{n,T} = 0\} \to 1,$$

and the limit distributions of the number of trees of fixed sizes in a random graph from $\mathcal{A}_{n,T}$ coincide with the corresponding limit distributions in a random forest and are described in Theorems 1.5.1 and 1.5.2; the limit distribution of the maximum size of trees in a random graph from $\mathcal{A}_{n,T}$ is given in Theorem 1.6.1.

Now let $n, T \to \infty$ such that $\theta = 2T/n \to \lambda$, $0 < \lambda < 1$. According to Theorem.1.4.3, under these conditions,

$$F_{n,N} = \frac{n^{2T}\sqrt{1 - \lambda}}{2^T T!}(1 + o(1)).$$

(1.8.6)

If $n, T \to \infty$, $2T/n \to \lambda$, $0 < \lambda < 1$, and $m = o(n)$, then by Theorem 1.4.3,

$$F_{n-m,N} = \frac{(n - m)^{2(T-m)}\sqrt{1 - \lambda}}{2^{T-m}(T - m)!}(1 + o(1)).$$

(1.8.7)

Since $\theta = 2T/n \to \lambda$, $0 < \lambda < 1$, implies $2(T - m)/(n - m) \le \theta$, there exists a constant c such that

$$F_{n-m,N} \le \frac{c(n - m)^{2(T-m)}}{2^{T-m}(T - m)!}.$$

(1.8.8)

In subsequent proofs, we will use a cumbersome technical estimate given in the following lemma.

Lemma 1.8.1. *Let* $n, T \to \infty$ *and let there be constants* λ_0 *and* λ_1 *such that* $0 < \lambda_0 \le \theta = 2T/n \le \lambda_1 < 1$. *Then*

$$
c_{n,T}(m) = e^{2Tm/n}\left(1 - \frac{1}{T}\right)\cdots\left(1 - \frac{m-1}{T}\right)\left(1 - \frac{m}{n}\right)^{2(T-m)}
$$

$$
\times \left(1 - \frac{1}{n}\right)\cdots\left(1 - \frac{m-1}{n}\right) \le 1, \tag{1.8.9}
$$

where $m_0 \le m \le T$ *and* m_0 *is sufficiently large.*

Proof. Write the logarithm of $c_{n,T}(m)$ as

$$
\log c_{n,T}(m) = \frac{2Tm}{n} + \sum_{i=1}^{m-1}\log\left(1 - \frac{i}{T}\right) + 2(T-m)\log\left(1 - \frac{m}{n}\right)
$$

$$
+ \sum_{i=1}^{m-1}\log\left(1 - \frac{i}{n}\right)
$$

$$
= -\sum_{k=1}^{\infty}\frac{1}{kT^k}\sum_{i=1}^{m-1}i^k - \sum_{k=2}^{\infty}\frac{2T}{k}\left(\frac{m}{n}\right)^k
$$

$$
+ \sum_{k=1}^{\infty}\frac{2m}{k}\left(\frac{m}{n}\right)^k - \sum_{k=1}^{\infty}\frac{1}{kn^k}\sum_{i=1}^{m-1}i^k.
$$

Using

$$
\sum_{i=1}^{m-1}i^k \ge \frac{(m-1)^{k+1}}{k+1},
$$

we obtain the estimate

$$
\log c_{n,T}(m) \le \sum_{k=1}^{\infty}\frac{2m}{k}\left(\frac{m}{n}\right)^k - \sum_{k=1}^{\infty}\frac{2T}{k+1}\left(\frac{m}{n}\right)^{k+1}
$$

$$
- \sum_{k=1}^{\infty}\frac{(m-1)^{k+1}}{k(k+1)T^k} - \sum_{k=1}^{\infty}\frac{(m-1)^{k+1}}{k(k+1)n^k}
$$

$$
= \sum_{k=1}^{\infty}\left(\frac{m^{k+1}}{k(k+1)n^k}\left(2(k+1) - \frac{2Tk}{n}\right)\right.
$$

$$
\left. - \left(1 - \frac{1}{m}\right)^{k+1}\frac{n^k}{T^k} - \left(1 - \frac{1}{m}\right)^{k+1}\right).
$$

To prove the assertion of the lemma, we note that for sufficiently large m,

$$c_k = 2(k+1) - \theta k - \left(1 - \frac{1}{m}\right)^{k+1} \frac{2^k}{\theta^k} - \left(1 - \frac{1}{m}\right)^{k+1} \leq 0$$

for all k. Indeed, since $0 < \lambda_0 \leq a \leq \lambda_1 < 1$, for sufficiently large m,

$$\left(1 - \frac{1}{m}\right)^{k+1} \frac{2^k}{\theta^k} \geq 2^k, \quad k \geq 1,$$

and therefore

$$c_k \leq 2(k+1) - 2^k,$$

which implies that $c_k \leq 0$ for all $k \geq 3$ and sufficiently large m. In addition,

$$c_1 = 4 - \theta - \left(1 - \frac{1}{m}\right)^2 \frac{2}{\theta} - \left(1 - \frac{1}{m}\right)^2 \leq 3 - \theta - \frac{2}{\theta} + \frac{4}{\theta m} + \frac{2}{m},$$

$$c_2 = 6 - 2\theta - \left(1 - \frac{1}{m}\right)^3 \frac{4}{\theta^2} - \left(1 - \frac{1}{m}\right)^3 \leq 5 - 2\theta - \frac{4}{\theta^2} + \frac{4}{\theta^2 m} + \frac{4}{m},$$

and $c_1 < 0$, $c_2 < 0$ for sufficiently large m, since for $0 < \lambda_0 \leq \theta \leq \lambda_1 < 1$,

$$3 - \theta - \frac{2}{\theta} < 0, \qquad 5 - 2\theta - \frac{4}{\theta^2} < 0.$$

∎

Let $b_{n,i}$ be the number of connected unicyclic graphs with n vertices that belong to $A_{n,T}^{(i)}$, $i = 1, 2, 3$. If this specification is of no significance, we write b_n for the number of connected unicyclic graphs. Let $a_{n,T}(k)$ be the number of graphs in $A_{n,T}$ with exactly k cycles. It is clear that

$$a_{n,T}(k) = \frac{1}{k!} \sum_{m=k}^{\infty} \binom{n}{m} F_{n-m,N} \sum_{m_1 + \cdots + m_k = m} \frac{m! \, b_{m_1} \cdots b_{m_k}}{m_1! \cdots m_k!}. \qquad (1.8.10)$$

As in Section 1.7, let

$$B(x) = \sum_{n=1}^{\infty} \frac{b_n x^n}{n!},$$

$$(1.8.11)$$

$$B_i(x) = \sum_{n=1}^{\infty} \frac{b_{n,i} x^n}{n!}, \quad i = 1, 2, 3,$$

and set

$$x = \theta e^{-\theta}.$$

For such x, according to (1.7.1) and (1.7.2),

$$B_1(x) = -\tfrac{1}{2}\log(1-\theta) + \theta - \tfrac{1}{4}(1-(1-\theta)^2)$$
$$= -\tfrac{1}{2}\log(1-\theta) + \tfrac{1}{2}\theta + \tfrac{1}{4}\theta^2,$$

$$B_2(x) = -\tfrac{1}{2}\log(1-\theta) - \tfrac{1}{4}(1-(1-\theta)^2)$$
$$= -\tfrac{1}{2}\log(1-\theta) - \tfrac{1}{2}\theta + \tfrac{1}{4}\theta^2,$$

$$B_3(x) = -\tfrac{1}{2}\log(1-\theta) - \theta + \tfrac{1}{4}(1-(1-\theta)^2)$$
$$= -\tfrac{1}{2}\log(1-\theta) - \tfrac{1}{2}\theta - \tfrac{1}{4}\theta^2.$$

Theorem 1.8.2. *If $n, T \to \infty$ such that $\theta = 2T/n \to \lambda$, $0 < \lambda < 1$, then for any $i = 1, 2, 3$ and any fixed $k = 0, 1, \ldots,$*

$$a_{n,T}^{(i)}(k) = \frac{n^{2T}\sqrt{1-\lambda}\Lambda_i^k}{2^T T! k!}(1+o(1)),$$

where $a_{n,T}^{(i)}(k)$ is the number of graphs in $A_{n,T}^{(i)}$ with exactly k cycles, and

$$\Lambda_1 = -\frac{1}{2}\log(1-\lambda) + \frac{\lambda}{2} + \frac{\lambda^2}{4},$$

$$\Lambda_2 = -\frac{1}{2}\log(1-\lambda) - \frac{\lambda}{2} + \frac{\lambda^2}{4},$$

$$\Lambda_3 = -\frac{1}{2}\log(1-\lambda) - \frac{\lambda}{2} - \frac{\lambda^2}{4}.$$

Proof. We partition the first sum of (1.8.10) into two parts, S_1 and S_2. We set $M = T^{1/4}$ and include in S_1 the summands with $m < M$. For any x from the convergence domain of the series (1.8.11), the estimate

$$\sum_{m \geq M} \sum_{m_1 + \cdots + m_k = m} \frac{b_{m_1} x^{m_1} \cdots b_{m_k} x^{m_k}}{m_1! \cdots m_k!} \leq (B(x))^{k-1} \sum_{m \geq M/k} \frac{b_m x^m}{m!} \quad (1.8.12)$$

holds. As in Section 1.7, let d_n be the number of connected graphs of single-valued mappings of a set with n elements into itself and let

$$d(x) = \sum_{m=1}^{\infty} \frac{d_m x^m}{m!}.$$

Since

$$b_m \leq d_m = (m-1)! \sum_{k=0}^{m-1} \frac{m^k}{k!} \leq (m-1)! \, e^m$$

(see the proof of Lemma 1.7.5), the estimate

$$\sum_{m \geq M/k} \frac{b_m x^m}{m!} \leq \sum_{m \geq M/k} (ex)^m$$

holds. Recall that we chose $x = \theta e^{-\theta}$. According to the hypothesis of the theorem, $\theta = 2T/n \to \lambda, 0 < \lambda < 1$, and there exists $q < 1$ such that $ex = \theta e^{1-\theta} \leq q < 1$ beginning with some n. Therefore

$$\sum_{m \geq M/k} \frac{b_m x^m}{m!} \leq \frac{1}{1-q} q^{M/k}. \tag{1.8.13}$$

Taking into account estimates (1.8.8), (1.8.9), and Lemma 1.8.1, we find that

$$S_2 = \frac{1}{k!} \sum_{m \geq M} \sum_{m_1 + \cdots + m_k = m} \binom{n}{m} F_{n-m,N} \frac{m! \, b_{m_1} \cdots b_{m_k}}{m_1! \cdots m_k!}$$

$$\leq \frac{c}{k!} \sum_{m \geq M} \sum_{m_1 + \cdots + m_k = m} \frac{n! \, (n-m)^{2(T-m)} b_{m_1} \cdots b_{m_k}}{(n-m)! \, 2^{T-m} (T-m)! \, m_1! \cdots m_k!}$$

$$\leq \frac{cn^{2T}}{k! \, 2^T T!} \sum_{m \geq M} \sum_{m_1 + \cdots + m_k = m} \theta^m \left(1 - \frac{1}{n}\right) \cdots \left(1 - \frac{m-1}{n}\right)$$

$$\times \left(1 - \frac{m}{n}\right)^{2T-2m} \left(1 - \frac{1}{T}\right) \cdots \left(1 - \frac{m-1}{T}\right) \frac{b_{m_1} \cdots b_{m_k}}{m_1! \cdots m_k!}$$

$$\leq \frac{cn^{2T}}{k! \, 2^T T!} \sum_{m \geq M} \sum_{m_1 + \cdots + m_k = m} (\theta e^{-\theta})^m \frac{b_{m_1} \cdots b_{m_k}}{m_1! \cdots m_k!}$$

$$\leq \frac{c_1 n^{2T}}{k! \, 2^T T!} (B(x))^{k-1} \sum_{m \geq M/k} \frac{b_m x^m}{m!}$$

$$\leq \frac{c_2 n^{2T}}{k! \, 2^T T! \, (1-q)} q^{T^{1/4}/k},$$

where c_1, c_2 are some constants. Thus, under the conditions of the theorem,

$$S_2 = o\left(n^{2T}/(2^T T!)\right).$$

We now estimate the sum S_1. According to (1.8.8),

$$T! \, F_{n-m,N} = \frac{T! \, (n-m)^{2(T-m)} \sqrt{1-\lambda}}{2^{T-m} (T-m)!} (1 + o(1))$$

$$= \frac{n^{2T} x^m \sqrt{1-\lambda}}{2^T n^m} (1 + o(1))$$

uniformly in $m \leq M = T^{1/4}$.

Therefore, for any fixed $k = 1, 2, \ldots$,

$$S_1 = \frac{1}{k!} \sum_{m \le M} \sum_{m_1 + \cdots + m_k = m} m! \binom{n}{m} F_{n-m,N} \frac{b_{m_1} \cdots b_{m_k}}{m_1! \cdots m_k!}$$

$$= \frac{n^{2T} \sqrt{1-\lambda}}{k! \, 2^T \, T!} \sum_{m=k}^{M} \sum_{m_1 + \cdots + m_k = m} \frac{b_{m_1} x^{m_1} \cdots b_{m_k} x^{m_k}}{m_1! \cdots m_k!} (1 + o(1)).$$

Taking into account the estimate of S_2, we obtain

$$S_1 = \frac{n^{2T} \sqrt{1-\lambda}}{k! \, 2^T \, T!}$$

$$\times \left(\sum_{m=k}^{\infty} \sum_{m_1 + \cdots + m_k = m} \frac{b_{m_1} x^{m_1} \cdots b_{m_k} x^{m_k}}{m_1! \cdots m_k!} (1 + o(1)) + o(1) \right)$$

$$= \frac{n^{2T} \sqrt{1-\lambda}}{2^T \, T! \, k!} (B(x))^k (1 + o(1)).$$

Combining the estimates of S_1 and S_2 yields

$$a_{n,T}(k) = \frac{n^{2T} \sqrt{1-\lambda}}{k! \, 2^T \, T!} (B(x))^k (1 + o(1))$$

under the hypothesis of the theorem. Since $x = \theta e^{-\theta} \to \lambda e^{-\lambda}$, we also have

$$B_1(x) \to \Lambda_1, \qquad B_2(x) \to \Lambda_2, \qquad B_3(x) \to \Lambda_3.$$

∎

Theorem 1.8.3.   *If $n, T \to \infty$ such that $\theta = 2T/n \to \lambda$, $0 < \lambda < 1$, then*

$$a_{n,T}^{(i)} = \frac{n^{2T} \sqrt{1-\lambda}}{2^T \, T!} e^{\Lambda_i}, \quad i = 1, 2, 3,$$

where Λ_i, $i = 1, 2, 3$, as in Theorem 1.8.2.

Proof.   To obtain the asymptotics of $a_{n,T}$, we have to estimate the sum

$$a_{n,T} = \sum_{k=0}^{\infty} a_{n,T}(k). \tag{1.8.14}$$

After normalization, we have

$$\left(\frac{n^{2T}}{2^T \, T!} \right)^{-1} a_{n,T} = \sum_{k=0}^{\infty} \left(\frac{n^{2T}}{2^T \, T!} \right)^{-1} a_{n,T}(k), \tag{1.8.15}$$

where for any fixed $k = 0, 1, \ldots$,

$$\left(\frac{n^{2T}}{2^T \, T!} \right)^{-1} a_{n,T}(k) \to \frac{B^k(\lambda e^{-\lambda}) \sqrt{1-\lambda}}{k!}$$

as $n, T \to \infty$, $2T/n \to \lambda$, $0 < \lambda < 1$. We can pass to the limit under the sum in (1.8.15) if the series converges uniformly with respect to the parameters n, T. To see this, it suffices to obtain an estimate

$$\left(\frac{n^{2T}}{2^T T!}\right)^{-1} a_{n,T} \le A_k \qquad (1.8.16)$$

such that the series $\sum_{k=0}^{\infty} A_k$ converges. Using (1.8.8) and (1.8.9) and reasoning as we did in the proof of the estimate of S_2 give

$$a_{n,T}(k) = \frac{1}{k!} \sum_{m=1}^{n} \sum_{m_1+\cdots+m_k=m} m! \binom{n}{m} F_{n-m,N} \frac{b_{m_1} \cdots b_{m_k}}{m_1! \cdots m_k!}$$

$$\le \frac{cn^{2T}}{k! \, 2^T T!} \sum_{m=k}^{\infty} \sum_{m_1+\cdots+m_k=m} \frac{x^m b_{m_1} \cdots b_{m_k}}{m_1! \cdots m_k!}$$

$$= \frac{cn^{2T} (B(k))^k}{2^T T! \, k!}.$$

Thus we have an estimate of the form (1.8.16) and can pass to the limit under the sum in (1.8.15) to obtain

$$a_{n,T} = \frac{n^{2T} \sqrt{1-\lambda}}{2^T T!} \exp\left\{B\left(\lambda e^{-\lambda}\right)\right\}(1 + o(1)).$$

Depending on the set of graphs under consideration, replace $B(x)$ with $B_1(x)$, $B_2(x)$, or $B_3(x)$, and Theorem 1.8.3 is proved. ∎

A random graph from $\mathcal{A}_{n,T}$ has exactly $N = n - T$ trees and a random number of unicyclic components. We denote by $\varkappa_{n,T}^{(i)}$ the number of unicyclic components in a random graph from $\mathcal{A}_{n,T}^{(i)}$, $i = 1, 2, 3$.

Theorem 1.8.4. *If $n, T \to \infty$ such that $\theta = 2T/n \to \lambda$, $0 < \lambda < 1$, then for any $i = 1, 2, 3$ and for any fixed $k = 0, 1, \ldots$,*

$$\mathbf{P}\{\varkappa_{n,T}^{(i)} = k\} = \frac{\Lambda_i^k e^{-\Lambda_i}}{k!}(1 + o(1)),$$

where the Λ_i are as in Theorem 1.8.2.

Proof. The assertions of the theorem follow from Theorems 1.8.2 and 1.8.3, since

$$\mathbf{P}\{\varkappa_{n,T}^{(i)} = k\} = a_{n,T}^{(i)}(k) / a_{n,T}^{(i)}.$$

∎

Now we consider the case $\theta = 2T/n \to 1$. Let $\omega_{n,T}^{(i)}$ be the number of vertices that lie in the unicyclic components of a random graph from $\mathcal{A}_{n,T}^{(i)}$, $i = 1, 2, 3$. It is clear that if we know the distribution of a characteristic of the random graph under the condition $\{\omega_{n,T}^{(i)} = m\}$, the unconditional distribution can be obtained by averaging over the distribution of $\omega_{n,T}^{(i)}$.

Theorem 1.8.5. *If $n, T \to \infty$ such that $\varepsilon = 1 - 2T/n \to 0$ and $\varepsilon^3 n \to \infty$, then for any $i = 1, 2, 3$,*

$$\frac{2}{\varepsilon^2} P\{\omega_{n,T}^{(i)} = m\} = \frac{1}{\Gamma(1/4)} y^{-3/4} e^{-y}(1 + o(1))$$

uniformly with respect to m such that $y = \varepsilon^2 m/2$ lies in any fixed interval of the form $0 < y_0 \le y \le y_1 < \infty$, and there exists a constant A such that, for all m,

$$\frac{2}{\varepsilon^2} P\{\omega_{n,T}^{(i)} = m\} \le A y^{-3/4} e^{-y}.$$

Proof. We denote the number of graphs in $A_{n,T}^{(i)}$ by $a_{n,T}^{(i)}$ and the number of graphs for which $\omega_{n,T}^{(i)} = m$ by $a_{n,T,m}^{(i)}$. Clearly,

$$a_{n,T}^{(i)} = \sum_{m=0}^{\infty} a_{n,T,m}^{(i)}, \tag{1.8.17}$$

$$a_{n,T,m}^{(i)} = \binom{n}{m} u_m^{(i)} F_{n-m,N}. \tag{1.8.18}$$

We decompose the sum in (1.8.17) into two parts. Let $0 < y_0 < y_1 < \infty$, $y = \varepsilon^2 m/2$, and

$$S_1 = \sum_{m: y \in [y_0, y_1]} a_{n,T,m}^{(i)}, \qquad S_2 = \sum_{m=0}^{\infty} a_{n,T,m}^{(i)} - S_1.$$

By Theorem 1.7.1 and the equalities (1.7.21),

$$u_m^{(i)} = A_i m^{m-1/4}(1 + o(1)) \tag{1.8.19}$$

uniformly in m in the region $y_0 \le y \le y_1$, where A_i, $i = 1, 2, 3$, are defined in (1.7.22). There exists a constant c_1 such that, for all m,

$$u_m^{(i)} \le c_1 m^{m-1/4}. \tag{1.8.20}$$

To estimate $F_{n-m,N}$, it is convenient to use the intermediate formula (1.4.25). From (1.4.26) and the equality

$$\theta(2 - \theta) = 1 - \varepsilon^2,$$

we have

$$F_{n-m,N} = \frac{(n-m)!\,(1 - \varepsilon^2)^N}{N!\,x^{n-m}} P\{\zeta_N = n - m\}, \tag{1.8.21}$$

where, according to Theorem 1.4.1,

$$\mathbf{P}\{\zeta_N = k\} = \frac{1}{\sigma\sqrt{2\pi N}} e^{-u^2/2}(1 + o(1))$$

uniformly in k such that $u = (k - N\mu)/(\sigma\sqrt{N})$ lies in any finite interval,

$$\mu = \frac{2}{2 - \theta} = \frac{n}{N}, \qquad \sigma^2 = \frac{2(1 - \varepsilon)}{\varepsilon(1 + \varepsilon)^2}.$$

If $\varepsilon \to 0$, $\varepsilon^3 n \to \infty$, and $m\sqrt{\varepsilon/n} \to 0$, then for $k = n - m$,

$$u = \frac{(k - N\mu)}{\sigma\sqrt{N}} = -\frac{m}{\sigma\sqrt{N}} = -\frac{m\sqrt{\varepsilon}}{\sqrt{n}}(1 + o(1)).$$

Consequently,

$$\mathbf{P}\{\zeta_N = n - m\} = \frac{1}{\sigma\sqrt{2\pi N}}(1 + o(1)). \tag{1.8.22}$$

It follows from (1.8.21) and (1.8.22) that

$$F_{n-m,N} = \frac{(n - m)! \left(1 - \varepsilon^2\right)^N e^{n(1-\varepsilon)}\sqrt{\varepsilon}}{2^N N! (1 - \varepsilon)^n \sqrt{2\pi n}}(1 - \varepsilon)^m e^{-m(1-\varepsilon)}(1 + o(1)). \tag{1.8.23}$$

There exists a constant c such that

$$\sigma\sqrt{2\pi N}\mathbf{P}\{\zeta_N = k\} \le c;$$

therefore

$$F_{n-m,N} \le \frac{c(n - m)! \left(1 - \varepsilon^2\right)^N e^{n(1-\varepsilon)}\sqrt{\varepsilon}}{2^N N! (1 - \varepsilon)^n \sqrt{2\pi n}}(1 - \varepsilon)^m e^{-m(1-\varepsilon)} \tag{1.8.24}$$

for all m, $0 \le m \le T$. We note that as $\varepsilon \to 0$,

$$(1 - \varepsilon)^m e^{\varepsilon m} = e^{-y}(1 + o(1))$$

uniformly in m such that $y_0 \le y \le y_1$, and for all m,

$$(1 - \varepsilon)^m e^{\varepsilon m} \le e^{-y}.$$

Clearly, (1.8.23) holds uniformly in m such that $y = \varepsilon^2 m/2$ lies in the interval $[y_0, y_1]$. Therefore, if $n \to \infty$, $\varepsilon = 1 - 2T/n \to 0$, and $\varepsilon^3 n \to \infty$, then

$$F_{n-m,N} = f_n(n - m)! e^{-m-y}(1 + o(1)) \tag{1.8.25}$$

uniformly in m such that $y_0 \le y \le y_1$, where

$$f_n = \frac{\left(1 - \varepsilon^2\right)^N e^{n(1-\varepsilon)}\sqrt{\varepsilon}}{2^N N! (1 - \varepsilon)^n \sqrt{2\pi n}},$$

and there exists a constant A_0 such that for all m,

$$F_{n-m,N} \leq A_0 f_n (n-m)! \, e^{-m-y}. \tag{1.8.26}$$

Therefore, by (1.8.18), (1.8.19), and (1.8.25), we have the equality

$$a_{n,T,m}^{(i)} = n! \, A_i f_n \frac{m^{m-1/4} e^{-m-y}}{m!} (1 + o(1)) \tag{1.8.27}$$

$$= n! \, A_i f_n \frac{2^{1/4} \Gamma(1/4)}{\sqrt{\varepsilon}} \frac{1}{\Gamma(1/4)} y^{-3/4} e^{-y} \frac{\varepsilon^2}{2} (1 + o(1)),$$

which holds uniformly in m such that $y_0 \leq y \leq y_1$; and outside of this domain, by (1.8.18), (1.8.20), and (1.8.26), we have

$$a_{n,T,m}^{(i)} \leq \frac{An! \, f_n}{\sqrt{\varepsilon} \Gamma(1/4)} y^{-3/4} e^{-y} \frac{\varepsilon^2}{2}, \tag{1.8.28}$$

where A is a constant. The sum

$$\sum_{m : y \in [y_0, y_1]} \frac{1}{\Gamma(1/4)} y^{-3/4} e^{-y} \frac{\varepsilon^2}{2}$$

is the integral sum of the function $(\Gamma(1/4))^{-1} z^{-3/4} e^{-z}$ with step $\varepsilon^2/2$. Therefore, by choosing y_0 small enough and y_1 and n large enough, this sum can be made arbitrarily close to 1, and the sum for remaining values of m can be made arbitrarily small. Thus

$$a_{n,T}^{(i)} = n! \, A_i f_n \frac{2^{1/4} \Gamma(1/4)}{\sqrt{\varepsilon}} (1 + o(1)).$$

Now it follows from (1.8.27) and (1.8.28) that

$$P\{\omega_{n,T}^{(i)} = m\} = \frac{a_{n,T,m}^{(i)}}{a_{n,T}^{(i)}} = \frac{\varepsilon^2}{2\Gamma(1/4)} y^{-3/4} e^{-y} (1 + o(1))$$

uniformly in m such that $y_0 \leq y \leq y_1$ and that outside this domain,

$$P\{\omega_{n,T}^{(i)} = m\} \leq \frac{A\varepsilon^2}{2} y^{-3/4} e^{-y}.$$

This completes the proof of the theorem.   ∎

When we substitute the exact expressions for A_i and f_n, we obtain for $i = 1, 2, 3$,

$$a_{n,T}^{(i)} = C_i \frac{n! \left(1 - \varepsilon^2\right)^N e^{n(1-\varepsilon)}}{2^N N! \, (1 - \varepsilon)^n \sqrt{2\pi n}} (1 + o(1)), \tag{1.8.29}$$

where $C_1 = e^{3/4}$, $C_2 = e^{-1/4}$, and $C_3 = e^{-3/4}$. It is easy to confirm that if

$\varepsilon = 1 - 2T/n \to 0$, then

$$\frac{n! \left(1 - \varepsilon^2\right)^N e^{n(1-\varepsilon)}}{2^N N! (1 - \varepsilon)^n \sqrt{2\pi n}} = \frac{n^{2T}}{2^T T!} (1 + o(1)).$$

Thus, under the conditions of Theorem 1.8.5, the asymptotic formulas

$$a_{n,T}^{(i)} = \frac{C_i n^{2T}}{2^T T!} (1 + o(1)), \quad i = 1, 2, 3,$$

are valid.

Let $\kappa_{n,T}$ denote the number of unicyclic components in a random graph from $\mathcal{A}_{n,T}$ and use $\beta_{n,T}$ to denote the number of vertices in the maximal unicyclic component.

Theorem 1.8.6. *If $n, T \to \infty$ such that $\varepsilon = 1 - 2T/n \to 0$ and $\varepsilon^3 n \to \infty$, then for any fixed x,*

$$\mathbf{P}\left\{\kappa_{n,T} + \frac{1}{2} \log \varepsilon \leq x \sqrt{-\frac{1}{2} \log \varepsilon}\right\} \to \Phi(x) = \frac{1}{\sqrt{2\pi}} \int_{-\infty}^{x} e^{-u^2/2}\, du.$$

Proof. For any fixed x,

$$\mathbf{P}\left\{\kappa_{n,T} + \frac{1}{2} \log \varepsilon \leq x \sqrt{-\frac{1}{2} \log \varepsilon}\right\}$$

$$= \sum_{m=0}^{\infty} \mathbf{P}\{\omega_{n,T} = m\} \mathbf{P}\left\{\varkappa_m + \frac{1}{2} \log \varepsilon \leq x \sqrt{-\frac{1}{2} \log \varepsilon}\right\},$$

where $\varkappa_m$ is the number of components in a random graph from $\mathcal{U}_m$ discussed in Section 1.7. By Theorem 1.7.2, the random variable

$$\left(\varkappa_m - \frac{1}{4} \log m\right) \bigg/ \sqrt{\frac{1}{4} \log m}$$

is asymptotically normal with parameters $(0, 1)$.

Let $y = \varepsilon^2 m/2$ and $0 < y_0 \leq y \leq y_1 < \infty$. Then $\log m = \log(2y) - 2 \log \varepsilon$. Further, since $\varepsilon \to 0$,

$$\mathbf{P}\left\{\varkappa_m + \frac{1}{2} \log \varepsilon \leq x \sqrt{-\frac{1}{2} \log \varepsilon}\right\} \to \Phi(x)$$

uniformly in m such that $y \in [y_0, y_1]$ and does not depend on m asymptotically. In view of Theorem 1.8.5, by choosing y_0 small enough and y_1 and n large enough, the sum

$$\sum_{m:y\in[y_0,y_1]} \mathbf{P}\{\omega_{n,T} = m\}$$

can be made arbitrarily close to 1. Therefore

$$P\left\{\kappa_{n,T} + \frac{1}{2}\log\varepsilon \leq x\sqrt{-\frac{1}{2}\log\varepsilon}\right\} \to \Phi(x)$$

for any fixed x. ∎

Consider now the maximum size of the unicyclic components. Recall that in Section 1.7 we introduced $W_s(z, \gamma)$, setting $W_0(z, \gamma) = 1$, and

$$W_s(z, \gamma) = \int_{X_s(z,\gamma)} \frac{dx_1 \cdots dx_s}{x_1 \cdots x_s(z - x_1 - \cdots - x_s)^{3/4}},$$

where

$$X_s(z, \gamma) = \{x_i \geq \gamma,\ i = 1, \ldots, s,\ x_1 + \cdots + x_s \leq z\}, \quad s = 1, 2, \ldots.$$

Theorem 1.8.7. *If $n, T \to \infty$ such that $\varepsilon = 1 - 2T/n \to 0$ and $\varepsilon^3 n \to \infty$, then for any fixed $\gamma > 0$,*

$$P\{\varepsilon^2\beta_{n,T} \leq \gamma\} \to \sum_{s=0}^{\infty} \frac{(-1)^s}{4^s s!} Z_s(\gamma),$$

where

$$Z_s(\gamma) = \frac{1}{\Gamma(1/4)} \int_0^{\infty} y^{-3/4} e^{-y} W_s\left(1, \frac{\gamma}{2y}\right) dy, \quad s = 0, 1, \ldots.$$

Proof. For any fixed $\gamma > 0$,

$$P\{\varepsilon^2\beta_{n,T} \leq \gamma\} = \sum_{m=0}^{\infty} P\{\omega_{n,T} = m\} P\{\varepsilon^2\beta_m \leq \gamma\},$$

where β_m is the maximum size of the components in a random graph from $\mathcal{U}_m$ studied in Section 1.7. If $y = \varepsilon^2 m/2$ and $y \in [y_0, y_1]$, then

$$P\{\varepsilon^2\beta_m \leq \gamma\} = P\left\{\beta_m \leq \frac{\gamma}{2y}m\right\}.$$

By Theorem 1.7.3,

$$P\left\{\beta_m \leq \frac{\gamma}{2y}m\right\} = \sum_{s=0}^{\infty} \frac{(-1)^s}{4^s s!} W_s\left(1, \frac{\gamma}{2y}\right) + o(1).$$

It is clear that this holds uniformly in m such that $y \in [y_0, y_1]$. Choosing a small enough y_0 and a large enough y_1 and averaging over the distribution of $\omega_{n,T}$ prove Theorem 1.8.7. ∎

The number of trees in any graph of $\mathcal{A}_{n,T}$ is $N = n - T$. Let $\eta_{n,T}$ be the maximum size of trees in a random graph from $\mathcal{A}_{n,T}$.

Theorem 1.8.8.   *If n, $T \to \infty$ such that $\varepsilon = 1 - 2T/n \to 0$ and $\varepsilon^3 n \to \infty$, then*

$$P\{\beta\eta_{n,T} - u \le z\} \to e^{-e^{-z}},$$

where $\beta = -\log(\theta e^{-\theta})$, $\theta = 2T/n$, and u is the root of the equation

$$\left(\frac{2}{\pi}\right)^{1/2} N\beta^{3/2} = u^{5/2}e^u. \tag{1.8.30}$$

Proof.   It is clear that

$$P\{\beta\eta_{n,T} - u \le z\} = \sum_{m=0}^{\infty} P\{\omega_{n,T} = m\}P\{\beta\eta_{n-m,T-m} - u \le z\}. \tag{1.8.31}$$

Let $v = \varepsilon^3 n$. It is easily seen that, under the conditions of Theorem 1.8.8, the root of equation (1.8.30) can be written as

$$u = \log v - \tfrac{5}{2}\log\log v - \log 4\sqrt{\pi} + o(1). \tag{1.8.32}$$

Let $y = \varepsilon^2 m/2$ lie in a finite interval $0 < y_0 \le y \le y_1 < \infty$. Set

$$\theta_m = \frac{2(T - m)}{n - m}, \qquad \varepsilon_m = 1 - \frac{2(T - m)}{n - m},$$

$$v_m = \varepsilon_m^3 n, \qquad \beta(m) = -\log(\theta_m e^{\theta_m}).$$

Since $\varepsilon_m = \varepsilon(1 + o(1))$, it follows from (1.8.32) that the root of the equation

$$\left(\frac{2}{\pi}\right)^{1/2} N(\beta(m))^{3/2} = u^{5/2}e^u$$

can be written as

$$u_m = \log v_m - \tfrac{5}{2}\log\log v_m - \log 4\sqrt{\pi} + o(1) = u + o(1)$$

uniformly in y in any fixed interval $[y_0, y_1]$. Therefore, by applying Theorem 1.6.3, we obtain

$$P\{\beta\eta_{n-m,T-m} - u \le z\} \to e^{-e^{-z}} \tag{1.8.33}$$

uniformly in $y \in [y_0, y_1]$. In the main part of the sum in (1.8.31), this probability does not depend on m asymptotically. Therefore, averaging (1.8.33) over the distribution of $\omega_{n,T}$ proves Theorem 1.8.8. ∎

When we compare Theorems 1.8.7 and 1.8.8, we see that the maximum size of trees in a random graph from $\mathcal{A}_{n,T}$ is greater than the maximum size of the unicyclic components, since $\beta = \varepsilon^2/2(1 + o(1))$ and $u \to \infty$. Let $\alpha_{n,T}$ be the maximum size of components of a random graph from $\mathcal{A}_{n,T}$, that is,

$$\alpha_{n,T} = \max(\beta_{n,T}, \eta_{n,T}).$$

Averaging over the distribution of $\omega_{n,T}$ gives the following theorem.

Theorem 1.8.9. *If $n, T \to \infty$ such that $\varepsilon = 1 - 2T/n \to 0$ and $\varepsilon^3 n \to \infty$, then for any fixed z,*

$$\mathbf{P}\{\beta\alpha_{n,T} - u \leq z\} \to e^{-e^{-z}},$$

where $\beta = -\log(\theta e^{-\theta})$, $\theta = 2T/n$, and u is the root of the equation

$$\left(\frac{2}{\pi}\right)^{1/2} N\beta^{2/3} = u^{5/2}e^u.$$

To conclude this section, we consider the case where $n, T \to \infty$ such that $\varepsilon^3 n$ tends to a constant.

Theorem 1.8.10. *If $n, T \to \infty$ such that $\varepsilon n^{1/3} \to 2 \cdot 3^{-2/3}v$, where $\varepsilon = 1 - 2T/n$ and v is a constant, then for any $i = 1, 2, 3$,*

$$a_{n,T}^{(i)} = \frac{c_i n! \, e^n}{2^N N! \sqrt{N}} p(v)(1 + o(1)),$$

where

$$c_1 = \frac{\sqrt{3}e^{3/4}}{2\sqrt{2}\Gamma(1/4)}, \qquad c_2 = \frac{\sqrt{3}e^{-1/4}}{2\sqrt{2}\Gamma(1/4)}, \qquad c_3 = \frac{\sqrt{3}e^{-3/4}}{2\sqrt{2}\Gamma(1/4)},$$

$$p(v) = \int_0^\infty y^{-3/4} p(-v - y; 3/2, -1)\, dy,$$

and $p(u; 3/2, -1)$ is the density of the stable law defined by (1.4.18).

Proof. We again use

$$a_{n,T} = \sum_{m=0}^{T} \binom{n}{m} u_m F_{n-m,N}. \tag{1.8.34}$$

According to Theorem 1.7.1, as $m \to \infty$,

$$u_m = Am^{m-1/4}(1 + o(1)), \tag{1.8.35}$$

where the value of the coefficient A depends on the type of the unicyclic components in $\mathcal{A}_{n,T}$, and

$$A_1 = \frac{\sqrt{2\pi}e^{3/4}}{2^{1/4}\Gamma(1/4)}, \qquad A_2 = \frac{\sqrt{2\pi}e^{-1/4}}{2^{1/4}\Gamma(1/4)}, \qquad A_3 = \frac{\sqrt{2\pi}e^{-3/4}}{2^{1/4}\Gamma(1/4)}.$$

To estimate $F_{n-m,N}$, we use formula (1.4.25) with $\theta = 1$. Then

$$F_{n-m,N} = \frac{(n-m)!}{2^N N! \, e^{-n+m}} \mathbf{P}\{\zeta_N = n - m\}, \tag{1.8.36}$$

where $\zeta_N = \xi_1 + \cdots + \xi_N$ is a sum of independent random variables with distribution (1.4.19):

$$\mathbf{P}\{\xi_1 = k\} = \frac{2k^{k-2}e^{-1}}{k!}, \quad k = 1, 2, \ldots.$$

By Theorem 1.4.2,

$$bN^{2/3}\mathbf{P}\{\zeta_N = k\} = p(u; 3/2, -1)(1 + o(1))$$

uniformly in k such that $u = (k - 2N)/(bN^{2/3})$ lies in any fixed finite interval.

Under the conditions of Theorem 1.8.9,

$$(n - 2N)/\big(bN^{2/3}\big) \to -v.$$

Let $y = m/(bN^{2/3})$ and $0 < y_0 \le y \le y_1 < \infty$. Then, under the conditions of the theorem,

$$u = \frac{(n - m - 2N)}{bN^{2/3}} \to -v - y.$$

Thus, by (1.8.36),

$$F_{n-m,N} = \frac{(n-m)! \, p(-v - y; 3/2, -1)}{2^N N! \, e^{-n+m} bN^{2/3}}(1 + o(1)) \tag{1.8.37}$$

uniformly in m such that $y \in [y_0, y_1]$. Since $b = 2(2/3)^{2/3}$, from (1.8.35) and (1.8.37), we obtain

$$a_{n,T,m} = \binom{n}{m} u_m F_{n-m,N}$$

$$= \frac{An! \, m^{m-1/4} p(-v - y; 3/2, -1)}{m! \, 2^N N! \, e^{-n+m} bN^{2/3}}(1 + o(1))$$

$$= \frac{An! \, e^N p(-v - y; 3/2, -1)}{2NN! \, \sqrt{2\pi} m^{3/4} bN^{2/3}}(1 + o(1))$$

$$= \frac{An! \, e^n \sqrt{3}}{2^{3/4} 2^N N! \, \sqrt{2\pi N}} y^{-3/4} p(-v - y; 3/2, -1) \frac{1}{bN^{2/3}}(1 + o(1))$$

uniformly in m such that $y \in [y_0, y_1]$. To obtain $a_{n,T}$, we need to carry out the summation in (1.8.35). If we choose a small enough y_0 and a large enough y_1, substitute the expression of $a_{n,T,m}$ into (1.8.34), note that the obtained sum is the integral sum of the function $z^{-3/4} p(-v - y; 3/2, -1)$ with step $b^{-1}n^{-2/3}$, and omit the needed estimation of the tails, we have

$$a_{n,T} = \frac{cn! \, e^n}{2^N N! \, \sqrt{N}} \int_0^\infty y^{-3/4} p(-v - y; 3/2, -1) \, dy(1 + o(1)),$$

where

$$c = \frac{A\sqrt{3}}{2^{3/4}\sqrt{2\pi}}.$$

Recall our convention that if we consider the set $\mathcal{A}_{n,T}^{(i)}$, then A is replaced by A_i, $i = 1, 2, 3$.   ∎

It follows from Theorem 1.8.10 that the number $\omega_{n,T}$ of the vertices that form the unicyclic components in a random graph of $\mathcal{A}_{n,T}$ has the following limit distribution:

If $n, T \to \infty$ such that $\varepsilon = 1 - 2T/n \to 0$ and $\varepsilon^3 n \to v$, then

$$bN^{2/3}\mathbf{P}\{\omega_{n,T} = m\} = \frac{1}{p(v)}y^{-3/4}p(-v - y; 3/2, -1)(1 + o(1))$$

uniformly in m such that $y = m/(bN^{2/3})$ lies in any fixed interval of the form $0 < y_0 \le y \le y_1 < \infty$ and $p(v)$ is defined in Theorem 1.8.10.

1.9. Notes and references

In this book, we use a probabilistic approach to combinatorial problems. Section 1.1 provides the results from probability theory that suffice for the probabilistic analysis presented in the book. All of the results in Section 1.1 can be found in standard treatments of probability theory; however, we follow [76], where these results are given along with full proofs.

A detailed discussion of the saddle-point method can be found in [42]. Theorem 1.1.7 is a simplified version of the corresponding theorem that gives a full asymptotic expansion of $G(\lambda)$.

The proof of the local limit theorem (Theorem 1.1.11) was suggested by B. V. Gnedenko and is contained in the book [49], which remains one of the best textbooks on the limit theorems of probability theory (see also [43, 122, 60]). The approximation of the binomial distribution by the normal and Poisson laws was investigated by Yu. V. Prokhorov [125] (see also [90]). The inequality from Theorem 1.1.16 was proposed by Hoeffding [59] for sums of bounded random variables (see also [122]).

Section 1.2 is devoted to a description of the generalized scheme of allocation of particles, which is a generalization of the multinomial trials. It was introduced in [69] and now has a significant place in probabilistic combinatorics (see also [78]). Successful applications of the generalized scheme are mostly limited to the equiprobable cases; there are only a few examples where a nonequiprobable scheme has a natural combinatorial interpretation. Along with the nonequiprobable multinomial distribution, Example 1.2.3 is an example of a nonequiprobable scheme.

Example 1.2.4 concerns random forests with rooted trees and is related to branching processes. Indeed, the distribution (1.2.11) is that of the total progeny in the Galton–Watson process $\mu(t, G)$, which begins with one particle that has Poisson-distributed numbers of offspring of a particle. Therefore a random forest with N trees and n nonroot vertices can be represented by the same process that begins with N particles under the condition that the total progeny is $n + N$. We describe more precisely the correspondence between random trees and the branching process $\mu(t, G)$, whose distribution of the number of offspring of one particle is the Poisson distribution with parameter λ.

Let $\mu_r(t, G)$ be the number of particles at time t having exactly r direct descendants, and let $\nu(G)$ be the total progeny over the whole period of evolution of the process.

Consider the set T_n of all rooted trees whose nonroot vertices are labeled $1, 2, \ldots, n$, and whose root is labeled by 0. Assigning the probability $(n + 1)^{-n+1}$ to each tree of T_n gives the uniform distribution on T_n.

Any vertex of a tree is joined to the root by a unique path, whose number of edges is called the height of the corresponding vertex. We assume that all the edges of a tree are directed from the root and call the number of edges emanating from a vertex the degree of the vertex.

Let $\mu_r(t, T_n)$, $r, t = 0, 1, \ldots, n$, be the number of vertices of height t having degree r. Consider the matrices $\|\mu_r(t, T_n)\|$ and $\|\mu_r(t, G)\|$, $t, r = 0, 1, \ldots, n$, and a matrix $M = \|m_r(t)\|$ of the same dimension with nonnegative elements. Kolchin [73] showed that

$$\mathbf{P}\{\|\mu_r(t, T_n)\| = M\} = \mathbf{P}\{\|\mu_r(t, G)\| = M \mid \nu(G) = n + 1\}.$$

This relation means that the distribution of any random variable that can be expressed in terms of the random variables $\mu_r(t, T_n)$, $r, t = 0, 1, \ldots, n$, coincides with the conditional distribution of the corresponding random characteristic of the branching process under the condition that $\nu(G) = n + 1$.

This scheme has been used widely to obtain a complete description of the properties of random trees and forests [73, 74, 75, 111, 112, 113, 114, 116]. Recently Yu. L. Pavlov [118, 119] discovered that the branching process that has a geometric distribution of the number of offsprings corresponds – in the same sense as discussed above – to a random plane planted tree with unlabeled vertices. This representation of random plane planted trees is also mentioned in [4, 136, 138]. Note that we are aware of only these two branching processes that have the Poisson and the geometric distributions of the number of offspring, which lead to sets of trees with uniform distribution. Results on more general classes of forests with nonuniform distributions can be found in [120, 121].

The correspondence between random plane planted trees and a branching process that has a geometric distribution appears to be deep and can be considered as a correspondence of realizations, that is, there exists a one-to-one correspondence between the set of such trees and the realizations of the corresponding branching process. It seems that this fact was first pointed out in an explicit form by V. A. Vatutin [138].

The general approach to investigating connectivity and the sizes of components of random graphs of various types is presented in Section 1.3. This general approach was first outlined by Kolchin [78], but its particular forms had already been used to investigate other random graphs, such as random permutations, random mappings, and random forests of rooted trees [71, 72, 73, 74, 75].

Forests of nonrooted trees are investigated in Sections 1.4–1.6. Section 1.4 concerns the number of such forests. The number of forests of N labeled rooted trees with n nonroot vertices is $N(N + n)^{n-1}$. In contrast to the forests of rooted trees, the number $F_{n,N}$ of nonrooted forests cannot be expressed by a simple formula. A complete analysis of the random forests of nonrooted trees was conducted by V. E. Britikov, who used the generalized scheme of allocation. The possibility of using such an approach was pointed out in [78, 77]. When Britikov began investigating $F_{n,N}$, it was known only that for any fixed N as $n \to \infty$,

$$F_{n,N} = \frac{n^{n-2}}{2^{N-1}(N-1)!}(1 + o(1)). \tag{1.9.1}$$

A complete description of the asymptotic behavior of $F_{n,N}$ can be found in [29]. In particular, formula (1.9.1) is generalized for $N \to \infty$ and proves that if $n \to \infty$ and $(1 - 2T/n)^3 n \to -\infty$, then

$$F_{n,N} = \frac{n^{n-2}}{2^{N-1}(N-1)!}\left(\frac{2T}{n} - 1\right)^{-5/2}(1 + o(1)).$$

The cases in which $(1 - 2T/n)^3 n$ tends to a constant and $(1 - 2T/n)^3 n \to \infty$ are covered by Theorems 1.4.4 and 1.4.3, respectively.

Section 1.5 deals with the numbers μ_r of trees with r vertices, $r = 3, 4, \ldots$, in a random forest. A complete description of the limit distributions of these random variables was obtained by Britikov [30]. Theorems 1.5.1 and 1.5.2 summarize the results proved in [30], where, in addition, the behavior of μ_1 and μ_2 is analyzed.

The general approach used to investigate the order statistics in the generalized scheme was suggested in [70] and is also described in Lemma 1.2.2 in [78]. In Section 1.6, we apply this approach to the maximum size of trees in random unrooted forests. The results of this section were obtained by Britikov [28]. Theorems 1.6.1–1.6.5 cover all possible regular variations of the parameters n and N, but not the case where N is bounded. Clearly, for any fixed k, the size of the kth largest tree of the forest can be analyzed in the same way. Luczak and Pittel [101] realized this posibility and interpreted the results of their analysis as an evolution of a random forest (see also [31]).

It is pertinent to note here the results that concern the investigations of the ordered series of components of wide classes of random graphs [4, 7, 14, 15, 35, 36, 41, 56]. There are two natural ways of labeling the components. One way is to arrange them in decreasing order; the other is to use a particular random labeling called the size-biased permutation. For the first type of labeling, let $M_1 \geq M_2 \geq \cdots$ be the sequence of sizes of the components of a graph with n vertices numbered in decreasing order. Let C_1 be the size of the component that contains the vertex with label 1, let C_2 be the size of the component that contains the vertex with the smallest label among the vertices not included in the first component, and so on.

It is clear that the joint distribution of the random variables $C_1, C_2, \ldots$ normalized by n places unit mass on the set Δ of infinite sequences of nonnegative numbers such that

$$\Delta = \{(x_1, x_2, \ldots), x_1 + x_2 + \cdots = 1\},$$

and the joint distribution of $M_1, M_2, \ldots$ normalized by n is concentrated on the set

$$\nabla = \{(x_1, x_2, \ldots) \in \Delta, \ x_1 \geq x_2 \geq \cdots\}.$$

For some classes of graphs, the limit distributions of the sequences $C_1, C_2, \ldots$ and $M_1, M_2, \ldots$ are known. Let us describe a class of the limit distributions.

Let $Z_1, Z_2, \ldots$ be independent identically distributed random variables with density

$$\theta(1 - z)^{\theta - 1}, \quad 0 < z < 1, \quad \theta > 0.$$

Let

$$Y_1 = Z_1, \qquad Y_2 = Z_2(1 - Z_1), \qquad Y_3 = Z_3(1 - Z_1)(1 - Z_2), \ldots$$

and let $Y_{(1)}, Y_{(2)}, \ldots$ be the order statistics constructed from $Y_1, Y_2, \ldots$. The distribution of $Y_1, Y_2, \ldots$ on Δ is called the GEM distribution with parameter θ, and the distribution of $Y_{(1)}, Y_{(2)}, \ldots$ on ∇ is called the Poisson–Dirichlet distribution with parameter θ.

It is known that the distribution of the random variables $M_1, M_2, \ldots$ normalized by n for the cycle sizes of a random permutation of degree n converges, as $n \to \infty$, to the Poisson–Dirichlet distribution with parameter $\theta = 1$ and that the random variable C_1 is uniformly distributed on the set $\{1, \ldots, n\}$ (see, for example [78]). For random mappings, the distributions of the random variables $C_1, C_2, \ldots$ and $M_1, M_2, \ldots$ normalized by n converge, respectively, to the GEM distribution and the Poisson–Dirichlet distribution with parameter $\theta = 1/2$ [3].

As usual, let α_r denote the number of components of size r of a random graph with n vertices. The joint distribution of the random variables $\alpha_1, \ldots, \alpha_n$ of the form

$$\mathbf{P}\{\alpha_1 = a_1, \ldots, \alpha_n = a_n\} = \binom{\theta + n - 1}{n}^{-1} \frac{\theta^{a_1 + \cdots + a_n}}{1^{a_1} 2^{a_2} \cdots n^{a_n} a_1! \cdots a_n!},$$

where $a_1, \ldots, a_n$ are nonnegative integers such that $a_1 + 2a_2 + \cdots + na_n = n$ is similar to the joint distribution of the random variables $\alpha_1, \ldots, \alpha_n$ for a random permutation (see Lemma 1.3.7). This distribution arises frequently in population genetics and is known as the Ewens distribution [40, 67].

If the random variables $C_1, C_2, \ldots$ and $M_1, M_2, \ldots$ correspond to a graph with the Ewens distribution of $\alpha_1, \ldots, \alpha_n$ with parameter θ, then as $n \to \infty$, the distributions of the normalized random variables converge, respectively, to the GEM distribution and the Poisson–Dirichlet distribution with the same parameter θ [67]. See also [139, 140, 141].

Section 1.7 contains the results on unicyclic random graphs obtained in [77]. The analysis of random graphs with components of two types presented in Section 1.8 is also contained in [77]. The idea of considering a graph as a combination of connected components of certain types can be attributed to Agadzhanyan [1, 2]. The results of Section 1.8 can be found in [77].

2

Evolution of random graphs

2.1. Subcritical graphs

This chapter deals with several models of random graphs with n labeled vertices and T edges as $n, T \to \infty$. The parameter $\theta = 2T/n$ plays a decisive role in the behavior of random graphs, and it may be interpreted as time in the evolution of the graphs. It turns out that many of the characteristics change their behavior abruptly near the point $\theta = 1$. It is convenient to distinguish three domains of the variation of the parameter θ. We say that a random graph is subcritical if $n, T \to \infty$ in such a way that $(1 - \theta)^3 n \to \infty$. Thus, for a subcritical graph, θ may tend to unity, but not too fast. A critical graph is characterized by the conditions that $n, T \to \infty$ and $(1 - \theta)^3 n$ tends to a constant. And, finally, a graph is supercritical if $n, T \to \infty$ and $(1 - \theta)^3 n \to -\infty$.

In this section we consider three sets of graphs. Let $\mathcal{G}_{n,T}^{(1)}$ be the set of all graphs with n labeled vertices and T edges with loops and multiple edges, provided each vertex may have no more than one loop and each pair of vertices may be connected by no more than two edges. Let $\mathcal{G}_{n,T}^{(2)}$ be the set of all graphs with n labeled vertices and T edges that have no loops; however, each edge may occur twice, so that each pair of vertices may be connected by no more than two edges. And, finally, let $\mathcal{G}_{n,T}^{(3)}$ be the set of all graphs with n labeled vertices and T edges that have neither loops nor multiple edges.

Denote the number of graphs in $\mathcal{G}_{n,T}^{(i)}$ by $g_{n,T}^{(i)}, i = 1, 2, 3$. We introduce the uniform distribution on $\mathcal{G}_{n,T}^{(i)}, i = 1, 2, 3$, assigning equal probabilities to all elements of the corresponding set, and denote by $G_{n,T}^{(i)}$ a random graph such that

$$\mathbf{P}\{G_{n,T}^{(i)} = G\} = \left(g_{n,T}^{(i)}\right)^{-1}$$

for any $G \in \mathcal{G}_{n,T}^{(i)}, i = 1, 2, 3$.

Recall that in Section 1.8 we considered the sets $A_{n,T}^{(i)}, i = 1, 2, 3$, of all graphs with n labeled vertices and T edges with components of two types: trees and unicyclic components. In $A_{n,T}^{(3)}$, the unicyclic components have neither loops nor multiple edges; in $A_{n,T}^{(2)}$, the unicyclic components have no loops, but may contain cycles of length 2; and in $A_{n,T}^{(1)}$, the unicyclic components may contain loops and cycles of length 2. Thus,

$$A_{n,T}^{(i)} \subset G_{n,T}^{(i)}, \quad i = 1, 2, 3.$$

The results of Section 1.8 allow us to describe the limit distributions of various characteristics of subcritical random graphs $G_{n,T}^{(i)}, i = 1, 2, 3$.

Theorem 2.1.1. *If $n, T \to \infty$ such that $(1 - 2T/n)^3 n \to \infty$, then for any $i = 1, 2, 3$,*

$$\mathbf{P}\{G_{n,T}^{(i)} \in A_{n,T}^{(i)}\} \to 1.$$

Proof. It is clear that

$$\mathbf{P}\{G_{n,T}^{(i)} \in A_{n,T}^{(i)}\} = a_{n,T}^{(i)} / g_{n,T}^{(i)}.$$

We need to determine the asymptotics of $g_{n,T}^{(i)}, i = 1, 2, 3$, under the conditions of Theorem 2.1.1 to match the results on $a_{n,T}^{(i)}$ from Section 1.8.

Recall that if $\theta = 2T/n \to \lambda, 0 \le \lambda \le 1$, then by Theorems 1.8.1, 1.8.2, and assertion (1.8.29),

$$a_{n,T}^{(i)} = \frac{c_i(\lambda)n^{2T}}{2^T T!}(1 + o(1)) \qquad (2.1.1)$$

for any $i = 1, 2, 3$, where

$$c_1(\lambda) = e^{\lambda/2 + \lambda^2/4}, \qquad c_2(\lambda) = e^{-\lambda/2 + \lambda^2/4}, \qquad c_3(\lambda) = e^{-\lambda/2 - \lambda^2/4}.$$

If $n, T \to \infty$ and $T^3/n^4 \to 0$, then

$$
\begin{aligned}
g_{n,T}^{(3)} &= \binom{n(n-1)/2}{T} \\
&= \frac{(n(n-1))^T}{2^T T!}\left(1 - \frac{2}{n(n-1)}\right)\left(1 - \frac{4}{n(n-1)}\right)\cdots\left(1 - \frac{2(T-1)}{n(n-1)}\right) \\
&= \frac{n^{2T} e^{-T/n - T^2/n^2}}{2^T T!}(1 + o(1)),
\end{aligned}
\qquad (2.1.2)
$$

and Theorem 2.1.1 is proved for $i = 3$.

It is clear that each graph from $G_{n,T}^{(2)}$ can be obtained by a choice of T edges, which is equivalent to an allocation of T particles into $\binom{n}{2}$ cells, provided each cell contains no more than two particles. Therefore

$$g_{n,T}^{(2)} = \sum_{t_1+2t_2=T} \binom{S}{t_1}\binom{S-t_1}{t_2},$$

where $S = \binom{n}{2}$, t_1 cells have exactly one particle, and t_2 cells have two particles. Hence,

$$g_{n,T}^{(2)} = \sum_{t_1+2t_2=T} \frac{S!}{t_1!t_2!(S-t_1-t_2)!}$$

$$= \frac{1}{T!}\sum_{0\le t\le T/2} \frac{T!S!}{t!(T-2t)!(S-T+t)!}.$$

For any fixed t,

$$\frac{T!S!}{(T-2t)!(S-T+t)!} = T^{2t}S^{T-t}e^{-T^2/(2S)}(1+o(1))$$

$$= S^T \left(\frac{2T^2}{n^2}\right)^t e^{-T^2/n^2}(1+o(1)).$$

Therefore, under the conditions of Theorem 2.1.1,

$$g_{n,T}^{(2)} = \frac{S^T e^{-T^2/n^2}}{T!}\sum_{t=0}^{\infty}\frac{1}{t!}\left(\frac{2T^2}{n^2}\right)^t (1+o(1))$$

$$= \frac{n^{2T}e^{-T/n-T^2/n^2}}{2^T T!}e^{2T^2/n^2}(1+o(1))$$

$$= \frac{n^{2T}}{2^T T!}e^{-T/n+T^2/n^2}(1+o(1)). \tag{2.1.3}$$

Similarly, each graph from $G_{n,T}^{(3)}$ can be obtained by a choice of T edges, which is equivalent to an allocation of T particles into $n + \binom{n}{2}$ cells, provided that no more than two particles are allocated into each of $\binom{n}{2}$ cells and only one particle may be put into each of n cells. Therefore, putting $S = \binom{n}{2}$ yields

$$g_{n,T}^{(1)} = \sum_{t_1+t_2+2t_3=T} \binom{n}{t_1}\binom{S}{t_2}\binom{S-t_2}{t_3}.$$

By the same arguments under the conditions of Theorem 2.1.1,

$$g_{n,T}^{(1)} = \frac{n^{2T}}{2^T T!}e^{T/n+T^2/n^2}(1+o(1)). \tag{2.1.4}$$

Then, by comparing (2.1.1) to (2.1.2), (2.1.3), and (2.1.4), we obtain the assertion of the theorem. ∎

According to Theorem 2.1.1, each of the subcritical graphs $G_{n,T}^{(i)}$, $i = 1, 2, 3$, consists of trees and unicyclic components and, with probability tending to 1, does not contain more complicated components.

Given a random graph G, denote by $\mu_r(G)$ the number of trees of size r, by $\eta(G)$ the maximum size of trees, by $\omega(G)$ the total number of vertices in the unicyclic components, by $\varkappa(G)$ the number of unicyclic components, by $\beta(G)$ the maximum size of the unicyclic components, and by $\alpha(G)$ the maximum size of the components.

Let $\gamma(G_{n,T}^{(i)})$ be a characteristic of the random graph $G_{n,T}^{(i)}$ and let $\gamma_{n,T}^{(i)}$ be the corresponding characteristic of the random graph from $\mathcal{A}_{n,T}^{(i)}$. Then, by the formula of total probability,

$$\mathbf{P}\{\gamma(G_{n,T}^{(i)}) \leq x\} = \mathbf{P}\{G_{n,T}^{(i)} \in \mathcal{A}_{n,T}^{(i)}\}\mathbf{P}\{\gamma_{n,T}^{(i)} \leq x\}$$
$$+ \mathbf{P}\{G_{n,T}^{(i)} \notin \mathcal{A}_{n,T}^{(i)}\}\mathbf{P}\{\gamma(G_{n,T}^{(i)}) \leq x \mid G_{n,T}^{(i)} \notin \mathcal{A}_{n,T}^{(i)}\}$$

for any x. By Theorem 2.1.1,

$$\mathbf{P}\{G_{n,T}^{(i)} \in \mathcal{A}_{n,T}^{(i)}\} \to 1$$

if the graph $G_{n,T}^{(i)}$ is subcritical. Therefore, for any characteristic $\gamma(G_{n,T}^{(i)})$ of the subcritical graph,

$$\mathbf{P}\{\gamma(G_{n,T}^{(i)}) \leq x\} = \mathbf{P}\{\gamma_{n,T}^{(i)} \leq x\}(1 + o(1)) + o(1), \qquad (2.1.5)$$

and if $\mathbf{P}\{\gamma_{n,T}^{(i)} \leq x\}$ tends to a limit, then the probability $\mathbf{P}\{\gamma(G_{n,T}^{(i)}) \leq x\}$ has the same limit. Thus, many of the results of Section 1.8 can be reformulated for the corresponding characteristics of the random graphs $G_{n,T}^{(i)}$, $i = 1, 2, 3$. If $\gamma(G_{n,T}^{(i)})$ is an integer-valued characteristic, then for any fixed integer k,

$$\mathbf{P}\{\gamma(G_{n,T}^{(i)}) = k\} = \mathbf{P}\{\gamma_{n,T}^{(i)} = k\}(1 + o(1)) + o(1), \qquad (2.1.6)$$

and if $\mathbf{P}\{\gamma_{n,T}^{(i)} = k\}$ has a nonzero limit, then relation (2.1.6) allows us to obtain the limit of the probability $\mathbf{P}\{\gamma(G_{n,T}^{(i)}) = k\}$.

Theorem 2.1.2. *If $n, T \to \infty$ such that $T/n \to 0$, then for any $i = 1, 2, 3$,*

$$\mathbf{P}\{\omega(G_{n,T}^{(i)}) = 0\} \to 1.$$

If $n, T \to \infty$ such that $\varepsilon = 1 - 2T/n \to 0$ and $\varepsilon^3 n \to \infty$, then for any fixed $x > 0$ and any $i = 1, 2, 3$,

$$\mathbf{P}\{\omega(G_{n,T}^{(i)})\varepsilon^2/2 \leq x\} \to \frac{1}{\Gamma(1/4)} \int_0^x y^{-3/4} e^{-y}\, dy.$$

Proof. The assertions of the theorem follow from (2.1.5), (2.1.6), and Theorems 1.8.1 and 1.8.5. ∎

Theorem 2.1.3. *If the graph $G_{n,T}^{(i)}$ is subcritical, $i = 1, 2, 3$, and $r = r(n, T) \geq 3$ varies such that $N p_r(\theta) \to \infty$, then for any fixed x,*

$$\mathbf{P}\left\{\frac{\mu_r(G_{n,T}^{(i)}) - N p_r(\theta)}{\sigma_{rr}(\theta)\sqrt{N}} \leq x\right\} \to \frac{1}{\sqrt{2\pi}} \int_{-\infty}^{x} e^{-u^2/2} \, du,$$

where

$$N = n - T,$$

$$\theta = 2T/n,$$

$$p_r(\theta) = \frac{2k^{k-2}\theta^{k-1}e^{-k\theta}}{k!(2-\theta)}, \quad k = 1, 2, \ldots,$$

$$\sigma_{rr}(\theta) = \frac{p_r(\theta)(1 - p_r(\theta)) - (\mu - k)^2 p_r(\theta)}{\sigma^2},$$

$$\mu = \frac{2}{(2-\theta)},$$

$$\sigma^2 = \frac{2\theta}{(1-\theta)(2-\theta)^2}.$$

If $r = r(n, T) \geq 3$ varies such that $N p_r(\theta) \to \lambda$, $0 < \lambda < \infty$, then for any fixed $k = 0, 1, \ldots$,

$$\mathbf{P}\{\mu_r(G_{n,T}^{(i)}) = k\} = \frac{\lambda^k e^{-\lambda}}{k!}(1 + o(1)).$$

Proof. In view of (2.1.5) and (2.1.6), the assertion of the theorem follows from Theorems 1.5.1 and 1.5.2 because, by Theorem 2.1.2, the number $\omega(G_{n,T}^{(i)})$ of vertices in the unicyclic components for subcritical graphs is small compared with the total number of vertices; more precisely, $\mathbf{P}\{\omega(G_{n,T}^{(i)}) \leq n^{2/3}\} \to 1$. ∎

Theorem 2.1.4. *If $n, T \to \infty$ such that $T/n \to 0$, $r = r(n, T) > 1$ and $N p_r(\theta) \to \infty$, $N p_{r+1}(\theta) \to \lambda$, $0 \leq \lambda < \infty$, then for any $i = 1, 2, 3$,*

$$\mathbf{P}\{\alpha(G_{n,T}^{(i)}) = r\} = \mathbf{P}\{\eta(G_{n,T}^{(i)}) = r\} = e^{-\lambda} + o(1),$$

$$\mathbf{P}\{\alpha(G_{n,T}^{(i)}) = r + 1\} = \mathbf{P}\{\eta(G_{n,T}^{(i)}) = r + 1\} = 1 - e^{-\lambda} + o(1).$$

Proof. In view of (2.1.5) and (2.1.6), the assertions of the theorem follow from Theorem 1.6.1. ∎

Theorem 2.1.5. *If $i = 1, 2, 3$ and $n, T \to \infty$ such that $\theta = 2T/n \to \lambda$, $0 < \lambda < 1$, then for any fixed $k = 0, 1, \ldots$,*

$$\mathbf{P}\{x(G_{n,T}^{(i)}) = k\} = \frac{\Lambda_i^k e^{-\Lambda_i}}{k!}(1 + o(1)),$$

where

$$\Lambda_1 = -\frac{1}{2}\log(1-\lambda) + \frac{\lambda}{2} + \frac{\lambda^2}{4},$$

$$\Lambda_2 = -\frac{1}{2}\log(1-\lambda) - \frac{\lambda}{2} + \frac{\lambda^2}{4},$$

$$\Lambda_3 = -\frac{1}{2}\log(1-\lambda) - \frac{\lambda}{2} - \frac{\lambda^2}{4}.$$

For any fixed $k = 0, \pm 1, \ldots,$

$$\mathbf{P}\{\alpha(G_{n,T}^{(i)}) - [a] \le k\} = \mathbf{P}\{\eta(G_{n,T}^{(i)}) - [a] \le k\}(1 + o(1))$$

$$= \exp\left\{-\frac{(\lambda - 1 - \log\lambda)^{5/2}}{(e^{\lambda-1} - \lambda)}e^{(k+\{a\})(\lambda-1-\log\lambda)}\right\}(1 + o(1)),$$

where

$$a = \frac{\log n - (5/2)\log\log n}{\theta - 1 - \log\theta},$$

$[a]$ and $\{a\}$ are, respectively, the integer and fractional parts of a.

Proof. The assertions of the theorem follow from (2.1.5), (2.1.6), and Theorems 1.8.4 and 1.6.2.   ∎

Theorem 2.1.6. *If $i = 1, 2, 3$ and $n, T \to \infty$ such that $\varepsilon = 1 - 2T/n \to 0$ and $\varepsilon^3 n \to \infty$, then for any fixed x,*

$$\mathbf{P}\left\{x(G_{n,T}^{(i)}) + \frac{1}{2}\log\varepsilon \le x\sqrt{-\frac{1}{2}\log\varepsilon}\right\} \to \frac{1}{\sqrt{2\pi}}\int_{-\infty}^{x} e^{-u^2/2}\,du,$$

and for any fixed $x > 0$,

$$\mathbf{P}\{\varepsilon^2\beta(G_{n,T}^{(i)}) \le x\} = \sum_{s=0}^{\infty}\frac{(-1)^s}{4^s s!}Z_s(x)(1 + o(1)),$$

where $Z_s(x)$ is defined in Theorem 1.8.7. Finally, for any fixed z,

$$\mathbf{P}\{\beta\alpha(G_{n,T}^{(i)}) - u \le z\} = \mathbf{P}\{\beta\eta(G_{n,T}^{(i)}) - u \le z\}(1 + o(1)) = e^{-e^{-z}}(1 + o(1)),$$

where $\beta = -\log(\theta e^{-\theta})$, $\theta = 2T/n$, and u is the root of the equation

$$\left(\frac{2}{\pi}\right)^{1/2} N\beta^{3/2} = u^{5/2}e^u. \tag{2.1.7}$$

Proof. The results of the theorem are the consequences of (2.1.5), (2.1.6), and Theorems 1.8.6, 1.8.7, 1.8.8, and 1.8.9.   ∎

2.2. Critical graphs

Recall that a graph with n vertices and T edges is called critical if $n, T \to \infty$ such that $\varepsilon = 1 - 2T/n \to 0$ and $\varepsilon^3 n$ tends to a constant. We have seen that many of the characteristics of the random graphs $G_{n,T}^{(i)}$, $i = 1, 2, 3$, change their behavior if $\theta = 2T/n$ approaches the value 1. For example, the number of cycles, or the number of unicyclic components $\varkappa(G_{n,T}^{(i)})$, tends to zero in probability if $\theta \to 0$, has the Poisson distribution with parameter $\Lambda_i, i = 1, 2, 3$, respectively, if $\theta \to \lambda$, $0 < \lambda < 1$, where

$$\Lambda_1 = -\frac{1}{2} \log(1 - \lambda) + \frac{\lambda}{2} + \frac{\lambda^2}{4},$$

$$\Lambda_2 = -\frac{1}{2} \log(1 - \lambda) - \frac{\lambda}{2} + \frac{\lambda^2}{4},$$

$$\Lambda_3 = -\frac{1}{2} \log(1 - \lambda) - \frac{\lambda}{2} - \frac{\lambda^2}{4},$$

and is asymptotically normal with parameters $(-\frac{1}{4} \log \varepsilon, -\frac{1}{4} \log \varepsilon)$ if $\varepsilon \to 0$, $\varepsilon^3 n \to \infty$. Thus, $\theta = 1$ is a singular point and one can correctly suppose that the behavior of the graphs near this point is interesting but difficult to investigate. Indeed, not much is known about the properties of critical graphs. We present here only one assertion about this behavior.

Recall that $A_{n,T}^{(i)}$ is the set of graphs with n labeled vertices and T edges that consists of trees and unicyclic components with neither loops nor multiple edges for $i = 3$, without loops and with cycles of length 2 allowed for $i = 2$, and with cycles of lengths 1 and 2 allowed for $i = 1$.

Theorem 2.2.1. *If $n, T \to \infty$ such that $\varepsilon n^{1/3} \to 2 \cdot 3^{-2/3} v$, where v is a constant, then for any random graph $G_{n,T}^{(i)}$, $i = 1, 2, 3$,*

$$\mathbf{P}\{G_{n,T}^{(i)} \in A_{n,T}^{(i)}\} = \frac{\sqrt{3\pi}}{\sqrt{2}\Gamma(1/4)} e^{4v^3/27} p(v)(1 + o(1)),$$

where

$$p(v) = \int_0^\infty y^{-3/4} p(-v - y; 3/2, -1) \, dy$$

and $p(y; 3/2, -1)$ is the density of the stable law, introduced in Theorem 1.4.2, with the characteristic function

$$f(t) = \exp\left\{ -|t|^{3/2} e^{i\pi t/(4|t|)} \right\}.$$

Proof. It is clear that

$$\mathbf{P}\{G_{n,T}^{(i)} \in A_{n,T}^{(i)}\} = a_{n,T}^{(i)} / g_{n,T}^{(i)},$$

where $a_{n,T}^{(i)}$ is the number of graphs in $\mathcal{A}_{n,T}^{(i)}$, and $g_{n,T}^{(i)}$ is the number of graphs in $\mathcal{G}_{n,T}^{(i)}$, $i = 1, 2, 3$. In accordance with Theorem 1.8.10,

$$a_{n,T}^{(i)} = \frac{c_i n! \, e^n}{2^N N! \sqrt{N}} p(v)(1 + o(1)),$$

where $N = n - T$,

$$c_1 = \frac{\sqrt{3} e^{3/4}}{2\sqrt{2}\Gamma(1/4)}, \quad c_2 = \frac{\sqrt{3} e^{-1/4}}{2\sqrt{2}\Gamma(1/4)}, \quad c_1 = \frac{\sqrt{3} e^{-3/4}}{2\sqrt{2}\Gamma(1/4)}.$$

In the previous section, we proved that

$$g_{n,T}^{(i)} = \frac{n^{2T} c_i(1)}{2^T T!}(1 + o(1)),$$

where $c_1(1) = e^{3/4}$, $c_2(1) = e^{-1/4}$, $c_3(1) = e^{-3/4}$.

Since $T = n(1 - \varepsilon)/2$ and $\varepsilon^3 n \to 8v^3/9$, we easily find

$$\frac{n! \, e^n T! \, 2^T}{2^N N! \sqrt{N} n^{2T}} = 2\sqrt{\pi} e^{4v^3/27}(1 + o(1))$$

and, consequently,

$$\frac{a_{n,T}^{(i)}}{g_{n,T}^{(i)}} = \frac{\sqrt{3\pi}}{\sqrt{2}\Gamma(1/4)} e^{4v^3/27} p(v)(1 + o(1)).$$

■

The function $p(v)$ can be represented by a convergent power series. The function

$$g(v) = p(-v) = \int_0^\infty y^{-3/4} p(v - y; 3/2, -1) \, dy$$

can be thought of as the convolution of the function

$$g_1(y) = \begin{cases} y^{-3/4}, & y > 0, \\ 0, & y \le 0 \end{cases}$$

and the function $g_2(y) = p(y; 3/2, -1)$, so that

$$g(v) = \int_{-\infty}^\infty g_1(y) g_2(v - y) \, dy.$$

Therefore the Fourier transform $\hat{g}(t)$ of the function $g(v)$ is the product of the Fourier transforms of the functions $g_1(y)$ and $g_2(y)$. The Fourier transform $\hat{g}_1(t)$ of the function $g_1(y)$ is

$$\hat{g}_1(t) = \frac{2\pi e^{i\pi t/(8|t|)}}{\sqrt{2}\Gamma(3/4)|t|^{1/4}},$$

and the Fourier transform $\hat{g}_2(t)$ of the function $g_2(y) = p(y; 3/2, -1)$ is the characteristic function of this density:

$$\hat{g}_2(t) = \exp\left\{ -|t|^{3/2} e^{i\pi t/(4|t|)} \right\}.$$

Thus,

$$\hat{g}(t) = \frac{2\pi e^{i\pi t/(8|t|)}}{\sqrt{2}\Gamma(3/4)|t|^{1/4}} \exp\left\{ -|t|^{3/2} e^{i\pi t/(4|t|)} \right\}.$$

By the inversion formula,

$$g(v) = \frac{1}{2\pi} \int_{-\infty}^{\infty} e^{-itv} \hat{g}(t)\, dt$$

$$= \frac{1}{\sqrt{2}\Gamma(3/4)} \int_{-\infty}^{\infty} e^{-itv} |t|^{-1/4} e^{i\pi t/(8|t|)} \exp\left\{ -|t|^{3/2} e^{i\pi t/(4|t|)} \right\} dt,$$

and therefore, under the hypotheses of Theorem 2.2.1,

$$\mathbf{P}\{G_{n,T}^{(i)} \in A_{n,T}^{(i)}\} = \frac{\sqrt{3\pi}}{\sqrt{2}\Gamma(1/4)\sqrt{2}\Gamma(3/4)} h(v)(1 + o(1)),$$

where

$$h(v) = \int_{-\infty}^{\infty} e^{itv} |t|^{-1/4} e^{i\pi t/(8|t|)} \exp\left\{ -|t|^{3/2} e^{i\pi t/(4|t|)} \right\} dt.$$

Since $\Gamma(1/4)\Gamma(3/4) = \sqrt{2}\pi$, we obtain

$$\mathbf{P}\{G_{n,T}^{(i)} \in A_{n,T}^{(i)}\} = \frac{\sqrt{3}}{2\sqrt{2\pi}} h(v)(1 + o(1)). \tag{2.2.1}$$

The function $h(v)$ can be represented by a convergent power series.

Theorem 2.2.2. *If $n, T \to \infty$ such that $\varepsilon n^{1/3} \to 2 \cdot 3^{-2/3} v$, where v is a constant, then for any random graph $G_{n,T}^{(i)}$, $i = 1, 2, 3$,*

$$\mathbf{P}\{G_{n,T}^{(i)} \in A_{n,T}^{(i)}\} = P(v)(1 + o(1)),$$

where

$$P(v) = \sqrt{\frac{2}{3\pi}} e^{4v^3/27} \sum_{k=0}^{\infty} \frac{v^k}{k!} \Gamma\left(\frac{2k}{3} + \frac{1}{2}\right) \cos\frac{\pi k}{3}.$$

Proof. Let us represent $h(v)$ by a power series in v. Since the left-hand side of (2.2.1) is real,

$$h(v) = \Re \int_{-\infty}^{\infty} e^{itv} |t|^{-1/4} e^{i\pi t/(8|t|)} \exp\left\{ -|t|^{3/2} e^{i\pi t/(4|t|)} \right\} dt.$$

Consider first the integral

$$h_1(v) = \int_0^\infty e^{itv} t^{-1/4} e^{i\pi/8} \exp\left\{-t^{3/2} e^{i\pi/4}\right\} dt.$$

By expanding e^{itv}, we obtain

$$h_1(v) = e^{i\pi/8} \sum_{k=0}^\infty \frac{(iv)^k}{k!} \int_0^\infty t^{k-1/4} \exp\left\{-t^{3/2} e^{i\pi/4}\right\} dt.$$

After the change of variables $t^{3/2} e^{i\pi/4} = z$, we obtain

$$h_1(v) = \frac{2}{3} \sum_{k=0}^\infty \frac{v^k}{k!} \exp\left\{\frac{i\pi k}{3}\right\} \Gamma\left(\frac{2k}{3} + \frac{1}{2}\right).$$

Therefore

$$\Re h_1(v) = \frac{2}{3} \sum_{k=0}^\infty \frac{v^k}{k!} \cos\frac{\pi k}{3} \Gamma\left(\frac{2k}{3} + \frac{1}{2}\right). \tag{2.2.2}$$

Similarly, for

$$h_2(v) = \int_{-\infty}^0 e^{itv} |t|^{-1/4} e^{-i\pi/8} \exp\left\{-|t|^{3/2} e^{i\pi t/(4|t|)}\right\} dt$$

$$= \int_0^\infty e^{-itv} t^{-1/4} e^{-i\pi/8} \exp\left\{-t^{3/2} e^{-i\pi/4}\right\} dt,$$

we obtain

$$\Re h_2(v) = \frac{2}{3} \sum_{k=0}^\infty \frac{v^k}{k!} \cos\frac{\pi k}{3} \Gamma\left(\frac{2k}{3} + \frac{1}{2}\right). \tag{2.2.3}$$

The assertion of the theorem follows from (2.2.1), (2.2.2), and (2.2.3). ∎

Theorem 2.2.2 allows us to calculate the limit values of $\mathbf{P}\{G_{n,T}^{(i)} \in \mathcal{A}_{n,T}^{(i)}\}$. For example,

$$P(0) = \sqrt{2/3}.$$

Some values of $P(v)$ are given in Table 2.1.

2.3. Random graphs with independent edges

When we were determining the number of graphs in the classes $\mathcal{G}_{n,T}^{(i)}, i = 1, 2, 3,$ in Section 2.1, we associated each of the classes with the corresponding equiprobable scheme of allocating particles into cells. It is easily seen from these correspondences that the realizations of each of the random graphs $G_{n,T}^{(i)}, i = 1, 2, 3,$ could be obtained by a sequential allocation of particles, but these random allocations are dependent. For example, if a pair of vertices has been connected in the random

Table 2.1. Values of $P(v)$

v	$P(v)$	v	$P(v)$	v	$P(v)$
-3.0	0.0053	-1.0	0.4919	1.2	0.9563
-2.8	0.0118	-0.8	0.5727	1.4	0.9653
-2.6	0.0239	-0.6	0.6470	1.6	0.9722
-2.4	0.0443	-0.4	0.7128	1.8	0.9776
-2.2	0.0755	-0.2	0.7693	2.0	0.9819
-2.0	0.1196	0.2	0.8551	2.2	0.9852
-1.8	0.1768	0.4	0.8860	2.4	0.9878
-1.6	0.2461	0.6	0.9105	2.6	0.9899
-1.4	0.3244	0.8	0.9297	2.8	0.9915
-1.2	0.4078	1.0	0.9447	3.0	0.9929

graph $G_{n,T}^{(3)}$ after allocating some of the edges, then the outcomes of all subsequent allocations cannot be the edges connecting these two vertices.

The classes of random graphs whose edges are independent seem to be easier to investigate by using the methods of probability theory. The best-known random graph with this property is $G_{n,p}$ with n vertices such that each of the $\binom{n}{2}$ possible edges belongs to the edge set of $G_{n,p}$ with probability p independently of the behavior of the other edges. This graph has a random number of edges with the binomial distribution with n trials and the probability of success p.

In this section, we consider the random graph $G_{n,T}$ with n vertices labeled $1, \ldots, n$ and T edges that can be obtained by T independent trials. In each trial, the loop at any point i occurs with probability n^{-2} and the edge connecting the vertices i and j, $i \neq j$, occurs with probability $2n^{-2}$. In other words, if the edge set of $G_{n,T}$ consists of T edges $((i(1), j(1)), \ldots, (i(T), j(T))$, then $i(1), j(1), \ldots, i(T), j(T)$ are independent identically distributed random variables taking the values $1, 2, \ldots, n$ with equal probabilities. It is clear that the realizations of the random graph $G_{n,T}$ are not equiprobable. For example, for $n = 2$ and $T = 1$, the graphs with a loop and an isolated vertex have the probabilities $1/4$ each, and the connected graph has the probability $1/2$. Nevertheless, this model has some advantages and is conducive to treatment by probabilistic methods.

Since $i(1), j(1), \ldots, i(T), j(T)$ are independent identically distributed random variables, we can associate to the random graph $G_{n,T}$ the classical scheme of allocating particles where $2T$ particles are allocated into n cells such that each particle falls into any of n cells with probability $1/n$ independently of the allocations of the other particles. By using this relationship, we can, for example, easily find the distribution of the number of loops in $G_{n,T}$. Indeed, we have T trials, corresponding to T edges, and in each of these trials a loop appears with probability $1/n$. Thus, the total number of loops α_1 in $G_{n,T}$ has the binomial distribution with parameters $(T, 1/n)$. The mean number of loops is $\mathbf{E}\alpha_1 = T/n$. If $2T/n \to \lambda, 0 < \lambda < \infty$, then the Poisson distribution with parameter $\lambda/2$ is the limit distribution for α_1.

Under the condition $\alpha_1 = m$, the other edges may be considered as the result of $T - m$ independent allocations into $\binom{n}{2}$ cells corresponding to $\binom{n}{2}$ possible edges of the complete graph with n vertices. Therefore, with $\alpha_1 = m$, the number α_2 of cycles of length 2 in $G_{n,T}$ can be thought of as the number of cells with exactly two particles in the classical (equiprobable) scheme of allocation of $T - m$ particles into $\binom{n}{2}$ cells. The classical scheme of allocation has been well studied. In particular, if $n, T \to \infty$ such that $2T/n \to \lambda, 0 < \lambda < \infty$, then the distribution of the number of cells, occupied by exactly two particles each, converges to the Poisson distribution with parameter $\lambda^2/4$. Since the limit distribution does not depend on m for $m = o(n)$, averaging over the distribution of α_1 shows that α_1 and α_2 are asymptotically independent and their distributions approach the Poisson distributions.

Theorem 2.3.1. *If $n, T \to \infty$ such that $2T/n \to \lambda, 0 < \lambda < \infty$, then for any fixed nonnegative integers k_1 and k_2,*

$$\mathbf{P}\{\alpha_1 = k_1, \alpha_2 = k_2\} = \left(\frac{\lambda}{2}\right)^{k_1} \left(\frac{\lambda^2}{4}\right)^{k_2} e^{-\lambda/2 - \lambda^2/4}(1 + o(1)).$$

Because the edges of $G_{n,T}$ are independent, we can apply direct probabilistic approaches to investigations of the structure of $G_{n,T}$.

Theorem 2.3.2. *If $n, T \to \infty$ such that $T/n \to 0$, then in $G_{n,T}$, with probability tending to 1, there are no cycles and all the components are trees.*

Proof. Denote the number of cycles of length r with r distinct vertices by α_r, and let $\nu(G_{n,T}) = \alpha_1 + \cdots + \alpha_n$ be the total number of cycles considered as induced subgraphs of $G_{n,T}$. We can represent α_r as a sum of indicators. The edges of $G_{n,T}$ appear sequentially in T trials. We assign the numbers $1, 2, \ldots, T$ to the trials and arrange (in some order) all $\binom{T}{r}$ possible subsets of cardinality r of the trial numbers. We define the random variable ξ_i to be equal to 1 if the subset of trial numbers labeled with i forms a cycle in $G_{n,T}$, and $\xi_i = 0$ otherwise. It is clear that

$$\alpha_r = \xi_1 + \cdots + \xi_{\binom{T}{r}}.$$

In turn, each of the random variables $\xi_1, \ldots, \xi_{\binom{T}{r}}$ can be represented as a sum of indicators. The cycle corresponding to the subset with label i can be constructed from r different vertices and r different edges. There exist $\binom{n}{r}$ possibilities to choose these r vertices and $(r-1)!/2$ possibilities to construct a cycle from these r vertices for $r \geq 3$. Each construction fixes r edges that must occur. These r edges can occur at r fixed places of the subset labeled i, and there exist $r!$ possibilities to assign these r edges to r places. Thus the event $\{\xi_i = 1\}$ can be realized by one of the $\binom{n}{r}(r-1)!\, r!/2$ variants.

For $r \geq 3$, each of these variants has the probability $(2/n^2)^r$. Thus,

$$\mathbf{E}\alpha_r = \binom{T}{r}\binom{n}{r}\frac{(r-1)!\, r!}{2}\left(\frac{2}{n^2}\right)^r. \tag{2.3.1}$$

It is not difficult to check that this formula is also valid for $r = 1$ and $r = 2$.

It follows from (2.3.1) that

$$\mathbf{E}\alpha_r \leq \frac{T^r n^r (r-1)!\, r!}{r!\, r!\, 2}\left(\frac{2}{n^2}\right)^r = \left(\frac{2T}{n}\right)^r \frac{1}{2r}.$$

Therefore,

$$\mathbf{E}v(G_{n,T}) = \sum_{r=1}^{n} \mathbf{E}\alpha_r$$

has the upper bound

$$\mathbf{E}v(G_{n,T}) \leq \sum_{r=1}^{\infty} \left(\frac{2T}{n}\right)^r \frac{1}{2r}.$$

Under the conditions of the theorem, $\mathbf{E}v(G_{n,T})$ tends to zero and the number of cycles in $G_{n,T}$ is zero with probability approaching 1. ∎

We denote by $\mathcal{A}_{n,T}$ the set of all graphs with n labeled vertices and T edges whose components are trees and unicyclic components. Note that loops and cycles of length 2 are permitted. As before, $\theta = 2T/n$, $\varepsilon = 1 - 2T/n$.

Theorem 2.3.3. *If $n, T \to \infty$ such that $\varepsilon^3 n \to \infty$, then*

$$\mathbf{P}\{G_{n,T} \notin \mathcal{A}_{n,T}\} \leq \frac{4}{\varepsilon^3 n}.$$

Proof. We have to prove that under the conditions of the theorem, the graph $G_{n,T}$ has no component with more than one cycle with probability less than $4/(\varepsilon^3 n)$. If in $G_{n,T}$ there exists such a component, then in $G_{n,T}$ there either exists a subgraph that consists of two cycles connected by a chain (pince-nez) or there exist two cycles that have a common sequence of edges (a cycle with a bridge). We use $\xi_{r,s}^{(t)}$ to denote the number of subgraphs of $G_{n,T}$ that consist of cycles of lengths r and s connected by a chain of t edges, and denote by $\xi_r^{(t)}$ the number of subgraphs of $G_{n,T}$ that consist of a cycle of length r with two vertices connected by a sequence of t edges. To prove the assertion of the theorem, it is sufficient to show that the mean number of such subgraphs tends to zero. It is clear that

$$P\{G_{n,T} \notin \mathcal{A}_{n,T}\} = P\left\{\sum_{r,s,t} \xi_{r,s}^{(t)} + \sum_{r,t} \xi_r^{(t)} > 0\right\} \leq E\left(\sum_{r,s,t} \xi_{r,s}^{(t)} + \sum_{r,t} \xi_r^{(t)}\right).$$

By reasoning in the same way as in the proof of formula (2.3.1), we obtain the estimates

$$E\xi_r^{(t)} \leq \binom{n}{r+t-1}\binom{r+t}{r}(r-1)!\,(r-1)(t-1)!$$

$$\times \binom{T}{r+t}(r+t)!\left(\frac{2}{n^2}\right)^{r+t} \leq \frac{2r}{n}\left(\frac{2T}{n}\right)^{r+t},$$

$$E\xi_{r,s}^{(t)} \leq \binom{n}{r+s+t-1}\frac{(r+s+t-1)!}{r!\,s!\,(t-1)!}r!\,s!\,(t-1)!$$

$$\times \binom{T}{r+s+t}(r+s+t)!\left(\frac{2}{n^2}\right)^{r+s+t} \leq \frac{2}{n}\left(\frac{2T}{n}\right)^{r+s+t}.$$

Thus, the mathematical expectation of the total number of pince-nez and cycles with a bridge can be estimated as follows:

$$\sum_{r,t=0}^{\infty} E\xi_r^{(t)} + \sum_{r,s,t=0}^{\infty} E\xi_{r,s}^{(t)}$$

$$\leq \frac{2}{n}\sum_{r,t=0}^{(t)} r\left(\frac{2T}{n}\right)^{r+t} + \frac{2}{n}\sum_{r,s,t=0}^{(t)}\left(\frac{2T}{n}\right)^{r+s+t} \leq \frac{4}{n(1-2T/n)^3}.\qquad\blacksquare$$

Theorem 2.3.4. *If $n, T \to \infty$ such that $\theta = 2T/n \to \lambda$, $0 < \lambda < 1$, then the distribution of the number of cycles $\nu(G_{n,T})$ in $G_{n,T}$ converges to the Poisson distribution with parameter*

$$\Lambda = -\frac{1}{2}\log(1-\lambda).$$

Proof. In view of Theorems 2.3.1 and 2.3.3, we can reduce the proof to the application of Theorem 2.1.5 concerning the random graph $G_{n,T}^{(3)}$ without loops and multiple edges. Indeed, by the formula of total probability,

$$P\{\nu(G_{n,T}) = k\} = \sum_{k_1+k_2 \leq k} P\{\alpha_1 = k_1,\ \alpha_2 = k_2,\ G_{n,T} \in \mathcal{A}_{n,T}\}$$

$$\times P\{\nu(G_{n,T}) = k \mid \alpha_1 = k_1,\ \alpha_2 = k_2,\ G_{n,T} \in \mathcal{A}_{n,T}\}$$

$$+ \sum_{k_1+k_2 \leq k} P\{\alpha_1 = k_1,\ \alpha_2 = k_2,\ G_{n,T} \notin \mathcal{A}_{n,T}\}$$

$$\times P\{\nu(G_{n,T}) = k \mid \alpha_1 = k_1,\ \alpha_2 = k_2,\ G_{n,T} \notin \mathcal{A}_{n,T}\}.$$

According to Theorem 2.3.3, $\mathbf{P}\{G_{n,T} \notin \mathcal{A}_{n,T}\} \to 0$, and it is not difficult to see that

$$\mathbf{P}\{G_{n,T} \in \mathcal{A} \mid \alpha_1 = k_1,\ \alpha_2 = k_2\} = \mathbf{P}\{G^{(3)}_{n,T-k_1-k_2} \in \mathcal{A}_{n,T}\},$$

$$\mathbf{P}\{\nu(G_{n,T}) = k \mid \alpha_k = k_1,\ \alpha_2 = k_2,\ G_{n,T} \in \mathcal{A}_{n,T}\}$$
$$= \mathbf{P}\{\varkappa(G^{(3)}_{n,T-k_1-k_2}) = k - k_1 - k_2\}.$$

Thus

$$\mathbf{P}\{\nu(G_{n,T}) = k\} = \sum_{k_1+k_2 \le k} \mathbf{P}\{\alpha_1 = k_1,\ \alpha_2 = k_2\} \tag{2.3.2}$$
$$\times \mathbf{P}\{\varkappa(G^{(3)}_{n,T-k_1-k_2}) = k - k_1 - k_2\}(1 + o(1)) + o(1).$$

According to Theorem 2.1.5, under the conditions of Theorem 2.3.4, for any fixed $k_1, k_2 = 0, 1, \ldots$, and $k \ge k_1 + k_2$,

$$\mathbf{P}\{\varkappa(G^{(3)}_{n,T}) = k - k_1 - k_2\} \to \frac{\Lambda_3^{k-k_1-k_2} e^{-\Lambda_3}}{(k - k_1 - k_2)!},$$

where

$$\Lambda_3 = -\frac{1}{2}\log(1 - \lambda) - \frac{\lambda}{2} - \frac{\lambda^2}{4}.$$

Now it follows from (2.3.2) and Theorem 2.3.1 that

$$\mathbf{P}\{\nu(G_{n,T}) = k\} = \sum_{k_1+k_2 \le k} \frac{1}{k_1!}\left(\frac{\lambda}{2}\right)^{k_1} e^{-\lambda/2} \frac{1}{k_2!}\left(\frac{\lambda^2}{4}\right)^{k_2} e^{-\lambda^2/4}$$

$$\times \frac{\Lambda_3^{k-k_1-k_2}}{(k - k_1 - k_2)!} e^{-\Lambda_3}(1 + o(1)) + o(1)$$

$$= \frac{e^{-\Lambda}}{k!} \sum_{k_1+k_2 \le k} \frac{k!}{k_1! k_2! (k - k_1 - k_2)!}$$

$$\times \left(\frac{\lambda}{2}\right)^{k_1} \left(\frac{\lambda^2}{4}\right)^{k_2} \Lambda_3^{k-k_1-k_2}(1 + o(1)) + o(1)$$

$$= \frac{\Lambda^k}{k!} e^{-\Lambda}(1 + o(1)),$$

where

$$\Lambda = \Lambda_3 + \frac{\lambda}{2} + \frac{\lambda^2}{4} = -\frac{1}{2}\log(1 - \lambda).$$

∎

By reasoning in the same way, we can reformulate the theorems proved for $G^{(3)}_{n,T}$ so that they can also be applied to subcritical and critical graphs $G_{n,T}$. As an example, we give an analogue of Theorem 2.1.6 on the number $\varkappa(G_{n,T})$ and on the maximum sizes $\eta(G_{n,T})$, $\beta(G_{n,T})$, and $\alpha(G_{n,T})$ of trees, unicyclic components, and all components in $G_{n,T}$, respectively.

Theorem 2.3.5. *If $n, T \to \infty$ such that $\varepsilon = 1 - 2T/n \to 0$ and $\varepsilon^3 n \to \infty$, then for any fixed x,*

$$\mathbf{P}\left\{\varkappa(G_{n,T}) + \frac{1}{2}\log\varepsilon \leq x\sqrt{-\frac{1}{2}\log\varepsilon}\right\} \to \frac{1}{\sqrt{2\pi}}\int_{-\infty}^{x} e^{-u^2/2}\,du;$$

for any fixed $x > 0$,

$$\mathbf{P}\{\varepsilon^2\beta(G_{n,T}) \leq x\} = \sum_{s=0}^{\infty}\frac{(-1)^s}{4^s s!}Z_s(x)(1 + o(1)),$$

where $Z_s(x)$ is defined in Theorem 1.8.7; and

$$\mathbf{P}\{\beta\alpha(G_{n,T}) - u \leq z\} = \mathbf{P}\{\beta\eta(G_{n,T}) - u \leq z\}(1 + o(1)) = e^{-e^{-z}}(1 + o(1)),$$

where $\beta = -\log(\theta e^{-\theta})$, $\theta = 2T/n$, and u is the root of the equation

$$\left(\frac{2}{\pi}\right)^{1/2}(n - T)\beta^{3/2} = u^{5/2}e^u.$$

For the same reasons, Theorem 2.2.2 can be extended to the critical graph $G_{n,T}$.

Theorem 2.3.6. *If $n, T \to \infty$ such that $\varepsilon n^{1/3} \to 2 \cdot 3^{-2/3}v$, where v is a constant, then*

$$\mathbf{P}\{G_{n,T} \in \mathcal{A}_{n,T}\} = \sqrt{\frac{2}{3\pi}}e^{4v^3/27}\sum_{k=0}^{\infty}\frac{v^k}{k!}\Gamma\left(\frac{2k}{3} + \frac{1}{2}\right)\cos\frac{\pi k}{3}(1 + o(1)).$$

For the supercritical case where $n, T \to \infty$ such that $\varepsilon^3 n \to -\infty$, we present here only the simplest results. In the final section of this chapter, we will give a short review of what is known about the supercritical graphs.

It is known that if $\theta = 2T/n \to \lambda$, $\lambda > 1$, a giant component appears in the graph $G_{n,T}^{(3)}$ and, with probability tending to 1, $G_{n,T}^{(3)}$ consists of trees, unicyclic components, and this giant component formed by all the vertices that are not contained in trees and unicyclic components. As $2T/n$ increases, the size of the giant component increases and the number of unicyclic components decreases.

If $\theta = 2T/n \to \lambda$, $1 < \lambda < \infty$, then the number of unicyclic components has a Poisson distribution. For $\theta \to \infty$, we have the following result.

Theorem 2.3.7. *If $n, T \to \infty$ such that $\theta = 2T/n \to \infty$, then with probability tending to 1, there are no unicyclic components in $G_{n,T}$.*

Proof. The number of unicyclic component with r vertices is not greater than $cr^{r-1/2}$, where c is a constant (see, e.g., [16]). Denote by $\varkappa_r(G_{n,T})$ the number of unicyclic components of size r in $G_{n,T}$. By reasoning as in the proof of (2.3.1), we find that

$$\mathbf{E}\varkappa_r(G_{n,T}) \leq c\binom{n}{r}\binom{T}{r}r^{r-1/2}r!\left(\frac{2}{n^2}\right)^r\left(1 - \frac{2r(n-r)}{n^2} - \frac{r(r-1)}{n^2}\right)^{T-r},$$

(2.3.3)

where the last factor is the probability that the $T - r$ edges, which were not used for the construction of unicyclic components, neither connect the vertices in the component with the vertices outside the component nor connect any pair of vertices in the component.

It is sufficient to prove that

$$\sum_{1\leq r\leq n}\mathbf{E}\varkappa_r(G_{n,T}) \to 0.$$

With the help of estimate (2.3.3), we find that

$$\sum_{1\leq r\leq n}\mathbf{E}\varkappa_r(G_{n,T}) \leq \sum_{1\leq r\leq n}(\theta e)^r e^{-2r(n-(r+1)/2)(T-r)/n^2}.$$

For sufficiently large n and $1 \leq r \leq n$,

$$e^{-2r(n-(r+1)/2)(T-r)/n^2} \leq e^{-r\theta/4}$$

and $q = \theta e^{1-\theta/4} < 1$. Therefore

$$\sum_{1\leq r\leq n}\mathbf{E}\varkappa(G_{n,T}) \leq \sum_{r=1}^{\infty}q^r = \frac{q}{1-q}.$$

Since $q = \theta e^{1-\theta/4} \to 0$ as $\theta \to \infty$, we conclude that a unicyclic component exists in $G_{n,T}$ with a probability that tends to zero. ■

Finally, we consider the behavior of the random graph $G_{n,T}$ near the point where the graph becomes connected. Denote the number of components in $G_{n,T}$ by $\varkappa_{n,T}$.

Theorem 2.3.8. *If $n \to \infty$ and $2T = n\log n + xn + o(n)$, where x is a constant, then with probability tending to 1, the graph consists of a giant connected component and isolated vertices. Also, for any fixed integer $k = 0, 1, \ldots,$*

$$\mathbf{P}\{\varkappa_{n,T} - 1 = k\} \to \frac{e^{-kx}}{k!}e^{-e^{-x}}.$$

Proof. We have to prove that, with probability tending to 1, $G_{n,T}$ consists of one giant component and isolated vertices, and that the distribution of the number of these isolated vertices converges to the Poisson distribution with parameter e^{-x}.

The edges of $G_{n,T}$ appear as a result of T independent trials, and these T trials can be considered as the allocation of $2T$ particles into n cells such that any particle is allocated independently of the other and, with equal probabilities, falls into any of n cells. Therefore the number of isolated vertices in $G_{n,T}$ has the same distribution as the number $\mu_0(2T, n)$ of empty cells in the well-studied classical scheme of allocating particles. Under the conditions of the theorem, the distribution of $\mu_0(2T, n)$ converges to the Poisson distribution with parameter e^{-x}.

To complete the proof, it suffices to show that, with probability tending to 1, the remaining vertices form one giant component. If, in addition to the isolated vertices, there were two other components, then the graph would contain a tree of size r, $2 \leq r \leq n/2$, such that any vertex of the tree would not be connected to any vertices outside the tree. A skeleton of one of the two components could play the role of such a tree.

By ξ_r we denote the number of trees of size r which are the skeletons of connected components of $G_{n,T}$. We will show that under the conditions of the theorem,

$$\sum_{2 \leq r \leq n/2} \mathbf{E}\xi_r \to 0,$$

and consequently, with probability tending to 1, such a tree does not occur in $G_{n,T}$. We can represent ξ_r as a sum of indicators and find that

$$\mathbf{E}\xi_r = \binom{T}{r-1}\binom{n}{r}r^{r-2}r!\left(\frac{2}{n^2}\right)^{r-1}\left(1 - \frac{2r(n-r)}{n^2}\right)^{T-r+1}. \qquad (2.3.4)$$

This formula is similar to (2.3.1): We choose r vertices and $r - 1$ edges that form the tree, and the last factor is the probability that none of the $T - r + 1$ edges that remain connects a vertex from the set of r selected vertices with a vertex from the set of $n - r$ remaining vertices.

By using formula (2.3.4), we can check, for example, that with probability tending to 1, there are no isolated edges in $G_{n,T}$. Indeed, for $r = 2$,

$$\mathbf{E}\xi_2 \leq 2T\left(1 - \frac{4(n-2)}{n^2}\right)^{T-1} \leq 2Te^{-4(n-2)(T-1)/n^2}, \qquad (2.3.5)$$

and the right-hand side of (2.3.5) tends to zero if $n \to \infty$ and $2T = n\log n + xn + o(n)$.

It follows from (2.3.4) that

$$\mathbf{E}\xi_r \leq \frac{t^{r-1}n^r r^{r-2}r!\,2^r}{(r-1)!\,r!\,n^{2(r-1)}}e^{-2r(n-r)(T-r+1)/n^2},$$

and for all sufficiently large n,

$$\mathsf{E}\xi_r \le \left(\frac{2T}{n}\right)^{r-1} ne^r \exp\left\{-\frac{2rT}{n}\cdot\frac{1}{2}\cdot\frac{8}{9}\right\}$$

$$= n\left(\frac{2T}{n}\right)^{r-1} e^r \exp\left\{-\frac{2T}{n}\cdot\frac{4r}{9}\right\}.$$

Therefore

$$\sum_{3\le r\le n/2} \mathsf{E}\xi_r \le \frac{n^2}{2T}\sum_{r=3}^{\infty}\left(\theta e^{1-4\theta/9}\right)^r$$

$$= \frac{n^2\left(\theta e^{1-4\theta/9}\right)^3}{2T\left(1-\theta e^{1-4\theta/9}\right)}.$$

If $n \to \infty$ and $2T = n\log n + xn + o(n)$, then

$$\theta e^{1-4\theta/9} = 2\log n e^{-4\log n/9+1-4x/9+o(1)},$$

and for all sufficiently large n,

$$\theta e^{1-4\theta/9} \le \frac{c\log n}{n^{4/9}},$$

where c is a constant.

Therefore, under the conditions of the theorem,

$$\sum_{3\le r\le n/2} \mathsf{E}\xi_r \to 0.$$

Taking into account that $\mathsf{E}\xi_2 \to 0$ also, we see that, with probability tending to 1, the graph $G_{n,T}$ has only one component besides the isolated vertices. ∎

2.4. Nonequiprobable graphs

The model of the random graph $G_{n,T}$ considered in the previous section can be easily extended to nonequiprobable graphs. However, the approach based on the generalized scheme of allocation, which reduces the investigations of equiprobable graphs to some problems concerning sums of independent random variables, does not apply to nonequiprobable graphs. In this case, few results have been obtained because of the lack of effective methods to investigate these objects.

In this section, we consider a generalization of the random graph $G_{n,T}$ of the previous section. We preserve the notation $G_{n,T}$ for this nonequiprobable graph with n vertices labeled with the numbers $1, 2, \ldots, n$ and T edges, which can be obtained by the following procedure. We consider T independent trials, in each of which one edge is drawn. The edge connects two different vertices or forms a loop;

the vertices with labels i and j are connected with the probability $2 p_i p_j$, and the loop at vertex i is formed with the probability p_i^2; $i, j = 1, \ldots, n$, $p_1, \ldots, p_n \geq 0$, $p_1 + \cdots + p_n = 1$. Thus, after T trials we have a realization of the random graph $G_{n,T}$, which may have loops and multiple edges.

The main result of this section is the following assertion.

Theorem 2.4.1. *Assume that* $p_i = a_i / n$, *where* $a_i = a_i(n)$, $0 < \varepsilon \leq a_i \leq E$, $i = 1, \ldots, n$, ε *and* E *are constants, and the limit*

$$a^2 = \lim_{n \to \infty} \frac{1}{n} \sum_{i=1}^{n} a_i^2$$

exists.

Then, if $n, T \to \infty$ *such that* $2T/n \to \lambda$, $0 < \lambda a^2 < 1$, *the distribution of the number of cycles* $\nu(G_{n,T})$ *in the graph* $G_{n,T}$ *converges to the Poisson distribution with parameter* $\Lambda = -\frac{1}{2} \ln(1 - \lambda a^2)$.

In proving the theorem, the limit distribution of the random variable α_r, the number of cycles of length r, and the joint limit distribution of $\alpha_{r_1}, \ldots, \alpha_{r_s}$ are obtained.

Theorem 2.4.2. *Under the conditions of Theorem 2.4.1, without the requirement* $\lambda a^2 < 1$, *the distribution of the random variable* α_r *for any fixed* r *tends to the Poisson distribution with parameter* $\lambda_r = \lambda^r a^{2r} / (2r)$.

Theorem 2.4.3. *Under the conditions of Theorem 2.4.1, without the requirement* $\lambda a^2 < 1$, *the joint distribution of* $\alpha_{r_1}, \ldots, \alpha_{r_s}$ *for any fixed* $1 \leq r_1 < \cdots < r_s$ *converges to the distribution of* s *independent random variables that have the Poisson distributions with parameters* $\lambda_{r_1}, \ldots, \lambda_{r_s}$, *respectively.*

The proof will be accomplished by the method of moments.

A cycle of length r has no self-intersections if it is composed of r vertices and exactly r edges of $G_{n,T}$. Denote by α_r the number of cycles without self-intersections of length r, $r \geq 3$, in the random graph $G_{n,T}$. For r distinct vertices $i_1, \ldots, i_r$, let $\xi_{i_1,\ldots,i_r} = 1$ if in $G_{n,T}$ there exists a cycle composed of these r vertices containing exactly r edges of $G_{n,T}$; in other cases, we set $\xi_{i_1,\ldots,i_r} = 0$. Then

$$\alpha_r = \sum_{i_1,\ldots,i_r} \xi_{i_1,\ldots,i_r}, \tag{2.4.1}$$

where the summation is taken over all $\binom{T}{r}$ distinct unordered sets of r distinct indices. In the complete graph with vertices $i_1, \ldots, i_r$, there exist $(r-1)!/2$ distinct cycles containing exactly r edges. We label these cycles in an arbitrary order with the numbers $j = 1, \ldots, (r-1)!/2$ and represent the random variable

$\xi_{i_1,\dots,i_r}$ as the sum of indicators:

$$\xi_{i_1,\dots,i_r} = \sum_{j=1}^{(r-1)!/2} \xi_{i_1,\dots,i_r}^{(j)}, \tag{2.4.2}$$

where $\xi_{i_1,\dots,i_r}^{(j)} = 1$ if the jth cycle exists in $G_{n,T}$, and $\xi_{i_1,\dots,i_r}^{(j)} = 0$ otherwise.

We now investigate the behavior of the random variable

$$\nu(G_{n,T}) = \alpha_1 + \cdots + \alpha_n,$$

where the variables α_r are defined by (2.4.1) for $r \geq 3$, α_1 is the number of loops, and α_2 is the number of pairs of parallel edges in $G_{n,T}$.

Each cycle in the graph $G_{n,T}$ may be thought of as the set of edges that form this cycle; therefore, the following assertion is needed for evaluating such probabilities as $\mathbf{P}\{\xi_{i_1,\dots,i_r}^{(j)} = 1\}$. Let $V_r = \{(i_1, j_1), \dots, (i_r, j_r)\}$ be the set of r distinct pairs of vertices in the graph $G_{n,T}$, where $i_k \neq j_k$, $k = 1, \dots, r$. Denote by $P(V_r)$ the probability of the event that all the edges from V_r occur in $G_{n,T}$.

Lemma 2.4.1. *If $n, T \to \infty$, $2T/n \to \lambda$, $0 < \lambda < \infty$, $0 < \varepsilon \leq a_i \leq E < \infty$, $i = 1, \dots, n$, then for arbitrary fixed ε, E, and r,*

$$P(V_r) = \frac{\lambda^r}{n^r} a_{i_1} a_{j_1} \dots a_{i_r} a_{j_r} \left(1 + O\left(\frac{1}{n}\right)\right) \tag{2.4.3}$$

uniformly with respect to $a_1, \dots, a_n$ and all sets V_r.

Moreover, for any $\delta > 0$, there exists a constant c such that, for all r and n,

$$P(V_r) \leq c \frac{(\lambda + \delta)^r}{n^r} a_{i_1} a_{j_1} \cdots a_{i_r} a_{j_r}. \tag{2.4.4}$$

Proof. Set $q_k = 2 p_{i_k} p_{j_k}$, $k = 1, \dots, r$. Then

$$
\begin{aligned}
P(V_r) &= \sum_{m_1,\dots,m_r \geq 1} \frac{T^{[m_1+\cdots+m_r]}}{m_1! \cdots m_r!} q_1^{m_1} \cdots q_r^{m_r} \\
&\quad \times (1 - q_1 - \cdots - q_r)^{T - m_1 - \cdots - m_r} \\
&= T^{[r]} q_1 \cdots q_r \bigg((1 - q_1 - \cdots - q_r)^{T-r} \\
&\quad + {\sum}' \frac{(T-r)^{[m_1+\cdots+m_r-r]}}{m_1! \cdots m_r!} q_1^{m_1-1} \cdots q_r^{m_r-1} \\
&\quad \times (1 - q_1 - \cdots - q_r)^{T - m_1 - \cdots - m_r} \bigg). \tag{2.4.5}
\end{aligned}
$$

Here $x^{[m]} = x(x-1)\cdots(x-m+1)$; the summation in $\sum'$ is taken over all sets $\{m_1, \dots, m_r\}$ in which $m_1, \dots, m_r \geq 1$ and there exists i, $1 \leq i \leq r$, such that

$m_i > 1$. It is clear that

$$(1 - q_1 - \cdots - q_r)^{T-r} \leq 1,$$

and for an arbitrary fixed r,

$$(1 - q_1 - \cdots - q_r)^{T-r} = 1 + O(1/n). \tag{2.4.6}$$

In addition,

$$\sideset{}{'}\sum \frac{(T-r)^{[m_1+\cdots+m_r-r]}}{m_1! \cdots m_r!} q_1^{m_1-1} \cdots q_r^{m_r-1}(1 - q_1 - \cdots - q_r)^{T-m_1-\cdots-m_r}$$

$$\leq \sum_{i=1}^{r} q_i(T-r)S_i, \tag{2.4.7}$$

where

$$S_i = \sum_{\substack{m_1,\ldots,m_r \geq 1 \\ m_i > 1}} \frac{(T-r-1)^{[m_1+\cdots+m_r-r-1]}}{m_1! \cdots m_r!}$$

$$\times \frac{q_1^{m_1-1} \cdots q_r^{m_r-1}}{q_i}(1 - q_1 - \cdots - q_r)^{T-m_1-\cdots-m_r}.$$

Let $l_i = m_i - 2$, $l_j = m_j - 1$, $j \neq i$ (recall that $m_i > 1$). Then

$$S_i = \sum_{l_1,\ldots,l_r \geq 0} \frac{(T-r-1)^{[l_1+\cdots+l_r]}}{(l_1+1)! \cdots (l_i+2)! \cdots (l_r+1)!}$$

$$\times q_1^{l_1} \cdots q_r^{l_r}(1 - q_1 - \cdots - q_r)^{T-r-1-l_1-\cdots-l_r}$$

$$\leq \sum_{l_1,\ldots,l_r \geq 0} \frac{(T-r-1)^{[l_1+\cdots+l_r]}}{l_1! \cdots l_r!}$$

$$\times q_1^{l_1} \cdots q_r^{l_r}(1 - q_1 - \cdots - q_r)^{T-r-1-l_1-\cdots-l_r} = 1. \tag{2.4.8}$$

Now assertion (2.4.3) follows from (2.4.5)–(2.4.8), and assertion (2.4.4) from (2.4.5), (2.4.7), and (2.4.8), since

$$\sum_{i=1}^{r} q_i(T-r) \leq \frac{2TrE^2}{n^2},$$

$$T^{[r]}q_1 \cdots q_r \leq \frac{(2T)^r}{n^{2r}} a_{i_1} a_{j_1} \cdots a_{i_r} a_{j_r}.$$

∎

Corollary 2.4.1. *If $n, T \to \infty$, $2T/n \to \lambda$, $0 < \lambda < \infty$, $0 < \varepsilon \le a_i \le E < \infty$, $i = 1, \dots, n$, then for arbitrary fixed ε, E, λ, and r,*

$$\mathbf{P}\{\xi^{(j)}_{i_1 \cdots i_r} = 1\} = \frac{\lambda^r}{n^r} a^2_{i_1} \cdots a^2_{i_r} \left(1 + O\left(\frac{1}{n}\right)\right)$$

uniformly with respect to j, $1 \le j \le (r-1)!/2$, all sets $\{i_1, \dots, i_r\}$ and $a_1, \dots, a_n$. Moreover, for any $\delta > 0$, there exists a constant c such that, for all r and n,

$$\mathbf{P}\{\xi^{(j)}_{i_1 \cdots i_r} = 1\} \le c \frac{(\lambda + \delta)^r}{n^r} a^2_{i_1} \cdots a^2_{i_r}.$$

Proof. The equality $\xi^{(j)}_{i_1 \cdots i_r} = 1$ holds if and only if in $G_{n,T}$ there exist r fixed edges, $\{(k_1, j_1), \dots, (k_r, j_r)\}$, $k_v \ne j_v$, $v = 1, \dots, r$, which form the jth cycle on the vertices $i_1, \dots, i_r$. For these edges, the sets $\{k_1, \dots, k_r\}$ and $\{j_1, \dots, j_r\}$ coincide with the set $\{i_1, \dots, i_r\}$. Therefore, the corollary follows from Lemma 2.4.1. ∎

The notation $\{i_1, \dots, i_r\}$ denotes an unordered set of distinct indices $i_1, \dots, i_r$; the number of such sets is $\binom{n}{r}$. For ordered sets of distinct indices $i_1, \dots, i_r$, we will use the notation $\langle i_1, \dots, i_r \rangle$; the number of such sets is $n^{[r]}$. By the symbols

$$\sum_{\{i_1,\dots,i_r\}}, \qquad \sum_{\langle i_1,\dots,i_r \rangle},$$

we will denote the summations over all distinct unordered and ordered sets of r distinct indices, respectively. It is clear that the summation over all unordered sets $\{i_1, \dots, i_r\}$ is well suited to summands $f_{i_1 \dots i_r}$ whose values are invariant with respect to the permutations of indices. For such summands,

$$\sum_{\langle i_1,\dots,i_r \rangle} f_{i_1 \dots i_r} = r! \sum_{\{i_1,\dots,i_r\}} f_{i_1 \dots i_r}, \tag{2.4.9}$$

and, moreover,

$$\sum_{\langle i^{(1)}_1,\dots,i^{(1)}_r \rangle,\dots,\langle i^{(k)}_1,\dots,i^{(k)}_r \rangle} f_{i^{(1)}_1 \dots i^{(1)}_r \dots i^{(k)}_1 \dots i^{(k)}_r} = \sum_{\langle j_1,\dots,j_{rk} \rangle} f_{j_1 \dots j_{rk}} \tag{2.4.10}$$

if the left-hand side summation is taken over all distinct ordered sets of distinct r-dimensional indices $i^{(1)}_1, \dots, i^{(k)}_r$.

Lemma 2.4.2. *If $0 < \varepsilon \le a_i \le E < \infty$, $i = 1, \dots, n$, then for any fixed r, as $n \to \infty$,*

$$\left(\sum_{i=1}^n a^2_i\right)^r = \sum_{\langle i_1,\dots,i_r \rangle} a^2_{i_1} \cdots a^2_{i_r} \left(1 + O\left(\frac{1}{n}\right)\right). \tag{2.4.11}$$

Proof. The following representation is valid:

$$\left(\sum_{i=1}^{n} a_i^2\right)^r = \sum_{i_1,\dots,i_r=1}^{n} a_{i_1}^2 \cdots a_{i_r}^2 = \sum_{\langle i_1,\dots,i_r\rangle} a_{i_1}^2 \cdots a_{i_r}^2 + \sum_{\langle i_1,\dots,i_r\rangle}^{*} a_{i_1}^2 \cdots a_{i_r}^2,$$

where the summation in the first sum is taken over all distinct ordered sets of distinct indices, and in the asterisked sum, over all distinct ordered sets, each have at least two identical indices. The number of summands in the first sum is $n^{[r]}$; the number of summands in the second sum is equal to $n^r - n^{[r]}$ and does not exceed $c_r n^{r-1}$ where the constant c_r depends only on r. Therefore

$$\sum_{\langle i_1,\dots,i_r\rangle} a_{i_1}^2 \cdots a_{i_r}^2 \geq n^{[r]} \varepsilon^{2r}, \qquad \sum_{\langle i_1,\dots,i_r\rangle}^{*} a_{i_1}^2 \cdots a_{i_r}^2 \leq c_r n^{r-1} E^{2r},$$

and the proof is complete.   ∎

Corollary 2.4.2. *Under the conditions of Theorem 2.4.2, for any fixed $r \geq 3$,*

$$\mathbf{E}\alpha_r \to \frac{\lambda^r a^{2r}}{2r}.$$

Moreover, for any $\delta > 0$, there exists a constant c such that

$$\mathbf{E}\alpha_r \leq c\frac{(\lambda+\delta)^r (a^2+\delta)^r}{2r}.$$

Proof. Using representations (2.4.1) and (2.4.2), with the aid of (2.4.9), Corollary 2.4.1, and Lemma 2.4.2, we obtain

$$\mathbf{E}\alpha_r = \frac{(r-1)!\,\lambda^r}{2}\,\frac{1}{n^r}\,\sum_{\{i_1,\dots,i_r\}} a_{i_1}^2 \cdots a_{i_r}^2 \left(1 + O\left(\frac{1}{n}\right)\right)$$

$$= \frac{(r-1)!\,\lambda^r}{2n^r r!}\,\sum_{\langle i_1,\dots,i_r\rangle} a_{i_1}^2 \cdots a_{i_r}^2 \left(1 + O\left(\frac{1}{n}\right)\right)$$

$$= \frac{\lambda^r}{2r}\left(\frac{1}{n}\sum_{i=1}^{n} a_i^2\right)^r (1 + o(1)) = \frac{\lambda^r a^{2r}}{2r}(1 + o(1)).$$

The second assertion follows immediately from the inequality of Corollary 2.4.1.
                                                                                  ∎

We now evaluate the factorial moments of α_r. If $S_n = \xi_1 + \cdots + \xi_n$, where $\xi_1, \dots, \xi_n$ take the values 0 and 1 only, then according to Theorem 1.1.4,

$$S_n(S_n - 1)\cdots(S_n - m + 1) = \sum_{\langle k_1,\dots,k_m\rangle} \xi_{k_1}\cdots\xi_{k_m}, \qquad (2.4.12)$$

where the summation is taken over all distinct ordered sets of m distinct indices.

In our case, the indices have a composite structure because

$$\alpha_r = \sum_{\{i_1,\dots,i_r\}} \sum_{j=1}^{(r-1)!/2} \xi_{i_1\cdots i_r}^{(j)}.$$

The following representation is analogous to (2.4.12):

$$\alpha_r(\alpha_r - 1)\cdots(\alpha_r - m + 1) = \sum \xi_{i_1^{(1)}\cdots i_r^{(1)}}^{(j_1)} \cdots \xi_{i_1^{(m)}\cdots i_r^{(m)}}^{(j_m)}, \qquad (2.4.13)$$

where the summation is taken over all distinct ordered sets

$$\left(\left(\{i_1^{(1)},\dots,i_r^{(1)}\}, j_1 \right), \dots, \left(\{i_1^{(m)},\dots,i_r^{(m)}\}, j_m \right) \right)$$

of distinct indices of the form $(\{i_1,\dots,i_r\}, j)$; the set $\{i_1,\dots,i_r\}$ in the index is considered an unordered set of distinct indices, and j indicates the number of the cycle formed by the vertices $i_1,\dots,i_r$.

We show that under the conditions of Theorem 2.4.2, for any fixed r and any fixed $m \geq 1$,

$$\mathbf{E}\alpha_r^{[m]} \to \left(\frac{\lambda^r a^{2r}}{2r} \right)^m. \qquad (2.4.14)$$

This assertion for $m = 1$ follows from Corollary 2.4.2.

In order to become accustomed to the more complicated notation, we first consider the case $m = 2$. By (2.4.13),

$$\mathbf{E}\alpha_r^{[2]} = \sum_{\left(\left(\{i_1^{(1)},\dots,i_r^{(1)}\}, j_1 \right), \left(\{i_1^{(2)},\dots,i_r^{(2)}\}, j_2 \right) \right)} \mathbf{P}\left\{ \xi_{i_1^{(1)}\cdots i_r^{(1)}}^{(j_1)} = \xi_{i_1^{(2)}\cdots i_r^{(2)}}^{(j_2)} = 1 \right\}.$$

Decompose the right-hand side sum into two sums. Let the first sum Σ_1 include the summands with nonintersecting sets $\{i_1^{(1)},\dots,i_r^{(1)}\}$ and $\{i_1^{(2)},\dots,i_r^{(2)}\}$. When we take into account that in this case $2r$ edges must exist to guarantee

$$\xi_{i_1^{(1)}\cdots i_r^{(1)}}^{(j_1)} = \xi_{i_1^{(2)}\cdots i_r^{(2)}}^{(j_2)} = 1,$$

and by using Lemma 2.4.1, we obtain

$$\mathbf{P}\left\{ \xi_{i_1^{(1)}\cdots i_r^{(1)}}^{(j_1)} = \xi_{i_1^{(2)}\cdots i_r^{(2)}}^{(j_2)} = 1 \right\}$$

$$= \left(\frac{\lambda}{n} \right)^{2r} a_{i_1^{(1)}}^2 \cdots a_{i_r^{(1)}}^2 a_{i_1^{(2)}}^2 \cdots a_{i_r^{(2)}}^2 \left(1 + O\left(\frac{1}{n} \right) \right).$$

Therefore

$$\Sigma_1 = \left(\frac{\lambda}{n} \right)^{2r} \left(\frac{(r-1)!}{2} \right)^2 \sum_{\left(\{i_1^{(1)},\dots,i_r^{(1)}\}, \{i_1^{(2)},\dots,i_r^{(2)}\} \right)} a_{i_1^{(1)}}^2 \cdots a_{i_r^{(1)}}^2 a_{i_1^{(2)}}^2 \cdots a_{i_r^{(2)}}^2$$

$$\times (1 + O(1/n)).$$

It is clear by virtue of (2.4.9) and (2.4.10) that

$$\sum_{(\{i_1^{(1)},...,i_r^{(1)}\},\{i_1^{(2)},...,i_r^{(2)}\})} a_{i_1^{(1)}}^2 \cdots a_{i_r^{(1)}}^2 a_{i_1^{(2)}}^2 \cdots a_{i_r^{(2)}}^2 = \frac{1}{(r!)^2} \sum_{(i_1,...,i_{2r})} a_{i_1}^2 \cdots a_{i_{2r}}^2.$$

Therefore, by virtue of Lemma 2.4.2,

$$\Sigma_1 = \left(\frac{(r-1)!}{2}\right)^2 \frac{\lambda^{2r}}{(r!)^2} \left(\frac{1}{n}\sum_{i=1}^n a_i^2\right)^{2r} (1+o(1)) = \left(\frac{\lambda^r a^{2r}}{2r}\right)^2 (1+o(1)).$$

$$(2.4.15)$$

We now show that the remaining sum Σ_2 tends to zero.

The summation in Σ_2 is taken over the pairs of composite indices in which the sets $\{i_1^{(1)},\ldots,i_r^{(1)}\}$ and $\{i_1^{(2)},\ldots,i_r^{(2)}\}$ have at least one common element. Each composite index $(\{i_1,\ldots,i_r\},j)$ corresponds to a cycle in the complete graph with n vertices; the cycle consists of r edges and the vertices $i_1,\ldots,i_r$. Two cycles corresponding to the indices $(\{i_1^{(1)},\ldots,i_r^{(1)}\},j_1)$ and $(\{i_1^{(2)},\ldots,i_r^{(2)}\},j_2)$ can have $M < 2r$ distinct vertices and L distinct edges. We decompose the sum Σ_2 into the sums $\Sigma_{M,L}$ containing summands with fixed values of the parameters M and L. The number of such sums does not exceed $(2r)^2$; therefore it is sufficient to prove that any sum $\Sigma_{M,L}$ tends to zero. It is easy to see that in the case $M < 2r$, the inequality $L \geq M+1$ is valid. The number of summands in the sum $\Sigma_{M,L}$ does not exceed n^M, and the probability that L fixed edges appear in $G_{n,T}$ does not exceed, by virtue of (2.4.4), the value cn^{-L}. This implies

$$\Sigma_{M,L} \leq \frac{c}{n^{L-M}} \leq \frac{c}{n}. \qquad (2.4.16)$$

Therefore, as $n \to \infty$,

$$\Sigma_2 \to 0. \qquad (2.4.17)$$

The assertion (2.4.14) for $m = 2$ follows from (2.4.15) and (2.4.17).

Now let us consider the factorial moment of an arbitrary order m. By (2.4.13),

$$\mathbf{E}\alpha_r^{[m]} = \Sigma_1 + \Sigma_2,$$

where the sum Σ_1 includes only summands that do not have a pair of sets from $\{i_1^{(1)},\ldots,i_r^{(1)}\},\ldots,\{i_1^{(m)},\ldots,i_r^{(m)}\}$ with common elements. In this case, rm edges must occur in the graph $G_{n,T}$ to guarantee that the corresponding random variables equal 1. From this and Lemma 2.4.1, it follows that

$$\mathbf{P}\left\{\xi_{i_1^{(1)}\cdots i_r^{(1)}}^{(j_1)} = 1,\ldots,\xi_{i_1^{(m)}\cdots i_r^{(m)}}^{(j_m)} = 1\right\}$$

$$= \left(\frac{\lambda}{n}\right)^{mr} a_{i_1^{(1)}}^2 \cdots a_{i_r^{(1)}}^2 \cdots a_{i_1^{(m)}}^2 \cdots a_{i_r^{(m)}}^2 (1+o(1)),$$

and, by (2.4.9) and Lemma 2.4.2,

$$\Sigma_1 = \left(\frac{\lambda^r a^{2r}}{2r}\right)^m (1 + o(1)). \tag{2.4.18}$$

It remains to prove that the sum Σ_2 taken over the remaining sets of indices tends to zero. The summation in Σ_2 is taken over m sets of composite indices that have at least one common element in at least one pair of the sets $\{i_1^{(p)}, \ldots, i_r^{(p)}\}$, $\{i_1^{(q)}, \ldots, i_r^{(q)}\}$, $p \neq q$. Recall that each composite index corresponds to a cycle in the complete graph with n vertices. The cycles corresponding to m indices can contain M distinct vertices and L distinct edges. We decompose the sum Σ_2 into the sums $\Sigma_{M,L}$ containing summands with fixed values of the parameters M and L. The number of such sums does not exceed $(rm)^2$; therefore it is sufficient to prove that any sum $\Sigma_{M,L}$ tends to zero. It is clear that if $M < rm$, then $L \geq M+1$. Thus, since the number of summands in the sum $\Sigma_{M,L}$ does not exceed n^M, and by (2.4.3) the probability of L fixed edges occurring in $G_{n,T}$ does not exceed cn^{-L},

$$\Sigma_{M,L} \leq \frac{c}{n^{L-M}} \leq \frac{c}{n}.$$

Therefore, as $n \to \infty$,

$$\Sigma_2 \to 0. \tag{2.4.19}$$

The assertion (2.4.14) follows from (2.4.18) and (2.4.19).

By (2.4.14), the limit distribution for α_r, $r \geq 3$, is the Poisson distribution with parameter $\lambda_r = \lambda^r a^{2r}/(2r)$. It is easy to see that in the current situation the number of loops α_1 and the number of pairs of parallel edges α_2 approach the Poisson distributions with parameters $\lambda_1 = \lambda a^2/2$ and $\lambda_2 = \lambda^2 a^4/4$, respectively.

This proves Theorem 2.4.2.

The more general Theorem 2.4.3 can be proved analogously. It is sufficient to verify that under the conditions of the theorem,

$$\mathbf{E}\alpha_{r_1}^{[m_1]} \cdots \alpha_{r_s}^{[m_s]} \to \lambda_{r_1}^{m_1} \cdots \lambda_{r_s}^{m_s}$$

for arbitrary fixed integers $m_1, \ldots, m_s$, where

$$\lambda_r = \frac{\lambda^r a^{2r}}{2r}.$$

By (2.4.13),

$$\alpha_{r_k}^{[m_k]} = \sum_{\langle (I_1^{(k)}, j_1^{(k)}), \ldots, (I_{m_k}^{(k)}, j_{m_k}^{(k)}) \rangle} \xi_{I_1^{(k)}}^{(j_1^{(k)})} \cdots \xi_{I_{m_k}^{(k)}}^{(j_{m_k}^{(k)})},$$

where

$$I_l^{(k)} = \{i_1^{(l,k)}, \ldots, i_{r_k}^{(l,k)}\}, \quad l = 1, \ldots, m_k, \quad k = 1, \ldots, s,$$

are unordered sets of r_k vertices, and $j_l^{(k)}$, $l = 1, \ldots, m_k$, $k = 1, \ldots, s$, are the numbers of cycles of length r_k under the labeling chosen.

Therefore

$$\mathbf{E}\alpha_{r_1}^{[m_1]} \cdots \alpha_{r_s}^{[m_s]}$$

$$= \sum_I \mathbf{P}\left\{\xi_{I_1^{(1)}}^{\left(j_1^{(1)}\right)} = 1, \ldots, \xi_{I_{m_1}^{(1)}}^{\left(j_{m_1}^{(1)}\right)} = 1, \ldots, \xi_{I_1^{(s)}}^{\left(j_1^{(s)}\right)} = 1, \ldots, \xi_{I_{m_s}^{(s)}}^{\left(j_{m_s}^{(s)}\right)} = 1\right\},$$

where

$$I = \left\{\left\{\left(I_1^{(1)}, j_1^{(1)}\right), \ldots, \left(I_{m_1}^{(1)}, j_{m_1}^{(1)}\right)\right\}, \ldots, \left\{\left(I_1^{(s)}, j_1^{(s)}\right), \ldots, \left(I_{m_s}^{(s)}, j_{m_s}^{(s)}\right)\right\}\right\}.$$

We decompose the sum on the right-hand side of this representation into two parts; let the sum Σ_1 include only summands with the distinct elements in all $I_l^{(k)}$, $l = 1, \ldots, m_k$, $k = 1, \ldots, s$; and let the sum Σ_2 include all the remaining summands. For the summands of the first sum, the corresponding random variables equal 1 only if there exist $m_1 r_1 + \cdots + m_s r_s$ fixed edges in $G_{n,T}$. Therefore, by Lemma 2.4.1,

$$\mathbf{P}\left\{\xi_{I_1^{(1)}}^{\left(j_1^{(1)}\right)} = 1, \ldots, \xi_{I_{m_s}^{(s)}}^{\left(j_{m_s}^{(s)}\right)} = 1\right\}$$

$$= \left(\frac{\lambda}{n}\right)^{m_1 r_1 + \cdots + m_s r_s} a_{i_1^{(1,1)}}^2 \cdots a_{i_{r_1}^{(1,1)}}^2 \cdots a_{i_1^{(m_s,s)}}^2 \cdots a_{i_{r_s}^{(m_s,s)}}^2 (1 + o(1)),$$

and, by (2.4.9), (2.4.10), and Lemma 2.4.2,

$$\Sigma_1 = \left(\frac{\lambda^{r_1} a^{2r_1}}{2r_1}\right)^{m_1} \cdots \left(\frac{\lambda^{r_s} a^{2r_s}}{2r_s}\right)^{m_s} (1 + o(1)).$$

It remains to prove that Σ_2 tends to zero. The summation in Σ_2 is taken over sets of composite indices in which at least one of the elements $1, 2, \ldots, n$ is encountered at least twice. A cycle corresponds to each of the composite indices. The existence of a common element in the cycles implies that the number M of distinct vertices contained in the cycles and the number L of distinct edges involved in the cycles satisfy $L \geq M + 1$. We decompose the sum Σ_2 into a finite number of sums $\Sigma_{M,L}$ containing summands with fixed values of the parameters M and L. By virtue of (2.4.3), for each of these sums, the estimate

$$\Sigma_{M,L} \leq \frac{c}{n^{L-M}} \leq \frac{c}{n}$$

holds because the number of summands does not exceed n^M, and the probability of L fixed edges occurring in $G_{n,T}$ does not exceed cn^{-L}. This proves Theorem 2.4.3.

To prove Theorem 2.4.1, we need the following auxiliary assertion.

Lemma 2.4.3. *Let* $\xi_1^{(n)}, \ldots, \xi_n^{(n)}$ *be nonnegative integer-valued random variables such that for an arbitrary fixed s and arbitrary nonnegative integers $k_1, \ldots, k_s$,*

$$\mathbf{P}\{\xi_1^{(n)} = k_1, \ldots, \xi_s^{(n)} = k_s\} \to \frac{a_1^{k_1} \cdots a_s^{k_s}}{k_1! \cdots k_s!} e^{-a_1 - \cdots - a_s}$$

as $n \to \infty$, where $a_1, a_2, \ldots$ is a fixed sequence of nonnegative numbers. Moreover, suppose

$$\mathbf{E}(\xi_{s+1}^{(n)} + \cdots + \xi_n^{(n)}) \to 0 \tag{2.4.20}$$

as $s \to \infty$, uniformly in n, and let

$$\sum_{k=1}^{\infty} a_k = A < \infty.$$

Then the distribution of the random variable $\zeta_n^{(n)} = \xi_1^{(n)} + \cdots + \xi_n^{(n)}$ converges to the Poisson distribution with parameter A.

Proof. We show that for an arbitrary fixed $\varepsilon > 0$ and an arbitrary fixed m,

$$\left| \mathbf{P}\{\zeta_n^{(n)} = m\} - \frac{A^m e^{-A}}{m!} \right| \le \varepsilon$$

for sufficiently large n. For fixed ε and m, there exists s such that

$$\left| \frac{A_s^m e^{-A_s}}{m!} - \frac{A^m e^{-A}}{m!} \right| \le \frac{\varepsilon}{3},$$

where $A_s = a_1 + \cdots + a_s$.

It is not hard to see that

$$\left| \mathbf{P}\{\zeta_n^{(n)} = m\} - \mathbf{P}\{\zeta_s^{(n)} = m\} \right| \le \mathbf{P}\{\xi_{s+1}^{(n)} + \cdots + \xi_n^{(n)} > 0\}.$$

Therefore, by (2.4.20), $|\mathbf{P}\{\zeta_n^{(n)} = m\} - \mathbf{P}\{\zeta_s^{(n)} = m\}| \le \varepsilon/3$ for sufficiently large s. Finally, the conditions of the lemma yield the convergence of the distribution of $\zeta_s^{(n)} = \xi_1^{(n)} + \cdots + \xi_s^{(n)}$ (for any fixed s) to the Poisson distribution with parameter $A_s = a_1 + \cdots + a_s$. Therefore

$$\left| \mathbf{P}\{\zeta_s^{(n)} = m\} - \frac{A_s^m e^{-A_s}}{m!} \right| \le \frac{\varepsilon}{3}$$

for sufficiently large s. ∎

Theorem 2.4.1 follows from Theorem 2.4.3 and Lemma 2.4.3, whose conditions are satisfied when $\lambda a^2 < 1$.

2.5. Notes and references

The investigation of the evolution of random graphs began when P. Erdős and A. Rényi published the results of their study [37] in 1960. Along with the basic properties of the random graph $G_{n,T}^{(3)}$, they discovered the effect known as a phase transition. At about the same time, V. E. Stepanov studied the graph $G_{n,p}$, as documented later [133, 134, 135]. Until recently, Stepanov's results had not seemed to receive wide recognition. In particular, Stepanov proved that if $p = c/n$, where c is a constant, $c > 1$, then the size of the giant component is asymptotically normal with mean $n\alpha(c)$ and variance $n\beta(c)$, where

$$\alpha(c) = 1 - \frac{\gamma}{c}, \qquad \beta(c) = \frac{\gamma(1 - \gamma/c)}{c(1 - \gamma)^2},$$

and $\gamma < 1$ is the root of the equation

$$\gamma e^{-\gamma} = ce^{-c}.$$

A similar assertion for the graph $G_{n,T}^{(3)}$ was proved by B. Pittel [123] about twenty years later. He found that the size of the giant component of $G_{n,T}^{(3)}$ is asymptotically normal with parameters $n\alpha(c)$ and $n\beta(c)(1 - 2\gamma + 2\gamma^2/c)$ as $n, T \to \infty$ and $2T/n \to c > 1$.

Many open questions concerning the evolution of random graphs remain. The main goal of this chapter is to demonstrate the approach based on the generalized scheme of allocation in investigations of the evolution of random graphs. Section 2.1 shows that fine properties of subcritical graphs can be obtained in a rather simple and natural way, especially as concerns the behavior of subcritical graphs near the critical point. The transition phenomena for the graph $G_{n,T}^{(3)}$ were first considered by B. Bollobas [20]. The results presented in Section 2.1 can be found in [77]. The approach based on the generalized scheme of allocation allowed us to prove asymptotic normality of the number of unicyclic components and find the limit distribution of the maximum sizes of trees and unicyclic components.

Section 2.2 is devoted to critical graphs. The behavior of random graphs near the critical point, and especially in the critical domain where the giant component appears, is very complicated and difficult to investigate. The investigations of the behavior are far from complete, but even now the results obtained could fill another book. Much information about random graphs can be found in the fundamental work by Bollobas [21] and in the book [105], which is devoted to the evolution of random graphs. A detailed investigation of the birth of the giant component is given in [63]. Supercritical graphs are considered by Luczak [99], who, in particular, proved that the right-hand bound of the critical domain is determined by the conditions $n, T \to \infty$, $(1 - 2T/n)^3 n \to -\infty$.

Formally, to analyze supercritical random graphs, we can use the representation of almost all such graphs as a combination of components of three types: one giant component, trees, and unicyclic components. However, this approach is hampered by the absence of a simple formula for the number of connected graphs with n vertices and T edges with $k = T - n > 0$. Note that $k = T - n$ is equal to the number of independent cycles in the graph and is called the cyclomatic number of the graph. Denote by $c(n, k)$ the number of connected graphs with n labeled vertices and a cyclomatic number k. It is clear that $c(n, -1)$ is the number of trees, and by the Cayley formula, $c(n, -1) = n^{n-2}$, whereas $c(n, 0)$ is the number u_n of unicyclic graphs considered in Section 1.7. The numbers $c(n, k)$ were investigated by Stepanov (see [10, 142, 143]) and E. M. Wright [151, 152] and are known as the Stepanov–Wright numbers (see [143]). As $n \to \infty$ and $k^3/n \to 0$,

$$c(n, k) = d(3\pi)^{1/2} \left(\frac{e}{12k} \right)^{k/2} n^{n+(k-1)/2}(1 + o(1)),$$

where, as it was proved by Meertens, $d = 1/(2\pi)$ (see Bender, Canfield, and McKay [16]).

We hope that the results of the study by Bender et al. [17], who give the asymptotics of $c(n, k)$ for all regular variations of the parameters n and k, can be used in the application of the generalized scheme to random graphs and help to bring the investigations of supercritical graphs to the level attained for the subcritical case in Section 2.1. Note that obtaining the limit distributions of numerical characteristics of supercritical graphs would be merely a problem of averaging if the joint distribution of the size of the giant component and the number of its edges were known.

The parameter $\theta = 2T/n$ plays the role of time in the evolution of random graphs. Therefore, each numerical characteristic of a random graph can be considered not only as a random variable, but also as a random process with the time parameter θ. Of significant interest is the approach using the convergence of such processes. This approach is used in the recent papers [34, 62, 127]. Note that the investigations of convergence of such random processes in combinatorial problems were started by B. A. Sevastyanov [132] and Yu. V. Bolotnikov [22, 23, 24].

The random graph $G_{n,T}$ discussed in Section 2.3 was investigated by Kolchin [79, 83]. This graph provides an appropriate model of the graph corresponding to the left-hand side of a system of random congruences modulo 2 considered in the next chapter. An analogy of Theorem 2.3.8 for bipartite graphs was proved by Saltykov [131].

The nonequiprobable version of the graph $G_{n,T}$ is considered in Section 2.4, where the results of the papers [88, 66, 65] are presented. Here we use the method of moments. The lack of regular methods for an asymptotic analysis of nonequiprobable graphs makes it impossible to carry out anything approaching a complete investigation of such graphs. It seems to us that developing the methods appropriate for the analysis of nonequiprobable combinatorial structures is a problem of great importance.

3

Systems of random linear equations in GF(2)

3.1. Rank of a matrix and critical sets

In this section, we consider systems of linear equations in GF(2), the field with elements 0 and 1. Let us begin with two examples where such systems appear.

Consider first a simple classification problem. Suppose we have a set of n objects of two sorts, for example, of two different weights. We may sequentially sample pairs of the objects from the set at random, compare the weights of the objects from the chosen pair, and determine whether the weights are identical or different. The problem is to identify the objects that have the same weight – actually, to estimate the probability of finding that solution. For a formal description of the situation, let $\{1, 2, \ldots, n\}$ be the set of objects under consideration and let x_j be the unknown type of the object j, $j = 1, \ldots, n$. We may assume that $x_1, \ldots, x_n$ take the values 0 and 1, depending on the class to which the object belongs. We choose a pair of objects $i(t)$ and $j(t)$ in the trial with number t, $t = 1, \ldots, T$, and let b_t be the result of their comparison: $b_t = 0$ if their weights are identical, and $b_t = 1$ otherwise. Thus, the results of the comparisons can be written as the following system of linear equations in GF(2):

$$x_{i(t)} + x_{j(t)} = b_t, \quad t = 1, \ldots, T. \tag{3.1.1}$$

It is clear that the system can be rewritten in the matrix form

$$AX = B,$$

where $X = (x_1, \ldots, x_n)$ and $B = (b_1, \ldots, b_T)$ are column-vectors, and the elements a_{tj} of the matrix $A = \|a_{tj}\|$, $t = 1, \ldots, T$, $j = 1, \ldots, n$, are random variables whose distribution is determined by the sampling procedure. It is convenient to associate the system, or more precisely, the matrix A, with the random graph $G_{n,T}$ with n vertices that correspond to the variables $x_1, \ldots, x_n$. The graph has T edges $(i(t), j(t))$, $t = 1, \ldots, T$. Therefore the graph can have loops and multiple edges, depending on the sampling procedure.

In this chapter, we consider the characteristics of the graph $G_{n,T}$ that are related to some of the properties of the system (3.1.1). It is clear that the connectedness of the graph is an important characteristic for the classification problem. Indeed, in the case where the graph is connected, we can determine all values of the variables $x_1, \ldots, x_n$ if we set one of them equal to 0 or 1. In both cases, the partitions of the set are the same, but the system has two different solutions. In the case where the graph $G_{n,T}$ is disconnected, the system has more than two solutions; therefore a complete classification is impossible.

Now let the vector B consist of independent random variables that take the values 0 and 1. If the balance is out of order, the weighings can sometimes be wrong, and the variables $b_1, \ldots, b_T$ can differ from the true values. In this case, we obtain a system with distorted entries on the right-hand side that sometimes has no solution. If the balance is completely wrong, we may assume that the variables $b_1 \ldots, b_T$ do not depend on the left-hand side of the system and take the values 0 and 1 with equal probabilities. In this situation, several natural problems arise. Does the right-hand side $b_1, \ldots, b_T$ depend on the left-hand side of the system or are the sides independent? Can we reconstruct the real values of $x_1, \ldots, x_n$ in the case where the right-hand parts $b_1, \ldots, b_T$ are distorted?

Let us turn to the second example. Let a vector $(c_1, \ldots, c_n)$ in GF(2) be given. If we take an initial vector $x_1, \ldots, x_n$, then we can develop the recurring sequence $x_{n+t}, t = 1, 2, \ldots$, by the following recurrence relation:

$$x_{n+t} = c_1 x_t + \cdots + c_n x_{n+t-1}, \quad t = 1, 2, \ldots. \tag{3.1.2}$$

This recurrence relation can be realized with the help of a device called a shift register, presented in Figure 3.1.1. A shift register consists of n cells or stages with labels $1, 2, \ldots, n$. The n-dimensional $(0, 1)$ vector of the contents of these stages is called the state of the shift register. At an initial moment, the state of the shift register under consideration is the vector $(x_1, \ldots, x_n)$. The choice of the vector $(c_1, \ldots, c_n)$ means that we choose the stages with numbers corresponding to the ones in the sequence $c_1, \ldots, c_n$ and form the mod 2 sum $x_{n+1} = c_1 x_1 + \cdots + c_n x_n$. At the next moment, the contents of all stages are shifted to the left so that x_n transfers to the stage numbered $n-1$, x_{n-1} transfers to the stage $n-2$, and so on, x_1 leaves the register, and the sum $x_{n+1} = c_1 x_1 + \cdots + c_n x_n$ is placed into the stage with label n. Thus the state $(x_1, \ldots, x_n)$ transfers to the state $(x_2, \ldots, x_{n+1})$.

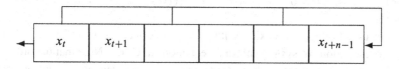

Figure 3.1.1. Shift register

The process is repeated. Thus, if $c_1, \ldots, c_n$ are given, then for any initial state $x_1, \ldots, x_n$, the recurring sequence (3.1.2) satisfies

$$x_{n+1} = c_1 x_1 + \cdots + c_n x_n,$$

$$x_{n+2} = c_1 x_2 + \cdots + c_n x_{n+1},$$

$$\cdots$$

$$x_{n+T} = c_1 x_T + \cdots + c_n x_{n+T-1}.$$

Let us change the notations and put $b_t = x_{n+t}$, $t = 1, 2, \ldots, T$, and $a_{11} = c_1, \ldots, a_{1n} = c_n$. Then the first relation becomes

$$a_{11} x_1 + \cdots + a_{1n} x_n = b_1.$$

It is clear that we can substitute $c_1 x_1 + \cdots + c_n x_n$ for x_{n+1} in the second relation and obtain

$$a_{21} x_1 + \cdots + a_{2n} x_n = b_2.$$

In the same way, we obtain

$$a_{11} x_1 + \cdots + a_{1n} x_n = b_1,$$

$$\cdots \tag{3.1.3}$$

$$a_{T1} x_1 + \cdots + a_{Tn} x_n = b_T.$$

Suppose that the initial state $(x_1, \ldots, x_n)$ is unknown and we observe the sequence $b_1, \ldots, b_T$. Then we can regard relations (3.1.3) as a system of linear equations with respect to the unknowns $x_1, \ldots, x_n$. A natural question is how many observations are needed to reconstruct the initial state and to obtain all elements of the sequence $b_t, t = T + 1, \ldots$.

The other situation concerns the feedback points $c_1, \ldots, c_n$. Suppose we observe the sequence $b_1, \ldots, b_T$, but the vector $(c_1, \ldots, c_n)$ determining the shift register is unknown. If the number of 1's in $(c_1, \ldots, c_n)$ is k, then there are $\binom{n}{k}$ possibilities for this vector. If we use an exhaustive search to find the true vector that corresponds to the observed sequence, we have the following situation. If the chosen vector is true, then system (3.1.3) is consistent for any T, but if the vector $(c_1, \ldots, c_n)$ is wrong, then the system becomes inconsistent for some T. Therefore the consistency of the system (3.1.3) serves as a test for selecting the true vector.

Let us introduce the auxiliary notions of a critical set and a hypercycle for our investigations of systems of linear equations in GF(2). Note that the ordinary notions of linear algebra, such as the notion of linear independence of vectors, rank of a matrix, Cramer's rule for finding the solutions of linear systems of equations, and so on, are extended in the obvious way to the n-dimensional vector space over GF(2). For example, if the rank of a $T \times n$ matrix $A = \|a_{tj}\|$ in GF(2) is r, then the homogeneous system of equations

$$AX = 0,$$

where $X = (x_1, \ldots, x_n)$ is the column-vector of unknowns, has exactly $n - r$ linearly independent solutions.

Denote by

$$a_t = (a_{t1}, \ldots, a_{tn}), \quad t = 1, \ldots, T,$$

the rows of the matrix A. If the coordinate-wise sum

$$a_{t_1} + \cdots + a_{t_m} = 0,$$

then the set $C = \{t_1, \ldots, t_m\}$ of row indices is called a critical set.

If C_1 and C_2 are critical sets and $C_1 \neq C_2$, then

$$C_1 \vartriangle C_2 = (C_1 \cup C_2) \setminus (C_1 \cap C_2)$$

is also a critical set.

Let $\varepsilon_1, \ldots, \varepsilon_s$ take the values 0 and 1. Critical sets $C_1, \ldots, C_s$ are called independent if

$$\varepsilon_1 C_1 \vartriangle \varepsilon_2 C_2 \vartriangle \cdots \vartriangle \varepsilon_s C_s = \varnothing,$$

if and only if $\varepsilon_1 = \cdots = \varepsilon_s = 0$.

Denote by $s(A)$ the maximum number of independent critical sets and by $r(A)$ the rank of the matrix A.

Theorem 3.1.1. *For any $T \times n$ matrix A in* GF(2),

$$s(A) + r(A) = T.$$

Proof. We consider the homogeneous system of equations

$$A'Y = 0 \tag{3.1.4}$$

in GF(2), where A' is the transpose of A. There is a one-to-one correspondence between the solutions of the system (3.1.4) and the critical sets: The solution $Y_{t_1,\ldots,t_m} = (y_1, \ldots, y_T)$, whose components $y_{t_1}, \ldots, y_{t_m}$ are 1 and the other components are zero, corresponds to the critical set $C = \{t_1, \ldots, t_m\}$. The linear independence of solutions corresponds to the independence of critical sets. Therefore the maximum number of critical sets $s(A)$ equals the maximum number of linearly independent solutions of system (3.1.4), which we know is $T - r(A)$.

∎

In addition to the critical sets of a $T \times n$ matrix $A = \|a_{tj}\|$, we consider a hypergraph G_A that is also defined by the matrix A. The set of vertices of the hypergraph G_A is the set $\{1, \ldots, n\}$ of column indices and the set of enumerated hyperedges is the set $\{e_1, \ldots, e_T\}$, where

$$e_t = \{j: a_{tj} = 1\}, \quad t = 1, \ldots, T.$$

Thus there exists a correspondence between a row $a_t = (a_{t1}, \ldots, a_{tn})$ and the hyperedge e_t, $t = 1, \ldots, T$. Note that the empty set corresponds to a row consisting of zeros.

The multiplicity of a vertex j in a set of hyperedges $C = \{e_{t_1}, \ldots, e_{t_m}\}$ is the number of hyperedges in C that contain this vertex.

A set of hyperedges $C = \{e_{t_1}, \ldots, e_{t_m}\}$ is called a hypercycle if each vertex of the hypergraph G_A has an even multiplicity in C, in other words, if the coordinate-wise sum of rows $a_{t_1} + \cdots + a_{t_m}$ in GF(2) equals the zero vector.

If each row of the matrix A contains exactly two 1's, then the hypergraph G_A is an ordinary graph, perhaps with multiple edges, and a hypercycle is an ordinary cycle or a union of cycles.

The set of the indices of hyperedges that form a hypercycle is a critical set for the matrix A. Let $\varepsilon_1, \ldots, \varepsilon_s$ take the values 0 and 1. Hypercycles $C_1, \ldots, C_s$ are independent, if

$$\varepsilon_1 C_1 \,\triangle\, \varepsilon_2 C_2 \,\triangle \cdots \triangle\, \varepsilon_s C_s = \varnothing,$$

if and only if $\varepsilon_1 = \cdots = \varepsilon_s = 0$. Therefore the maximum number $s(A)$ of critical sets of the matrix A equals the maximum number of independent hypercycles in G_A.

3.2. Matrices with independent elements

This section deals with random matrices with independent elements. Let $A = \|\alpha_{tj}\|$ be a $T \times n$ matrix whose elements are independent random variables taking the values 0 and 1 with equal probabilities, and let $\rho_n(T)$ be the rank of the matrix A in GF(2). The following theorem is the main result of this section.

Theorem 3.2.1. *Let $s \geq 0$ and m be fixed integers, $m + s \geq 0$. If $n \to \infty$ and $T = n + m$, then*

$$\mathbf{P}\{\rho_n(T) = n - s\} \to 2^{-s(m+s)} \prod_{i=s+1}^{\infty} \left(1 - \frac{1}{2^i}\right) \prod_{i=1}^{m+s} \left(1 - \frac{1}{2^i}\right)^{-1},$$

where the last product equals 1 for $m + s = 0$.

Proof. The limit theorem will be proved by using an explicit formula for $\mathbf{P}\{\rho_n(T) = n - s\}$. Denote by $\rho_n(t)$ the rank of the submatrix of A which consists of the first t rows of the matrix A. We interpret the parameter t as time and consider the process of sequential growth of the number of rows. Let $\xi_t = 1$ if the rank $\rho_n(t-1)$ increases after joining the tth row, and $\xi_t = 0$ if the rank preserves the previous value. It is clear that

$$\rho_n(t) = \xi_1 + \cdots + \xi_t.$$

It is not difficult to describe the probabilistic properties of the random variables $\xi_1, \ldots, \xi_T$. The event $\{\xi_t = 1\}$ means that the tth row is linearly independent with respect to the set of the rows with numbers $1, \ldots, t-1$, and the event $\{\xi_t = 0\}$ means that the row with number t is a linear combination of the preceding rows. If among the preceding $t-1$ rows there are exactly k linearly independent n-dimensional vectors, then the linear span of these k vectors contains 2^k vectors (all linear combinations of these k vectors). The matrix A is constructed in such a way that each row can be obtained by sampling with replacement from a box containing all 2^n distinct n-dimensional vectors. In other words, any row of the matrix A is independent of all other rows and is equal to any n-dimensional vector with probability 2^{-n}. Therefore

$$\mathbf{P}\{\xi_t = 0 \mid \rho_n(t-1) = k\} = \frac{2^k}{2^n},$$

$$\mathbf{P}\{\xi_t = 1 \mid \rho_n(t-1) = k\} = 1 - \frac{2^k}{2^n}.$$

(3.2.1)

Thus the process $\rho_n(t)$ is a Markov chain with stationary transition probabilities that are given by (3.2.1). To find $\mathbf{P}\{\rho_n(T) = n - s\}$, we can sum the probabilities of all trajectories of the Markov chain that lead from the origin to the point with coordinates $(n + m, n - s)$, that is, the trajectories such that $\rho_n(0) = 0$, $\rho_n(n + m) = n - s$. If we represent a trajectory as a "broken line" with intervals of growth and horizontal intervals, we see that any such a broken line has exactly $n + m - (n - s) = m + s$ horizontal intervals corresponding to $m + s$ zeros among the values of $\xi_1, \ldots, \xi_{n+m}$. The graph of the trajectory with $\xi_{t_1} = 0, \ldots, \xi_{t_{m+s}} = 0$ is illustrated in Figure 3.2.1.

By using (3.2.1) and Figure 3.2.1, we can easily write an explicit formula for the probability of a particular trajectory and for the total probability. The derivation of this probability is quite simple if $m + s = 0$. Indeed, the only trajectory with $\rho_n(0) = 0$ and $\rho_n(n+m) = n+m$ has no horizontal intervals, and at each interval the broken line increases; therefore

$$\mathbf{P}\{\rho_n(n+m) = n - s\} = \left(1 - \frac{1}{2^n}\right)\left(1 - \frac{2}{2^n}\right)\cdots\left(1 - \frac{2^{n+m-1}}{2^n}\right)$$

$$= \prod_{i=-m+1}^{n}\left(1 - \frac{1}{2^i}\right),$$

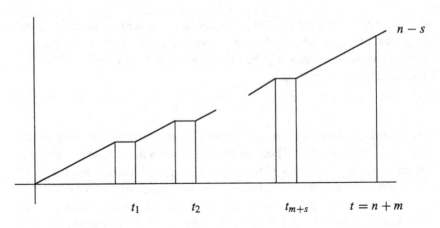

Figure 3.2.1. Graph of the trajectory with $\xi_{t_1} = \cdots = \xi_{t_{m+s}} = 0$

and in the case $m + s = 0$, as $n \to \infty$,

$$\mathbf{P}\{\rho_n(n+m) = n - s\} \to \prod_{i=s+1}^{\infty} \left(1 - \frac{1}{2^i}\right).$$

This coincides with the assertion of the theorem for $m + s = 0$ because the last product equals 1.

In the general case, for $m + s \leq 0$,

$\mathbf{P}\{\rho_n(n+m) = n - s\}$

$$= \sum_{1 \leq t_1 < \cdots < t_{m+s} \leq n+m} \mathbf{P}\{\xi_1 = 1, \ldots, \xi_{t_1-1} = 1, \xi_{t_1} = 0, \xi_{t_1+1} = 1, \ldots\}$$

$$= \sum_{1 \leq t_1 < \cdots < t_{m+s} \leq n+m} \left(1 - \frac{1}{2^n}\right) \cdots \left(1 - \frac{2^{t-(m+s)-1}}{2^n}\right)$$

$$\times \frac{2^{t_1-1+t_2-2+\cdots+t_{m+s}-m-s}}{2^{n(m+s)}}$$

$$= 2^{-s(m+s)} \prod_{k=0}^{t-(m+s)-1} \left(1 - \frac{2^k}{2^n}\right)$$

$$\times \sum_{1 \leq t_1 < \cdots < t_{m+s} \leq n+m} 2^{t_1-1+t_2-2+\cdots+t_{m+s}-m-s}.$$

Taking the factor $2^{(n-s)(m+s)}$ out of the sum yields

$$\mathbf{P}\{\rho_n(n+m) = n-s\}$$

$$= 2^{-s(m+s)} \prod_{i=s+1}^{n} \left(1 - \frac{1}{2^i}\right)$$

$$\times \sum_{1 \le t_1 < \cdots < t_{m+s} \le n+m} 2^{-(n-s)(m+s)+t_1-1+\cdots+t_{m+s}-m-s}.$$

As will be seen from the following evaluations, the moments $t_1, \ldots, t_{m+s}$ are concentrated at the end of the trajectory; therefore, in the sum of the formula, it is convenient to switch to the variables

$$i_l = -(t_l - l + s - n), \quad l = 1, \ldots, m+s.$$

It follows from $1 \le t_1 < \cdots < t_{m+s} \le n+m$ that

$$0 \le t_1 - 1 \le t_2 - 2 \le \cdots \le t_{m+s} - m - s \le n - s,$$

and by subtracting $n - s$ from each term, we obtain

$$-n+s \le t_1 - 1 - n + s \cdots \le t_{m+s} - m - s - n + s \le 0.$$

If we change the sign, we see that the domain $1 \le t_1 < \cdots < t_{m+s} \le n+m$ in terms of the new variables is

$$0 \le i_{m+s} \le \cdots \le i_1 \le n - s.$$

Thus

$$\mathbf{P}\{\rho_n(n+m) = n-s\} \tag{3.2.2}$$

$$= 2^{-s(m+s)} \prod_{i=s+1}^{n} \left(1 - \frac{1}{2^i}\right) \sum_{0 \le i_{m+s} \le \cdots \le i_1 \le n-s} 2^{-i_{m+s}-\cdots-i_1}.$$

It is easily seen that, as $n \to \infty$,

$$\prod_{i=s+1}^{n} \left(1 - \frac{1}{2^i}\right) \to \prod_{i=s+1}^{\infty} \left(1 - \frac{1}{2^i}\right), \tag{3.2.3}$$

and

$$\sum_{0 \le i_{m+s} \le \cdots \le i_1 \le n-s} 2^{-i_{m+s}-\cdots-i_1} \to \sum_{0 \le i_{m+s} \le \cdots \le i_1} 2^{-i_{m+s}-\cdots-i_1}. \tag{3.2.4}$$

To complete the proof it remains to transform the right-hand side of (3.2.4). It is not difficult to see that

$$\sum_{0 \le i_r \le \cdots \le i_2 \le i_1} 2^{-i_r - \cdots - i_2 - i_1}$$

$$= \sum_{0 \le i_r \le \cdots \le i_2} 2^{-i_r - \cdots - i_2} \sum_{i_2 \le i_1} 2^{-i_1}$$

$$= \left(1 - \frac{1}{2}\right)^{-1} \sum_{0 \le i_r \le \cdots \le i_3} 2^{-i_r - \cdots - i_3} \sum_{i_3 \le i_2} 2^{-2i_2}$$

$$= \left(1 - \frac{1}{2}\right)^{-1} \left(1 - \frac{1}{2^2}\right)^{-1} \sum_{0 \le i_r \le \cdots \le i_3} 2^{-i_r - \cdots - i_4 - 3i_3}$$

$$= \left(1 - \frac{1}{2}\right)^{-1} \left(1 - \frac{1}{2^2}\right)^{-1} \cdots \left(1 - \frac{1}{2^{r-1}}\right)^{-1} \sum_{0 \le i_r} 2^{-ri_r}$$

$$= \prod_{i=1}^{r} \left(1 - \frac{1}{2^i}\right)^{-1}. \tag{3.2.5}$$

Passing to the limit in (3.2.2) and taking into account (3.2.3), (3.2.4), and (3.2.5) provide the assertion of the theorem. ∎

Let the elements of a $T \times n$ matrix $A = \|\alpha_{tj}\|$ be independent and take the values 0 and 1 with equal probabilities. We consider the system of equations

$$AX = 0 \tag{3.2.6}$$

with respect to unknowns $X = (x_1, \ldots, x_n)$ in GF(2). Denote by $\nu_{n,T}$ the number of linearly independent solutions of this system of equations. If the rank $\rho_n(T)$ of the matrix A equals r, then $\nu_{n,T} = n - r$. Therefore Theorem 3.2.1 yields the following assertion.

Theorem 3.2.2. *Let $s \ge 0$ and m be fixed integers, $m + s \ge 0$. If $n \to \infty$, then*

$$P\{\nu_{n,n+m} = s\} \to 2^{-s(m+s)} \prod_{i=s+1}^{\infty} \left(1 - \frac{1}{2^i}\right) \prod_{i=1}^{m+s} \left(1 - \frac{1}{2^i}\right)^{-1},$$

where the last product equals 1 for $m + s = 0$.

In particular, for $m = s = 0$,

$$P\{\nu_{n,n} = 0\} \to \prod_{i=1}^{\infty} \left(1 - \frac{1}{2^i}\right) = 0.28878816\ldots.$$

The results of Theorems 3.2.1 and 3.2.2 are of special interest because they are stable in the sense that the limit distribution of the rank of a matrix is invariant with respect to deviations of the distributions of its elements from the equiprobable distribution.

Theorem 3.2.3. *Let the elements of a $T \times n$ matrix $A = \|\alpha_{tj}\|$ be independent and suppose there is a positive constant δ such that, for the probabilities $p_{tj}^{(n)} = P\{\alpha_{tj} = 1\}$, the inequalities*

$$\delta \leq p_{tj}^{(n)} \leq 1 - \delta, \quad t = 1, \ldots, T, \quad j = 1, \ldots, n,$$

hold. Let $s \geq 0$ and m be fixed integers, $m + s \geq 0$. Then, as $n \to \infty$,

$$P\{\rho_n(n + m) = n - s\} \to 2^{-s(m+s)} \prod_{i=s+1}^{\infty} \left(1 - \frac{1}{2^i}\right) \prod_{i=1}^{m+s} \left(1 - \frac{1}{2^i}\right)^{-1},$$

where the last product equals 1 for $m + s = 0$.

Because these results are outside of the main combinatorial direction of this book, we will omit the complicated proof of this theorem (see, e.g., [93]).

We illustrate the situation by proving that, under the conditions of Theorem 3.2.3, the mean value of the number of nontrivial solutions of system (3.2.6) is invariant to deviations of the distributions of elements of A from the equiprobable distribution.

Let $\mu_{n,T}$ be the number of nontrivial (i.e., nonzero) solutions of system (3.2.6). If we associate to the vector X an indicator that is 1 if X satisfies the system, then

$$E\mu_{n,T} = \sum_{X \neq 0} P\{AX = 0\}.$$

We will evaluate $E\mu_{n,T}$ by using the following lemma on summation of independent random variables in GF(2).

Lemma 3.2.1. *Let $\xi_1, \ldots, \xi_n$ be independent random variables that take the values 0 and 1 with probabilities*

$$P\{\xi_i = 1\} = \frac{1 - \Delta_i}{2}, \qquad P\{\xi_i = 0\} = \frac{1 + \Delta_i}{2}, \qquad i = 1, \ldots, n.$$

Then, in GF(2),

$$P\{\xi_1 + \cdots + \xi_n = 1\} = \frac{1 - \Delta_1 \cdots \Delta_n}{2}.$$

Proof. It is clear that it suffices to prove the assertion of the lemma for $n = 2$. In that case,

$$P\{\xi_1 + \xi_2 = 1\} = P\{\xi_1 = 1, \; \xi_2 = 0\} + P\{\xi_1 = 0, \; \xi_2 = 1\}$$

$$= \frac{(1 - \Delta_1)(1 + \Delta_2)}{4} + \frac{(1 + \Delta_1)(1 - \Delta_2)}{4}$$

$$= \frac{1 - \Delta_1 \Delta_2}{2}.$$

∎

If the elements of A are independent and take the values 0 and 1 with equal probabilities, then by Lemma 3.2.1, for any $X \neq 0$,

$$P\{AX = 0\} = (P\{\alpha_{11}x_1 + \cdots + \alpha_{1n}x_n = 0\})^T = 2^{-T}.$$

Therefore $E\mu_{n,T} = (2^n - 1)2^{-T}$, and for $T = n + m$, where m is a fixed integer,

$$E\mu_{n,n+m} = \frac{1}{2^m} - \frac{1}{2^{n+m}},$$

and as $n \to \infty$,

$$E\mu_{n,n+m} \to \frac{1}{2^m}.$$

Under some conditions on the nonequiprobable distribution of the matrix A, the last result still holds. Let

$$p_{tj}^{(n)} = P\{\alpha_{tj} = 1\},$$

and, as before, denote by $\mu_{n,T}$ the number of nontrivial solutions of system (3.2.6).

Theorem 3.2.4. *Under the conditions of Theorem 3.2.3,*

$$E\mu_{n,T} \to 2^{-m}.$$

Proof. By using the indicators as in the calculation of the mean number of solutions in the equiprobable case, we find that

$$E\mu_{n,T} = \sum_{X \neq 0} P\{AX = 0\} = \sum_{k=1}^{n} \sum_{1 \leq j_1 < \cdots < j_k \leq n} P_{j_1, \ldots, j_k}, \qquad (3.2.7)$$

where, for any fixed set $\{j_1, \ldots, j_k\}$ from the domain of summation, the term $P_{j_1, \ldots, j_k} = P\{AX = 0\}$ corresponds to the vector $X = (x_1, \ldots, x_n)$ whose elements with indices $j_1, \ldots, j_k$ are 1 and the remaining elements are zero.

We represent the probabilities $p_{tj}^{(n)}$ as

$$p_{tj}^{(n)} = \frac{1 - \Delta_{tj}}{2}.$$

According to the conditions of the theorem, there exists $\Delta < 1$ such that $|\Delta_{tj}| \le \Delta$ for all t and j. Since the rows of A are independent,

$$P_{j_1,\dots,j_k} = \prod_{i=1}^{T} P_{j_1,\dots,j_k}^{(t)},$$

where

$$P_{j_1,\dots,j_k}^{(t)} = \mathbf{P}\{\alpha_{tj_1} + \cdots + \alpha_{tj_k} = 0\}.$$

By Lemma 3.2.1,

$$\mathbf{P}\{\alpha_{tj_1} + \cdots + \alpha_{tj_k} = 0\} = \frac{1 + \Delta_{tj_1} \cdots \Delta_{tj_k}}{2},$$

and for all t and $1 \le j_1 < \cdots < j_k \le n$,

$$\frac{1 - \Delta^k}{2} \le P_{j_1,\dots,j_k}^{(t)} \le \frac{1 + \Delta^k}{2}.$$

Hence, for $P_{j_1,\dots,j_k}$, we obtain the bounds

$$\left(\frac{i - \Delta^k}{2}\right)^T \le P_{j_1,\dots,j_k}^{(t)} \le \left(\frac{1 + \Delta^k}{2}\right)^T.$$

By using these inequalities, we find from (3.2.7) that

$$\sum_{k=1}^{n} \binom{n}{k} \left(\frac{1 - \Delta^k}{2}\right)^T \le \mathbf{E}\mu_{n,T} \le \sum_{k=1}^{n} \binom{n}{k} \left(\frac{1 + \Delta^k}{2}\right)^T. \qquad (3.2.8)$$

Now let $T = n + m$, where m is a fixed integer. The left and the right sides of (3.2.8) can be estimated in the same way. Therefore we obtain only an estimate of the right-hand side. Let

$$S(\Delta) = \sum_{k=1}^{n} \binom{n}{k} \left(\frac{1 + \Delta^k}{2}\right)^{n+m}$$

and compare $S(\Delta)$ to

$$S(0) = \sum_{k=1}^{n} \binom{n}{k} \frac{1}{2^{n+m}}.$$

We have seen that $S(0) \to 2^{-m}$ as $n \to \infty$. We show that for any fixed Δ, $0 \le \Delta < 1$, the difference $S(\Delta) - S(0)$ tends to zero. We divide $S(\Delta)$ into two parts:

$$S_1(\Delta) = \sum_{1 \le k \le \varepsilon n} \binom{n}{k} \left(\frac{1+\Delta^k}{2}\right)^{n+m},$$

$$S_2(\Delta) = \sum_{\varepsilon n < k \le n} \binom{n}{k} \left(\frac{1+\Delta^k}{2}\right)^{n+m},$$

where ε, $0 < \varepsilon < 1/2$, will be chosen later. For the sake of simplicity, suppose that ε is such that εn is an integer; then for any ε and Δ, $0 \le \Delta < 1$, $0 < \varepsilon < 1/2$,

$$S_1(\Delta) = \sum_{1 \le k \le \varepsilon n} \binom{n}{k} \left(\frac{1+\Delta^k}{2}\right)^{n+m} \le \varepsilon n \left(\frac{1+\Delta}{2}\right)^{n+m} \binom{n}{\varepsilon n}$$

$$\le \left(\frac{1+\Delta}{2}\right)^{n+m} \frac{\varepsilon n n^{\varepsilon n}}{(\varepsilon n)^{\varepsilon n} \sqrt{\varepsilon n} e^{-\varepsilon n}}$$

by using the inequality $n! \ge n^n \sqrt{n} e^{-n}$. This bound for $S_1(\Delta)$ can be written as

$$S_1(\Delta) \le \sqrt{\varepsilon n} \left(\frac{1+\Delta}{2}\right)^m \left(\frac{1+\Delta}{2\varepsilon^\varepsilon e^{-\varepsilon}}\right)^n.$$

If we choose a sufficiently small ε, we can make the value $(1 + \Delta)/(2\varepsilon^\varepsilon e^{-\varepsilon})$ less than 1. For such ε, the bound tends to zero as $n \to \infty$. Thus, there exists a fixed ε, $0 < \varepsilon < 1/2$, such that the value $S_1(\Delta)$ and, consequently, $S_1(0)$ tend to zero, and $S_1(\Delta) - S_1(0) \to 0$.

We now estimate the difference $S_2(\Delta) - S_2(0)$. It is clear that

$$0 \le S_2(\Delta) - S_2(0) = \sum_{\varepsilon < k \le n} \binom{n}{k} \frac{1}{2^{n+m}}((1+\Delta^k) - 1)$$

$$\le \sum_{\varepsilon < k \le n} \binom{n}{k} \frac{1}{2^{n+m}}((1+\Delta^{\varepsilon n}) - 1)$$

$$= ((1+\Delta^{\varepsilon n})^{n+m} - 1) \sum_{\varepsilon n < k \le n} \binom{n}{k} \frac{1}{2^{n+m}}$$

$$\le \frac{1}{2^m}((1+\Delta^{\varepsilon n})^{n+m} - 1).$$

Since $(1 + \Delta^{\varepsilon n})^{n+m} \to 1$ as $n \to \infty$, it follows from the estimate obtained above that $S_2(\Delta) - S_2(0) \to 0$. Thus we have shown that $S(\Delta) - S(0) \to 0$ and $S(0) \to 2^{-m}$; hence, $S(\Delta) \to 2^{-m}$. Theorem 3.2.4 is thus proved. ∎

We can actually relax the hypotheses of Theorem 3.2.4. The result remains true if for $t = 1, \ldots, T$, $j = 1, \ldots, n$,

$$\frac{\log n + x_n}{n} \leq p_{tj}^{(t)} \leq 1 - \frac{\log n + x_n}{n},$$

where x_n tends to infinity arbitrarily slowly (see [93]). These bounds are exact in a sense because, as we will show in the next section, the limit distribution of the rank of a matrix A differs from the distribution given in Theorem 3.2.1 if the probability of 1's does not satisfy these inequalities.

3.3. Rank of sparse matrices

In Section 3.1, we introduced the notion of critical sets of a matrix. Recall that a set $\{t_1, \ldots, t_m\}$ of row indices of a matrix in GF(2) is called critical if the coordinate-wise sum of rows with indices $t_1, \ldots, t_m$ is the zero vector. The notion of independence of critical sets was also introduced, and $s(A)$ denoted the maximum number of independent critical sets of a matrix A. According to Theorem 3.1.1, the rank $r(A)$ of a matrix A is related to $s(A)$ by the equality $s(A) + r(A) = T$. Therefore, instead of the rank of a matrix, we can investigate the maximum number $s(A)$ of independent critical sets of the matrix.

In this section, critical sets are applied in the analysis of the rank of random sparse matrices. Let the elements of a $T \times n$ matrix $A = \|\alpha_{tj}\|$ be independent random variables such that

$$\mathbf{P}\{\alpha_{tj} = 1\} = \frac{\log n + x}{n}, \qquad \mathbf{P}\{\alpha_{tj} = 0\} = 1 - \frac{\log n + x}{n}, \qquad (3.3.1)$$

where x is a constant, $t = 1, \ldots, T$, $j = 1, \ldots, n$. We find the limit distribution of $s(A)$ for such a matrix.

Theorem 3.3.1. If $n, T \to \infty$ such that $T/n \to \alpha$, $0 < \alpha < 1$, and condition (3.3.1) is valid, then the distribution of the maximum number of independent critical sets $s(A)$ converges to the Poisson distribution with parameter $\lambda = \alpha e^{-x}$.

We show first that the distribution of the number of critical sets that correspond to zero rows of the matrix converges to a Poisson distribution. Denote the number of zero rows of the matrix A by $\xi_{n,T}$.

Lemma 3.3.1. If $n, T \to \infty$ such that $T/n \to \alpha$, $0 < \alpha < \infty$, and condition (3.3.1) is valid, then for any fixed $k = 0, 1, \ldots,$

$$\mathbf{P}\{\xi_{n,T} = k\} \to \frac{\lambda^k e^{-\lambda}}{k!},$$

where $\lambda = \alpha e^{-x}$.

Proof. The probability p_n that a fixed row consists entirely of zeros is

$$p_n = \left(1 - \frac{\log n + x}{n}\right)^n,$$

and under the conditions of the lemma,

$$p_n = \frac{1}{n} e^{-x}(1 + o(1)).$$

The random variable $\xi_{n,T}$ has the binomial distribution with parameters (T, p_n), where T is the number of trials and p_n is the probability of success. Under the conditions of the lemma, the mean number of successes $T p_n$ tends to αe^{-x}; hence, the binomial distribution converges to the Poisson distribution with parameter αe^{-x}. ∎

We now prove that if $\alpha < 1$, then with probability tending to 1, all critical sets consist of only zero rows.

Lemma 3.3.2. *If $n, T \to \infty$ such that $T/n \to \alpha$, $\alpha < 1$, and condition (3.3.1) is valid, then with probability tending to 1, the critical sets of A consist of only zero rows.*

Proof. We consider the total number of critical sets in which each contains at least one nonzero row. It is sufficient to prove that the mathematical expectation of this number tends to zero. Although the proof of this fact is straightforward, it involves many cumbersome estimations of sums containing the binomial coefficients.

An even number of successes among k independent trials with probability of success p occurs with probability $(1 + (q - p)^k)/2$.

Let us find the probability that k fixed rows form a critical set containing a nonzero row. The indices of these rows form a critical set if each column of the submatrix formed by these rows contains an even number of 1's. According to the remark on the probability that the number of successes is even, this probability equals

$$\frac{1}{2}\left(1 + \left(1 - \frac{2(\log n + x)}{n}\right)^k\right).$$

Therefore the probability that these k rows constitute a critical set equals

$$\frac{1}{2^n}\left(1 + \left(1 - \frac{2(\log n + x)}{n}\right)^k\right)^n.$$

Note that the probability that there is no 1 in all these k rows is equal to

$$\left(1 - \frac{\log n + x}{n}\right)^{kn}.$$

By using the corresponding indicators to represent the total number of nontrivial critical sets and the number of the critical sets that consist of zero rows, we obtain the following expression for the mean number of critical sets that do not consist of zero rows:

$$\sum_{k=0}^{T}\binom{T}{k}\frac{1}{2^n}r_k^n - \sum_{k=0}^{T}\binom{T}{k}\left(1-\frac{\log n + x}{n}\right)^{kn},$$

where

$$r_k = 1 + \left(1 - \frac{2(\log n + x)}{n}\right)^k.$$

We include the terms with $k = 0$ into these sums because they cancel each other. Note first that

$$\sum_{k=0}^{T}\binom{T}{k}\left(1-\frac{\log n + x}{n}\right)^{kn} = \left(1+\left(1-\frac{\log n + x}{n}\right)^n\right)^T,$$

and under the conditions of the lemma,

$$\left(1+\left(1-\frac{\log n + x}{n}\right)^n\right)^T = \left(1+\frac{1}{n}e^{-x}+o\left(\frac{1}{n}\right)\right)^T = e^{\alpha e^{-x}}(1+o(1)).$$

Now consider the sum

$$\sum_{k=0}^{T}\binom{T}{k}\frac{1}{2^n}r_k^n = \sum_{k=0}^{T}\binom{T}{k}\frac{1}{2^n}\left(1+\left(1-\frac{2(\log n)}{n}\right)^k\right)^n.$$

Set $a = 1 - 2(\log n + x)/n$ for now. The following equalities hold:

$$\sum_{l=0}^{T}\binom{T}{l}\frac{1}{2^n}r_k^n = \sum_{l=0}^{T}\binom{T}{l}\frac{1}{2^n}(1+a^l)^n = \sum_{l=0}^{T}\binom{T}{l}\frac{1}{2^n}\sum_{k=0}^{n}\binom{n}{k}a^{lk}$$

$$= \sum_{k=0}^{n}\binom{n}{k}\frac{1}{2^n}\sum_{l=0}^{T}\binom{T}{l}a^{kl} = \sum_{k=0}^{n}\binom{n}{k}\frac{1}{2^n}(1+a^k)^T.$$

Let

$$a_k = \binom{n}{k}\frac{1}{2^n}(1+a^k)^T,$$

and divide the sum

$$S(n,T) = \sum_{k=0}^{n}\binom{n}{k}\frac{1}{2^n}(1+a^k)^T$$

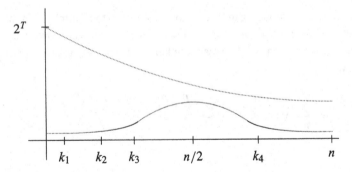

Figure 3.3.1. Graphs of the functions $\binom{n}{k}2^{-n}$ and r_k^T

into five parts so that

$$S(n, T) = S_1 + S_2 + S_3 + S_4 + S_5,$$

where

$$S_1 = \sum_{0 \le k \le k_1} a_k, \qquad S_2 = \sum_{k_1 < k \le k_2} a_k, \qquad S_3 = \sum_{k_2 < k \le k_3} a_k,$$

$$S_4 = \sum_{k_3 < k \le k_4} a_k, \qquad S_5 = \sum_{k_4 < k \le n} a_k,$$

$$k_1 = \varepsilon n, \quad k_2 = \frac{n}{2}(1 - \varepsilon), \quad k_3 = \frac{n}{2} - \frac{1}{2}n^{1/2+1/10}, \quad k_4 = \frac{n}{2} + \frac{1}{2}n^{1/2+1/10},$$

and the value of ε will be chosen later. For convenience we present the graphs of the functions $\binom{n}{k}2^{-n}$ and $r_k^T = (1 + (1 - 2(\log n + x)/n)^k)^T$ as functions of k in Figure 3.3.1.

The major contribution to $S(n, T)$ is made by the sum S_4. It is clear that

$$\left(1 - \frac{2(\log n + x)}{n}\right)^k = e^{2k(\log n + x)/n}(1 + o(1)) = \frac{1}{n}e^{-x}(1 + o(1))$$

uniformly in the integers $k = n/2 + u\sqrt{n}/2$ such that $|u| \le n^{1/10}$. These k form the domain of summation of S_4, which equals $\{k: |u| \le n^{1/10}\}$. Therefore

$$r_k^n = \left(1 + \left(1 - \frac{2(\log n + x)}{n}\right)^k\right)^T$$

$$= \left(1 + \frac{1}{n}e^{-x} + o\left(\frac{1}{n}\right)\right)^T$$

$$= e^{\alpha e^{-x}}(1 + o(1))$$

uniformly in k in the domain of summation of S_4. Thus

$$S_4 = \sum_{k:|u|\leq n^{1/10}} \binom{n}{k}\frac{1}{2^n}r_k^T = \sum_{k:|u|\leq n^{1/10}} \binom{n}{k}\frac{1}{2^n}e^{\alpha e^{-x}}(1+o(1))$$

$$= e^{\alpha e^{-x}} \sum_{k:|u|\leq n^{1/10}} \binom{n}{k}\frac{1}{2^n}(1+o(1)) = e^{\alpha e^{-x}}(1+o(1)),$$

since by the de Moivre–Laplace theorem,

$$\sum_{k:|u|\leq n^{1/10}} \binom{n}{k}\frac{1}{2^n} \to 1.$$

We now have to show that the remaining four sums tend to zero. We begin with

$$S_5 = \sum_{k\geq n/2+n^{1/10}\sqrt{n}/2} \binom{n}{k}\frac{1}{2^n}r_k^T.$$

Since r_k^T is monotone, we find that

$$S_5 \leq r_{k_4}^T \sum_{k:u\geq n^{1/10}} \binom{n}{k}\frac{1}{2^n}.$$

Under the conditions of the lemma $S_5 \to 0$, since, as was proved, $r_{k_4}^T \to e^{\alpha e^{-x}}$, and according to the de Moivre–Laplace theorem,

$$\sum_{k:u\geq n^{1/10}} \binom{n}{k}\frac{1}{2^n} \to 0.$$

Let us estimate

$$S_1 = \sum_{0\leq k\leq k_1} \binom{n}{k}\frac{1}{2^n}r_k^T.$$

By using the monotonicity of r_k^T, we find that, for sufficiently small ε such that εn is an integer,

$$S_1 \leq r_0^T \sum_{k\leq k_1} \binom{n}{k}\frac{1}{2^n} = 2^{T-n}\sum_{k\leq k_1}\binom{n}{k} \leq (1+k_1)\binom{n}{k}2^{T-n}$$

$$\leq \frac{(1+\varepsilon n)n^\varepsilon 2^T}{(\varepsilon n)^{\varepsilon n}\sqrt{\varepsilon n}e^{-\varepsilon n}2^n} \leq (1+\varepsilon n)\left(\frac{2^{T/n}}{2\varepsilon^\varepsilon e^{-\varepsilon}}\right)^n.$$

It is clear that $2^{T/n}(2\varepsilon^\varepsilon e^{-\varepsilon})^{-1} \leq q < 1$ for sufficiently small ε; therefore $S_1 \to 0$ as $n \to \infty$.

It remains to consider S_2 and S_3. Let us begin with

$$S_2 = \sum_{\varepsilon n<k\leq n(1-\varepsilon)/2} a_k.$$

We first show that a_k is a monotone increasing function for k such that $\varepsilon n \le k \le n(1 - \varepsilon)/2$. Indeed,

$$
\frac{a_{k+1}}{a_k} = \frac{\binom{n}{k+1} r_{k+1}^T}{\binom{n}{k} r_k^T}
$$

$$
= \frac{n - k}{k + 1} \left(\frac{1 + (1 - 2(\log n + x/n)^{k+1}}{1 + (1 - 2(\log n + x)/n)^k} \right)^T
$$

$$
\ge \frac{n - n(1 - \varepsilon)/2}{n(1 - \varepsilon)/2 - 1}
$$

$$
\times \left(1 - \frac{(1 - 2(\log n + x)/n)^k - (1 - 2(\log n + x)/n)^{k+1}}{1 + (1 - 2(\log n + x)/n)^k} \right)^T
$$

$$
\ge \frac{1 + \varepsilon}{1 - \varepsilon + 2/n}
$$

$$
\times \left(1 - \frac{(1 - 2(\log n + x)/n)^k - (1 - 2(\log n + x)/n)^{k+1}}{1 + (1 - 2(\log n + x)/n)^k} \right)^T .
$$

Since $1 + (1 - 2(\log n + x)/n)^k > 1$, we obtain

$$
\frac{a_{k+1}}{a_k} \ge \frac{1 + \varepsilon}{1 - \varepsilon + 2/n}
$$

$$
\times \left(1 - \left(1 - \frac{2(\log n + x)}{n} \right)^k \left(1 - 1 + \frac{2(\log n + x)}{n} \right) \right)^T
$$

$$
= \frac{1 + \varepsilon}{1 - \varepsilon + 2/n} \left(1 - \frac{2(\log n + x)}{n} \left(1 - \frac{2(\log n + x)}{n} \right)^k \right)^T .
$$

For sufficiently large n,

$$
(1 + \varepsilon)/(1 - \varepsilon + 2/n) \ge 1 + \varepsilon.
$$

Moreover, for k satisfying $\varepsilon n \le k \le n(1 - \varepsilon/2)$,

$$
\left(1 - \frac{2(\log n + x)}{n} \right)^k \le e^{-2k(\log n + x)/n} \le cn^{-2\varepsilon},
$$

where c is the constant $e^{-2\varepsilon x}$.

Thus, for sufficiently large n,

$$
\frac{a_{k+1}}{a_k} \ge (1 + \varepsilon) \left(1 - \frac{2c(\log n + x)}{n^{1+2\varepsilon}} \right)^T \ge (1 + \varepsilon)(1 - \varepsilon/2) \ge 1.
$$

If we estimate S_2, we can use the monotonicity of a_k to obtain the inequality

$$S_2 \le n a_{k_2}.$$

Let us estimate

$$a_{k_2} = \binom{n}{k_2} \frac{1}{2^n} \left(1 + \left(1 - \frac{2(\log n + x)}{n}\right)^{k_2}\right)^T.$$

Since a rough estimate is acceptable, we content ourselves with the bound

$$\binom{n}{k_2} \frac{1}{2^n} \le \sum_{k \le k_3} \binom{n}{k} \frac{1}{2^n} = \sum_{k: u \le -n^{1/10}} \binom{n}{k} \frac{1}{2^n}$$

$$= \frac{1}{\sqrt{2\pi}} \int_{-\infty}^{-n^{1/10}} e^{-u^2/2} \, du (1 + o(1))$$

$$= \frac{1}{\sqrt{2\pi} n^{1/10}} e^{-n^{1/5}/2} (1 + o(1)).$$

Here we used the well-known asymptotics

$$\int_{-\infty}^{-z} e^{-u^2/2} \, du = \frac{1}{z} e^{-z^2/2} (1 + o(1))$$

as $z \to \infty$. Thus, there exists a constant a such that

$$\sum_{k \le k_3} \binom{n}{k} \frac{1}{2^n} \le a e^{-n^{1/5}/2}.$$

Let us estimate the second factor of a_{k_2}. It is clear that

$$\left(1 + \left(1 - \frac{2(\log n + x)}{n}\right)^{k_2}\right)^T \le \left(1 + e^{-2k_2(\log n + x)/n}\right)^T$$

$$= \left(1 + e^{-(1-\varepsilon)(\log n + x)}\right)^T \le e^{T n^{-1+\varepsilon} e^{-x(1-\varepsilon)}} \le e^{bn^{\varepsilon}},$$

where b is a positive constant.

By combining the estimates of the two factors of a_{k_2}, we obtain the bound

$$S_2 \le n a_{k_2} \le n a e^{-n^{1/5}/2} e^{bn^{\varepsilon}},$$

and $S_2 \to 0$ if we choose $\varepsilon < 1/5$.

It remains to estimate

$$S_3 = \sum_{k_2 < k \le k_3} \binom{n}{k} \frac{1}{2^n} \left(1 + \left(1 - \frac{2(\log n + x)}{n}\right)^{k}\right)^T.$$

It is clear that

$$S_3 \leq \left(1 + \left(1 - \frac{2(\log n + x)}{n}\right)^{k_2}\right)^T \sum_{k \leq k_3} \binom{n}{k} \frac{1}{2^n}$$

$$\leq e^{bn^\varepsilon} a e^{-n^{1/5}/2},$$

and $S_3 \to 0$ if $\varepsilon < 1/5$. ∎

Proof of Theorem 3.3.1. The assertion of Theorem 3.3.1 follows from Lemmas 3.3.1 and 3.3.2 because by Lemma 3.3.2,

$$\mathbf{P}\{s(A) = \xi_{n,T}\} \to 1$$

under the conditions of the theorem. ∎

The following theorem is a corollary to Theorem 3.3.1. Suppose that

$$\frac{\log T + x}{T} \leq p_{tj}^{(n)} \leq 1 - \frac{\log T + x}{T}, \tag{3.3.2}$$

where x is a constant and $t = 1, \ldots, T$, $j = 1, \ldots, n$.

Theorem 3.3.2. *If $n, T \to \infty$ such that $T/n \to \alpha$, $1 < \alpha < \infty$, and condition (3.3.2) is valid, then the distribution of $s(A)$ converges to the Poisson distribution with parameter $\lambda = e^{-x}/\alpha$.*

Proof. Since the rank of a matrix is the maximum number of linearly independent rows or columns, we apply Theorem 3.3.1 to the transpose matrix and obtain the assertion of Theorem 3.3.2. ∎

Because we know the limit distribution of the rank of a matrix A, we can obtain some results for the behavior of the solutions to the system of linear equations with the matrix A. Let us consider the system

$$AX = B, \tag{3.3.3}$$

where the elements of the $T \times n$ matrix $A = \|\alpha_{tj}\|$ are independent, and for $t = 1, \ldots, T$, $j = 1, \ldots, n$,

$$\mathbf{P}\{\alpha_{tj} = 1\} = \frac{\log n + x}{n},$$

where x is a constant, the column-vector $B = (b_1, \ldots, b_T)$ is independent of A, and the random variables $b_1, \ldots, b_T$ are independent, taking the values 0 and 1 with equal probabilities.

Denote by $\mu_{n,T}^{(1)}$ the number of solutions of the system (3.3.3). The examples cited in Section 3.1 show that the consistency of linear systems plays a particular role in some of the problems related to such systems. The probability of consistency $P_{n,T}$ of system (3.3.3) is the probability that the system has at least one solution:

$$P_{n,T} = \mathbf{P}\{\mu_{n,T}^{(1)} > 0\}.$$

By using Theorem 3.3.1 we can easily prove the following assertion.

Theorem 3.3.3. *If $n, T \to \infty$ such that $T/n \to \alpha$, $0 < \alpha < 1$, and condition* (3.3.1) *is valid, then*

$$P_{n,T} \to e^{-\alpha e^{-x}/2}.$$

Proof. If the rank $r(A)$ of A equals r, then

$$\mathbf{P}\{\mu_{n,T}^{(1)} > 0 \,|\, r(A) = r\} = 2^{-T+r}. \tag{3.3.4}$$

Indeed, let the linearly independent rows have the indices $1, 2, \dots, r$. Then each of the rows with indices $r + 1, \dots, T$ is a linear combination of the first r rows, and for the system to be consistent, each of the right-hand parts $b_{r+1}, \dots, b_T$ must satisfy a linear relation of the form

$$\varepsilon_{1t} b_1 + \cdots + \varepsilon_{rt} b_r = b_t, \quad t = r + 1, \dots, T, \tag{3.3.5}$$

where $\varepsilon_{1t}, \dots, \varepsilon_{rt}$ are constants taking the values 0 and 1. The probability of the validity of any of the relations (3.3.5) is equal to $1/2$ and, hence, assertion (3.3.4) is true.

Since $\{r(A) = r\} = \{s(A) = T - r\}$, by the total probability formula,

$$P_{n,T} = \sum_{r=0}^{T} \mathbf{P}\{r(A) = r\} \frac{1}{2^{T-r}} = \sum_{s=0}^{T} \mathbf{P}\{s(A) = s\} \frac{1}{2^s}. \tag{3.3.6}$$

The last series from (3.3.6) is majorized by the series $\sum_{s=0}^{\infty} 2^{-s}$ and converges uniformly. Therefore it is possible to pass to the limit under the sum in (3.3.6). Passing to the limit with the help of Theorem 3.3.1 yields

$$P_{n,T} \to \sum_{s=0}^{\infty} \frac{\lambda^s e^{-\lambda}}{2^s s!} = e^{-\lambda/2},$$

where $\lambda = \alpha e^{-x}$.   ∎

3.4. Cycles and consistency of systems of random equations

In this section, we consider a system of T equations in GF(2):

$$x_{i(t)} + x_{j(t)} = \beta_t, \quad t = 1, \dots, T, \tag{3.4.1}$$

where $i(t)$, $j(t)$, $t = 1, \ldots, T$, are independent random variables that take the values $1, \ldots, n$ with equal probabilities, and the variables $\beta_1, \ldots, \beta_T$ take the values 0 and 1. We denote by $A_{n,T}$ the matrix of this system. As in Section 3.1, we associate the matrix $A_{n,T}$ to a graph $G_{n,T}$ with n labeled vertices that correspond to the variables $x_1, \ldots, x_n$. The graph has T edges $(i(t), j(t))$, $t = 1, \ldots, T$. Thus the edges of the graph $G_{n,T}$ may be considered an outcome of T independent trials: In each trial, an edge joins two different vertices i and j with probability $2n^{-2}$ and forms the loop at a vertex i with probability n^{-2}, $i, j = 1, \ldots, T$. Thus the graph $G_{n,T}'$ is the same as the graph considered in Section 2.3.

Denote by $\mu_{n,T}$ the number of solutions of the system (3.4.1) and consider the probability of consistency

$$P_{n,T} = \mathbf{P}\{\mu_{n,T} > 0\}.$$

We want to express $P_{n,T}$ in terms of the characteristics of $G_{n,T}$. Denote by $\varkappa_{n,T}$ the number of components of the graph $G_{n,T}$.

Theorem 3.4.1. *If $\beta_1 \ldots, \beta_T$ are independent random variables that take the values 0 and 1 with equal probabilities and do not depend on $A_{n,T}$, then*

$$P_{n,T} = \frac{1}{2^{T-n}} \sum_{k=1}^{n} \mathbf{P}\{\varkappa_{n,T} = k\} \frac{1}{2^k}.$$

Proof. We first assume that $G_{n,T}$ is a connected graph. We can then choose a tree that is a skeleton of the graph. This tree contains $n - 1$ edges that correspond to a subsystem containing $n - 1$ equations of the system. If we assign a fixed value to one of the unknowns, then with the help of the corresponding subsystem, we obtain the values of all other unknowns. Consequently, the right-hand sides of the remaining $T - n + 1$ equations must each take a fixed value for the system to be consistent. Since $\beta_1, \ldots, \beta_T$ are independent and take the values 0 and 1 with probabilities $1/2$, the probability of consistency is $(1/2)^{T-n+1}$ for $G_{n,T}$ connected.

Now assume the graph $G_{n,T}$ consists of k components with $n_1, \ldots, n_k$ vertices and $T_1, \ldots, T_k$ edges, respectively. The whole system is consistent if and only if each of its subsystem is consistent. Under the condition that the number of components $\varkappa_{n,T} = k$ and, consequently, that the system decomposes into k disjoint subsystems, the probability of consistency is

$$\frac{1}{2^{T_1 - n_1 + 1}} \frac{1}{2^{T_2 - n_2 + 1}} \cdots \frac{1}{2^{T_k - n_k + 1}} = \frac{1}{2^{T - n + k}}.$$

When we apply the formula of total probability, we obtain the assertion of the theorem. ∎

According to Theorem 3.4.1, the number of components of the graph $G_{n,T}$ can be used to investigate the system (3.4.1). Likewise, we can consider the maximum number of independent critical sets $s(A_{n,T})$ introduced in Section 3.1. According to Theorem 3.1.1, the maximum number of independent critical sets $s(A_{n,T})$ and the rank $r(A_{n,T})$ of the matrix $A_{n,T}$ are related by the equality

$$s(A_{n,T}) + r(A_{n,T}) = T.$$

It is not difficult to prove that

$$\varkappa_{n,T} = n - T + s(A_{n,T}),$$

and the rank $r(A_{n,T}) = n - \varkappa_{n,T}$. Thus, the assertion of Theorem 3.4.1 is equivalent to relation (3.3.6).

We remarked in Section 3.1 that a critical set of $A_{n,T}$ corresponds to a cycle or a union of cycles in the graph $G_{n,T}$, and the maximum number of critical sets $s(A_{n,T})$ equals the maximum number of independent cycles.

The graph $G_{n,T}$ was studied in Section 2.3. We have seen that if $n, T \to \infty$ such that $2T/n \to \lambda$, $0 < \lambda < 1$, then with probability tending to 1, the graph has no components with more than one cycle. Therefore, under these conditions, all cycles of $G_{n,T}$ are isolated and, consequently, independent. As in Section 3.1, we denote by $v(G_{n,T})$ the number of cycles in $G_{n,T}$.

It was proven (see Theorems 2.3.3 and 2.3.4) that if $2T/n \to \lambda$, $0 < \lambda < 1$, then

$$\mathbf{P}\{v(G_{n,T}) = s(A_{n,T})\} \to 1, \tag{3.4.2}$$

and for any fixed $k = 0, 1 \ldots$,

$$\mathbf{P}\{v(G_{n,T}) = k\} \to \frac{\Lambda^k e^{-\Lambda}}{k!}, \tag{3.4.3}$$

where

$$\Lambda = -\tfrac{1}{2} \log(1 - \lambda).$$

These results allow us to analyze the probability $P_{n,T}$ of consistency of the system (3.4.1).

Theorem 3.4.2. *If $n, T \to \infty$ such that $2T/n \to \lambda$, $0 < \lambda < 1$, and the right-hand sides $\beta_1 \ldots, \beta_T$ of the system (3.4.1) are independent random variables that take the values 0 and 1 with probabilities 1/2 and do not depend on $A_{n,T}$, then*

$$P_{n,T} \to (1 - \lambda)^{1/4}.$$

Proof. When we use Theorem 3.4.1 or the equivalent formula (3.3.6), we find that

$$P_{n,T} = \sum_{k=1}^{n} \mathbf{P}\{\varkappa_{n,T} = k\}\frac{1}{2^{T-n+k}} = \sum_{r=0}^{n} \mathbf{P}\{r(A_{n,T}) = r\}\frac{1}{2^{T-r}}$$

$$= \sum_{s=0}^{T} \mathbf{P}\{s(A_{n,T}) = s\}\frac{1}{2^s}.$$

Taking into account (3.4.2) and (3.4.3) and passing to the limit under the sum yield

$$P_{n,T} \to \sum_{s=0}^{\infty} \frac{\Lambda^s e^{-\Lambda}}{s!\, 2^s} = e^{-\Lambda/2} = (1 - \lambda)^{1/4}.$$

∎

In the same way, we can treat the nonequiprobable case, where the indices $(i(t), j(t))$, $t = 1, \ldots, T$, of the variables of system (3.4.1) are independent identically distributed random variables that take the value i with probability p_i, $i = 1, \ldots, n$, $p_1 + \cdots + p_n = 1$. As before, let the right-hand sides $\beta_1, \ldots, \beta_T$ be independent, take the values 0 and 1 with equal probabilities, and not depend on $A_{n,T}$. We retain the notation $P_{n,T}$ for the probability of consistency of such a system.

Theorem 3.4.3. *Let* $p_i = a_i/n$, *where* $a_i = a_i(n)$, $0 < \varepsilon_0 \le a_i \le \varepsilon_1 < \infty$, $i = 1, \ldots, n$, ε_0 *and* ε_1 *are constants, and let*

$$a^2 = \lim_{n\to\infty} \frac{1}{n} \sum_{i=1}^{n} a_i^2.$$

If $n, T \to \infty$ *such that* $2T/n \to \lambda$ *and* $a^2\lambda < 1$, *then*

$$P_{n,T} \to \left(1 - a^2\lambda\right)^{1/4}.$$

Proof. In Section 2.4, the nonequiprobable graph $G_{n,T}$ corresponding to the matrix $A_{n,T}$ was considered. The graph contains n labeled vertices and T edges that can be obtained by the following T independent trials. In each trial, one edge is drawn. The edge connects two different vertices i and j with probability $2p_i p_j$, and a loop at a vertex i is formed with probability p_i^2, $i, j = 1, \ldots, n$, $p_1 + \cdots + p_n = 1$.

According to Theorem 2.4.1, under the conditions of Theorem 3.4.3 for any fixed $k = 0, 1, \ldots$,

$$\mathbf{P}\{\nu(G_{n,T}) = k\} \to \frac{\Lambda^k e^{-\Lambda}}{k!},$$

where $\nu(G_{n,T})$ is the number of cycles in $G_{n,T}$, and

$$\Lambda = -\tfrac{1}{2}\log(1 - a^2\lambda).$$

If we reason as we did in the proof of Theorem 3.4.2, we obtain the assertion of Theorem 3.4.3. ∎

The proofs of Theorems 3.4.2 and 3.4.3 are mainly based on assertion (3.3.4) that

$$\mathbf{P}\{\mu_{n,T} > 0 \mid r(A_{n,T}) = r\} = 2^{-T+r}. \qquad (3.4.4)$$

The proof of this assertion in Section 3.3 used the fact that if r rows are linearly independent and $r(A_{n,T}) = r$, then each of the remaining rows is a linear combination of these r rows, and the system is consistent only if the corresponding right-hand sides satisfy a certain linear relation. If the right-hand sides $\beta_1, \ldots, \beta_T$ are independent, then such a relation is satisfied with probability $1/2$, and the events corresponding to different relations are independent. In other words, each cycle in $G_{n,T}$ imposes a restriction on the right-hand sides $\beta_1, \ldots, \beta_T$, these restrictions are independent, and each of them is satisfied with probability $1/2$.

If the right-hand sides $\beta_1, \ldots, \beta_T$ take the values 0 and 1 with unequal probabilities, then property (3.4.4) is not valid, and the corresponding formula for the probability $P_{n,T}$ of the consistency of the system becomes more complicated. In this section, we prove the following assertions.

Let

$$P_{n,T}(k) = \mathbf{P}\{\mu_{n,T} > 0, \ \nu(G_{n,T}) = k\}, \qquad P_{n,T} = \mathbf{P}\{\mu_{n,T} > 0\}.$$

Theorem 3.4.4. *Let the right-hand sides $\beta_1, \ldots, \beta_T$ of the system (3.4.1) be independent identically distributed random variables that take the values 0 and 1 with probabilities $1 - p$ and p, respectively, $0 < p < 1$, $\Delta = 1 - 2p$.*

If $n, T \to \infty$ such that $2T/n \to \lambda$, $0 < \lambda < 1$, then for any fixed $k = 0, 1, \ldots,$

$$P_{n,T}(k) \to \frac{1}{4^k k!}(-\log(1 - \lambda)(1 - \Delta\lambda))^k \sqrt{1 - \lambda},$$

$$P_{n,T} \to \left(\frac{1 - \lambda}{1 - \Delta\lambda}\right)^{1/4}.$$

Theorem 3.4.5. *Let the right-hand sides $\beta_1, \ldots, \beta_T$ of the system (3.4.1) take the values 0 and 1, and let $m = m(T)$ be the number of 1's in $\beta_1, \ldots, \beta_T$.*

If $n, T \to \infty$ such that $2T/n \to \lambda$, $0 < \lambda < 1$, and $m/T \to p$, $0 \le p \le 1$, then for any fixed $k = 0, 1, \ldots$,

$$P_{n,T}(k) \to \frac{1}{4^k k!}(-\log(1-\lambda)(1-\Delta\lambda))^k \sqrt{1-\lambda},$$

$$P_{n,T} \to \left(\frac{1-\lambda}{1-\Delta\lambda}\right)^{1/4},$$

where $\Delta = 1 - 2p$.

Before proceeding to the proof of these theorems, we will establish some auxiliary results. Let $\beta_1, \ldots, \beta_T$ be independent identically distributed random variables that take the values 0 and 1 with probabilities $1 - p$ and p, respectively; let $\Delta = 1 - 2p$; and let E be the set of the even numbers. Let $r_0 = 0$ and $r_1, \ldots, r_k$ be positive integers. We consider the random variables

$$\eta_i = \beta_{r_0 + \cdots + r_{i-1} + 1} + \cdots + \beta_{r_0 + \cdots + r_i}, \quad i = 1, \ldots, k.$$

Lemma 3.4.1.

$$\mathbf{P}\{\eta_i \in E, \ i = 1, \ldots, k\} = \frac{1}{2^k}(1 + \Delta^{r_1}) \cdots (1 + \Delta^{r_k}).$$

Proof. It suffices to note that the random variables $\eta_1, \ldots, \eta_k$ are independent and that the probability of the event of the sum $\beta_1 + \cdots + \beta_r$ being even equals $(1 + \Delta^r)/2$.   ∎

When the variables $\beta_1, \ldots, \beta_T$ are nonrandom, we need a similar assertion for the following scheme of allocating m particles into T cells. The cells are divided into $k+1$ groups of cells containing $r_1, \ldots, r_k$, $T - r_1 - \cdots - r_k$ cells, respectively. We assume that each cell can contain at most one particle, that $m \le T$, and that each of $\binom{T}{m}$ possible allocations are equiprobable. We introduce the random variables $\xi_1, \ldots, \xi_T$, setting $\xi_i = 0$ if the cell number i is empty, and $\xi_i = 1$ otherwise, for $i = 1, \ldots, T$. By analogy with the random variables $\eta_1, \ldots, \eta_k$, we define the random variables

$$\zeta_i = \xi_{r_0 + \cdots + r_{i-1} + 1} + \cdots + \xi_{r_0 + \cdots + r_i}, \quad i = 1, \ldots, k.$$

It is not difficult to verify the following assertions.

Lemma 3.4.2. *If $r_1, \ldots, r_k$ are fixed, $T \to \infty$, and $m/T \to 0$, then*

$$\mathbf{P}\{\zeta_i \in E, \ i = 1, \ldots, k\} \to 1.$$

Lemma 3.4.3. *If $r_1, \ldots, r_k$ are fixed, $T \to \infty$, and $m/T \to 1$, then*

$$\mathbf{P}\{\zeta_i \in E, \ i = 1, \ldots, k\} \to 1$$

if all $r_1, \ldots, r_k$ are even; and

$$P\{\zeta_i \in E, \ i = 1, \ldots, k\} \to 0$$

if at least one of $r_1, \ldots, r_k$ is odd.

Lemma 3.4.4. *If $r_1, \ldots, r_k$ are fixed, $T \to \infty$, and $m/T \to p$, $0 < p < 1$, then*

$$P\{\zeta_i \in E, \ i = 1, \ldots, k\} \to \frac{1}{2^k}\left(1 + \Delta^{r_1}\right) \cdots \left(1 + \Delta^{r_k}\right),$$

where $\Delta = 1 - 2p$.

We now consider the graph $G_{n,T}$ and mark the cycles in the graph by the following rule. Recall that $A_{n,T}$ is the set of all graphs with n labeled vertices and T edges whose components are trees and unicyclic components, allowing cycles of length 1 and 2. If a realization of the graph $G_{n,T}$ belongs to the set $A_{n,T}$, then every cycle of length r is marked with probability p_r independently of the others. If the graph contains a component with more than one cycle, then no cycle of the graph is marked. We denote by $p_{n,T}(k)$ the probability of the event that the number of cycles $\nu(G_{n,T})$ in the graph $G_{n,T}$ is equal to k and all cycles are marked. It is clear that the probability $p_{n,T}$ of the event that all cycles are marked equals

$$p_{n,T} = \sum_{k=0}^{\infty} p_{n,T}(k).$$

As in Section 1.7, we denote by d_m the number of mappings of the set $\{1, \ldots, m\}$ into itself whose graphs are connected, and by $d_m^{(r)}$ the number of mappings of the set $\{1, \ldots, m\}$ into itself whose graphs are connected and contain a cycle of length r. Let $F_{n,N}$ denote the number of forests with n labeled vertices and N trees, $T = n - N$.

Explicit expressions for d_m and $d_m^{(r)}$ are well known. By using the formula for the number of rooted trees, we obtain

$$d_m^{(r)} = \frac{m!}{(m-r)!} \, m^{m-r-1};$$

hence,

$$d_m = \sum_{r=1}^{m} d_m^{(r)} = (m-1)! \sum_{k=0}^{m-1} \frac{m^k}{k!}.$$

Lemma 3.4.5. *For any integer k, $1 \le k \le \min(n, T)$,*

$$p_{n,T}(k) = \frac{T!}{2^k k!}\left(\frac{2}{n^2}\right)^T \sum_{m=1}^{n} \binom{n}{m} F_{n-m,N} \sum_{m_1 + \cdots + m_k = m} \frac{m! \, D_{m_1} \cdots D_{m_k}}{m_1! \cdots m_k!},$$

where

$$D_m = \sum_{r=1}^{m} d_m^{(r)} p_r,$$

and for $k = 0$,

$$p_{n,T}(0) = F_{n,N} T! \left(\frac{2}{n^2}\right)^T.$$

Proof. For $k = 0$, the assertion is obvious. As in Section 1.7, let us denote by $b_n^{(r)}$ the number of connected graphs with n labeled vertices and one cycle of length r. It is clear that

$$b_n^{(1)} = d_n^{(1)}, \qquad b_n^{(2)} = d_n^{(2)}, \qquad b_n^{(r)} = d_n^{(r)}/2, \quad r \geq 3. \qquad (3.4.5)$$

Denote by $C_{n,T}$ the event that the graph $G_{n,T}$ contains no unmarked cycles. We represent the event

$$\{v(G_{n,T}) = k, \ G_{n,T} \in \mathcal{A}_{n,T}, \ C_{n,T}\}$$

as a union of the following disjoint events: In a specific order, T trials give T fixed edges that form a graph consisting of trees and k unicyclic components, including a marked cycle. It follows from this description that

$$p_{n,T}(k) = \mathbf{P}\{v(G_{n,T}) = k, \ G_{n,T} \in \mathcal{A}_{n,T}, \ C_{n,T}\}$$

$$= \sum_{m=k}^{n} \binom{n}{m} \sum_{m_1+\cdots+m_k=m} \frac{m!}{m_1!\cdots m_k! \, k!}$$

$$\times \sum_{r_1=1}^{m_1} b_{m_1}^{(r_1)} p_{r_1} \cdots \sum_{r_k=1}^{m_k} b_{m_k}^{(r_k)} p_{r_k} F_{n-m,T-m} T! \left(\frac{2}{n^2}\right)^T \frac{1}{2^{s_1+s_2}},$$

where $s_1 = s_1(r_1, \ldots, r_k)$ is the number of 1's among $r_1, \ldots, r_k$, and $s_2 = s_2(r_1, \ldots, r_k)$ is the number of 2's among $r_1, \ldots, r_k$. The factor 2^{-s_1} appears because the probability $2n^{-2}$ is replaced by n^{-2} in s_1 cases. The factor 2^{-s_2} reflects the fact that permuting trials in which two identical edges occur results in the same graph. The lemma follows from the relations (3.4.5). ∎

Theorem 3.4.6. *If $n, T \to \infty$ such that $2T/n \to \lambda$, $0 < \lambda < 1$, then for any fixed $k = 0, 1, \ldots$,*

$$p_{n,T}(k) = \frac{(D(\alpha))^k \sqrt{1-\lambda}}{2^k k!}(1 + o(1)),$$

where

$$D(x) = \sum_{m=1}^{\infty} \frac{D_m x^m}{m!}, \qquad \alpha = \lambda e^{-\lambda}.$$

Proof. The proof is similar to the proof of Theorem 1.8.2. We partition the sum from Lemma 3.4.5 into two parts. We put

$$M = T^{1/4}.$$

It is clear that for any x in the domain of convergence of the series

$$D(x) = \sum_{m=1}^{\infty} \frac{D_m x^m}{m!},$$

we have

$$\sum_{m \ge M} \sum_{m_1 + \cdots + m_k = m} \frac{D_{m_1} x^{m_1} \cdots D_{m_k} x^{m_k}}{m_1! \cdots m_k!} \le (D(x))^{k-1} \sum_{m \ge M/k} \frac{D_m x^m}{m!}.$$

$$(3.4.6)$$

Along with the function $D(x)$, let us introduce the generating function of the number of connected mappings

$$d(x) = \sum_{m=1}^{\infty} \frac{d_m x^m}{m!}.$$

The inequality

$$D(x) \le d(x) \qquad (3.4.7)$$

holds because

$$D_m = \sum_{r=1}^{m} d_m^{(r)} p_r \le \sum_{r=1}^{m} d_m^{(r)} = d_m.$$

Also,

$$d_m = (m-1)! \sum_{k=0}^{m-1} \frac{m^k}{k!} \le (m-1)! \, e^m,$$

which implies

$$\sum_{m \ge M/k} \frac{d_m x^m}{m!} \le \sum_{m \ge M/k} (ex)^m. \qquad (3.4.8)$$

Let

$$\theta(x) = \sum_{n=1}^{\infty} \frac{n^{n-1} x^n}{n!}, \qquad a(x) = \sum_{n=0}^{\infty} \frac{n^n x^n}{n!}.$$

By Example 1.3.2 and (1.4.8),

$$d(x) = \log a(x), \qquad a(x) = (1 - \theta(x))^{-1}.$$

We put $a = 2T/n$ and $x = ae^{-a}$ for $a < 1$. Then

$$\theta(x) = a, \qquad d(x) = -\log(1 - a).$$

Under the hypothesis of the theorem, $a = 2T/n \to \lambda$, $0 < \lambda < 1$, and for $x = ae^{-a}$, there exists $q < 1$ such that $ex = ae^{1-a} \le q < 1$ for sufficiently large n. Therefore

$$\sum_{m \ge M/k} \frac{d_m x^m}{m!} \le \frac{1}{1-q} q^{M/k}. \tag{3.4.9}$$

Using estimates (1.8.8), (1.8.9), and (3.4.6)–(3.4.9) yields

$$S_2 = \frac{T!}{2^k k!} \left(\frac{2}{n^2}\right)^T \sum_{m \ge M} \sum_{m_1 + \cdots + m_k = m} \binom{n}{m} F_{n-m,N} \frac{m! \, D_{m_1} \cdots D_{m_k}}{m_1! \cdots m_k!}$$

$$\le cT! \left(\frac{2}{n^2}\right)^T \sum_{m \ge M} \sum_{m_1 + \cdots + m_k = m} \frac{nn!(n-m)^{2(T-m)} d_{m_1} \cdots d_{m_k}}{(n-m)! \, 2^{T-m} (T-m)! \, m_1! \cdots m_k!}$$

$$\le c_1 \sum_{m \ge M} \sum_{m_1 + \cdots + m_k = m} n(ae^{-a})^m \frac{d_{m_1} \cdots d_{m_k}}{m_1! \cdots m_k!}$$

$$\le c_1 n (D(x))^{k-1} \sum_{m \ge M/k} \frac{d_m x^m}{m!} \le \frac{c_1 n}{1-q} q^{T^{1/4}/k} \le c_2 T q^{T^{1/4}/k},$$

where c_1, c_2 are constants. Thus, under the hypothesis of the theorem, $S_2 \to 0$. If $n, T \to \infty$, $2T/n \to \lambda$, $0 < \lambda < 1$, then by virtue of (1.8.7),

$$T! F_{n-m,N} = \frac{T! (n-m)^{2(T-m)} \sqrt{1-\lambda}}{2^{T-m} (T-m)!} (1 + o(1))$$

$$= \frac{n^{2T} x^m \sqrt{1-\lambda}}{2^T n^m} (1 + o(1))$$

uniformly in $m \le M = T^{1/4}$. Therefore, for any fixed $k = 1, 2, \ldots$,

$$S_1 = \frac{T!}{2^k k!} \left(\frac{2}{n^2}\right)^T \sum_{m \le M} \sum_{m_1 + \cdots + m_k = m} m! \binom{n}{m} F_{n-m,N} \frac{D_{m_1} \cdots D_{m_k}}{m_1! \cdots m_k!}$$

$$= \frac{\sqrt{1-\lambda}}{2^k k!} \sum_{m=k}^{M} \sum_{m_1 + \cdots + m_k = m} \frac{D_{m_1} x^{m_1} \cdots D_{m_k} x^{m_k}}{m_1! \cdots m_k!} (1 + o(1)).$$

By using the estimate of S_2, we obtain

$$S_1 = \frac{\sqrt{1-\lambda}}{2^k k!} \sum_{m=k}^{\infty} \sum_{m_1 + \cdots + m_k = m} \frac{D_{m_1} x^{m_1} \cdots D_{m_k} x^{m_k}}{m_1! \cdots m_k!} (1 + o(1)) + o(1)$$

$$= \frac{\sqrt{1-\lambda}}{2^k k!} (D(x))^k (1 + o(1)).$$

Combining the estimates of S_1 and S_2, we obtain, under the hypothesis of the theorem,

$$p_{n,T}(k) = \mathbf{P}\{\nu(G_{n,T}) = k,\ G_{n,T} \in \mathcal{A}_{n,T},\ C_{n,T}\}$$

$$= \frac{(D(x))^k \sqrt{1-\lambda}}{2^k k!}(1 + o(1)). \tag{3.4.10}$$

Hence the assertion of Theorem 3.4.6 for $k \geq 1$ follows, since $x = ae^{-a} = (2T/n)\,e^{-2T/n} \to \lambda e^{-\lambda} = \alpha$ and $D(x) \to D(\alpha)$. We use (1.8.6) and the representation from Lemma 3.4.5 and conclude that

$$p_{n,T}(0) = \sqrt{1-\lambda}(1 + o(1)).$$

∎

Corollary 3.4.1. *If $n, T \to \infty$ such that $2T/n \to \lambda$, $0 < \lambda < 1$, then the probability $p_{n,T}$ of the event that the graph $G_{n,T}$ contains no unmarked cycles satisfies the relation*

$$p_{n,T} = e^{D(\alpha)/2}\sqrt{1-\lambda}(1 + o(1)).$$

Proof. We denote by $p_{n,T}^{(1)}$ the probability of only marked cycles in the case where the graph has k unicyclic components and all the probabilities p_r are equal to 1, $r = 1, 2, \ldots$. In this case, $D(\alpha) = d(\alpha) = 2\Lambda = -\log(1-\lambda)$, and Theorem 3.4.6 gives

$$p_{n,T}^{(1)}(k) \to \frac{\Lambda^k e^{-\Lambda}}{k!}, \quad k = 0, 1, \ldots.$$

To prove the corollary, it suffices to show that in the sum

$$p_{n,T} = \sum_{k=0}^{n} p_{n,T}(k), \tag{3.4.11}$$

one can pass to the limit under the sum. Let us show that for any $\varepsilon > 0$, there exists K such that

$$\sum_{k=K+1}^{\infty} p_{n,T}(k) \leq \varepsilon. \tag{3.4.12}$$

We choose K such that

$$\sum_{k=K+1}^{\infty} \frac{\Lambda^k e^{-\Lambda}}{k!} \leq \frac{\varepsilon}{2},$$

and for fixed K, we choose n_0 so that for $n \geq n_0$,

$$\left| \sum_{k=0}^{K} p_{n,T}^{(1)}(k) - \sum_{k=0}^{K} \frac{\Lambda^k e^{-\Lambda}}{k!} \right| \leq \frac{\varepsilon}{2}.$$

Then, for $n \geq n_0$,

$$\left| \sum_{k=K+1}^{\infty} p_{n,T}^{(1)}(k) - \sum_{k=K+1}^{\infty} \frac{\Lambda^k e^{-\Lambda}}{k!} \right| = \left| \sum_{k=0}^{K} p_{n,T}^{(1)}(k) - \sum_{k=0}^{K} \frac{\Lambda^k e^{-\Lambda}}{k!} \right| \leq \frac{\varepsilon}{2},$$

and therefore

$$\sum_{k=K+1}^{\infty} p_{n,T}^{(1)}(k) \leq \varepsilon.$$

Since $p_{n,T}(k) \leq p_{n,T}^{(1)}(k)$, estimate (3.4.12) and the validity of passing to the limit under the sum are established. ∎

Proof of Theorems 3.4.4 and 3.4.5. A cycle leads to the inconsistency of system (3.4.1) if the sum of the right-hand sides of the subsystem corresponding to the cycle is odd. Let p_r be the probability that this sum is even for a cycle of length r. Then $P_{n,T}(k) = p_{n,T}(k)$ for any $k = 0, 1, \ldots$. Therefore Theorems 3.4.4 and 3.4.5 are direct corollaries to Theorem 3.4.6 and the fact proved above that one can pass to the limit under the sum in (3.4.11). To prove Theorem 3.4.4, we notice that in this case, according to Lemma 3.4.1,

$$p_r = (1 + \Delta^r)/2,$$

where $\Delta = 1 - 2p$; therefore

$$D(x) = \sum_{m=1}^{\infty} \frac{D_m x^m}{m!} = \sum_{m=1}^{\infty} \sum_{r=1}^{m} \frac{d_m^{(r)}(1 + \Delta^r)x^m}{2m!} \tag{3.4.13}$$

$$= \frac{1}{2} \sum_{m=1}^{\infty} \frac{d_m x^m}{m!} + \frac{1}{2} \sum_{m=1}^{\infty} \sum_{r=1}^{m} \frac{d_m^{(r)} \Delta^r x^m}{m!} = \frac{1}{2}(d(x) + d(x, \Delta)),$$

where

$$d(x, \Delta) = \sum_{m=1}^{\infty} \sum_{r=1}^{m} \frac{d_m^{(r)} \Delta^r x^m}{m!}.$$

For $x = ae^{-a}, 0 < a < 1$,

$$d(x) = -\log(1 - a),$$

and

$$d(x, \Delta) = -\log(1 - a\Delta).$$

Indeed,

$$d(x, \Delta) = \sum_{m=1}^{\infty} \sum_{r=1}^{m} \frac{d_m^{(r)} \Delta^r x^m}{m!} = \sum_{r=1}^{\infty} \sum_{m=1}^{\infty} \frac{d_m^{(r)} \Delta^r x^m}{m!}$$

$$= \sum_{r=1}^{\infty} \sum_{m=r}^{\infty} \frac{m^{m-r-1} \Delta^r x^m}{(m-r)!} = \sum_{r=1}^{\infty} \Delta^r x^r \sum_{t=0}^{\infty} \frac{(t+r)^{t-1} x^t}{t!}.$$

By using the well-known equality

$$\sum_{t=0}^{\infty} \frac{(t+r)^{t-1} x^t}{t!} = \frac{e^{ar}}{r}$$

from [124], Chapter 2, Problem 210 (see also [126]), we obtain

$$d(x, \Delta) = \sum_{r=1}^{\infty} \frac{\Delta^r x^r e^{ar}}{r} = \sum_{r=1}^{\infty} \frac{\Delta^r a^r}{r} = -\log(1 - a\Delta).$$

We conclude by noting that for $a = 2T/n \to \lambda, 0 < \lambda < 1$,

$$d(x) \to -\log(1 - \lambda), \qquad d(x, \Delta) = -\log(1 - a\Delta) \to \log(1 - \lambda\Delta).$$

Let us turn to the proof of Theorem 3.4.5. If $m/T \to 0$, then for any fixed k, all the cycles are marked with probability tending to 1. Therefore

$$P_{n,T}(k) = p_{n,T}^{(1)}(k)(1 + o(1)) = \frac{\Lambda^k e^{-\Lambda}}{k!}(1 + o(1)).$$

In the case where $m/T \to 1$, we have $p_r \to 0$ for odd r and $p_r \to 1$ for even r by Lemma 3.4.3. Therefore, in this case,

$$D_m \to D_m^{(2)} = \sum_{1 \le r \le m/2} d_m^{(2r)},$$

$$D(x) = \sum_{m=1}^{\infty} \frac{D_m x^m}{m!} \to B^{(2)}(\alpha) = \sum_{m=1}^{\infty} \frac{D_m^{(2)} \alpha^m}{m!}.$$

It is not difficult to see that

$$B^{(2)}(\alpha) = \sum_{r=1}^{\infty} \frac{\lambda^{2r}}{2r} = -\frac{1}{2}\log(1 - \lambda^2).$$

In the case where $m/T \to p, 0 < p < 1$, by Lemma 3.4.4,

$$p_r \to (1 + \Delta^r)/2,$$

and, as in (3.4.13),

$$D(x) \to D(\alpha) = (d(\alpha) + d(\alpha, \Delta))/2 = -\tfrac{1}{2} \log (1 - \lambda)(1 - \Delta\lambda).$$

■

3.5. Hypercycles and consistency of systems of random equations

In Section 3.2, we studied the rank of random matrices and found, in particular, that if the elements of a $T \times n$ matrix $A = \|\alpha_{tj}\|$ are independent identically distributed random variables taking the values 0 and 1 with equal probabilities, then the rank $r(A)$ of the matrix A has a threshold property: If $T/n \to \alpha$ and $\alpha < 1$, then $\mathbf{P}\{r(A) = T\} \to 1$, and if $T/n \to \alpha$ and $\alpha > 1$, then $\mathbf{P}\{r(A) = n\} \to 1$. In other words, the maximum number of independent critical sets $s(A)$ tends in probability to zero in the former case and to infinity in the latter case. A similar property apparently holds for the sparse matrices considered in Section 3.3: We proved only that if $\alpha < 1$, then $s(A)$ has in the limit a Poisson distribution, and $\mathbf{E}s(A) \to \infty$ for $\alpha > 1$.

In Section 3.4, we considered systems with at most two unknowns in each equation. It was shown that if $T/n \to \alpha$, $0 < \alpha < 1/2$, then the maximal number of independent critical sets or independent cycles in the corresponding graph approaches the Poisson distribution with parameter $\Lambda = -\tfrac{1}{2} \log(1 - 2\alpha)$. As follows from Theorem 2.1.6, if $\alpha > 1/2$, then $s(A)$ tends in probability to infinity.

The case of a matrix with independent and identically distributed random elements taking the values 0 and 1 with probabilities 1/2 and the case of a matrix with at most two elements in each row studied in Section 3.4 can be considered as the extreme cases in terms of the behavior of the rank and the maximum number of independent critical sets. In these cases, the threshold effect appears at the points $T/n = 1$ and $T/n = 1/2$, respectively.

In this section, we consider an intermediate case and obtain a weaker form of the threshold effect. We consider the system of random linear equations in GF(2):

$$x_{i_1(t)} + \cdots + x_{i_r(t)} = b_t, \quad t = 1, \ldots, T, \tag{3.5.1}$$

where $i_1(t), \ldots, i_r(t), t = 1, \ldots, T$, are independent identically distributed random variables taking the values $1, \ldots, n$ with equal probabilities, and the independent random variables $b_1, \ldots, b_T$ do not depend on the left-hand side of the system and take the values 0 and 1 with equal probabilities. If $r = 2$, we obtain the system considered in Section 3.4.

In Section 3.1, we introduced the notions of critical sets for a matrix and hypercycles for the hypergraph corresponding to a matrix. Denote by $A_{r,n,T}$ the matrix of system (3.5.1) and by $G_{r,n,T}$ the hypergraph with n vertices and T hyperedges $e_1, \ldots, e_T$ that corresponds to this matrix. Thus we consider a random hypergraph $G_{r,n,T}$, whose matrix $A = A_{r,n,T} = \|a_{tj}\|$ has the following structure. The elements of the matrix $a_{tj}, t = 1, \ldots, T, j = 1, \ldots, n$, are random variables and the rows of the matrix are independent. There are r ones allocated to each row: Each 1, independent of the others, is placed in each of n positions with probability $1/n$, and a_{tj} equals 1 if there are an odd number of 1's in position j of row t. Therefore, there are no more than r ones in each row.

For such regular hypergraphs, the following threshold property holds: If $n, T \to \infty$ such that $T/n \to \alpha$, then an abrupt change in the behavior of the rank of the matrix $A_{r,n,T}$ occurs while the parameter α passes the critical value α_r. This property can be expressed in terms of the total number of hypercycles in $G_{r,n,T}$. Let $s(A_{r,n,T})$ be the maximum number of independent critical sets of $A_{r,n,T}$ or independent hypercycles of the hypergraph $G_{r,n,T}$. Then

$$S(A_{r,n,T}) = 2^{s(A_{r,n,T})} - 1$$

is the total number of critical sets or hypergraphs.

In this section, we prove that the following threshold property is true for $S(A_{r,n,T})$.

Theorem 3.5.1. *Let $r \geq 3$ be fixed, $T, n \to \infty$ such that $T/n \to \alpha$. Then there exists a constant α_r such that $\mathbf{E}S(A_{r,n,T}) \to 0$ for $\alpha < \alpha_r$ and $\mathbf{E}S(A_{r,n,T}) \to \infty$ for $\alpha > \alpha_r$.*

The constant α_r is the first component of the vector that is the unique solution of the system of equations

$$e^{-x} \cosh \lambda \left(\frac{ar}{ar - x} \right)^a = 1,$$

$$\frac{x}{\lambda} \left(\frac{ar - x}{x} \right)^{1/r} = 1, \tag{3.5.2}$$

$$\lambda \tanh \lambda = x,$$

with respect to the variables a, x, and λ.

The numerical solution of the system of equations gives us the following values of the critical constants:

$$\alpha_3 = 0.8894 \ldots, \qquad \alpha_4 = 0.9671 \ldots, \qquad \alpha_5 = 0.9891 \ldots,$$

$$\alpha_6 = 0.9969 \ldots, \qquad \alpha_7 = 0.9986 \ldots, \qquad \alpha_8 = 0.9995 \ldots.$$

Expanding the solution of the system into powers of e^{-r} yields

$$\alpha_r \approx 1 - \frac{e^{-r}}{\log 2} - \frac{e^{-2r}}{\log 2}\left(\frac{r^2}{2} + \frac{r}{\log 2} - r - \frac{1}{2}\right),$$

which gives values close to the exact ones for $r \geq 4$.

Let us give some auxiliary results that will be needed for the proof of Theorem 3.5.1.

The total number of hypercycles $S(A_{r,n,T})$ in the hypergraph $G_{r,n,T}$ with the matrix $A_{r,n,T}$ can be represented as a sum of indicators. Let $\xi_{t_1,\dots,t_m} = 1$ if the hypercycle $C = \{e_{t_1}, \dots, e_{t_m}\}$ occurs in $G_{r,n,T}$, and $\xi_{t_1,\dots,t_m} = 0$ otherwise. It is clear that $\mathbf{P}\{\xi_{t_1,\dots,t_m} = 1\}$ does not depend on the indices $t_1, \dots, t_m$. Indeed, from the definition of the random hypergraph $G_{r,n,T}$, the indicator $\xi_{t_1,\dots,t_m} = 1$ if and only if there are an even number of 1's in each column of the submatrix consisting of the rows with indices $t_1, \dots, t_m$. The number of 1's in n columns of any m rows, before these numbers were reduced modulo 2, have the multinomial distribution with rm trials and n equiprobable outcomes.

Denote by $\eta_1(s, n), \dots, \eta_n(s, n)$ the contents of the cells in the equiprobable scheme of allocating s particles into n cells. In these notations, the number of 1's in the columns of any m rows, before those numbers have been reduced modulo 2, have a distribution that coincides with the distribution of the variables $\eta_1(rm, n), \dots, \eta_n(rm, n)$. Therefore

$$\mathbf{P}\{\xi_{t_1,\dots,t_m} = 1\} = \mathbf{P}\{\eta_1(rm, n) \in E, \dots, \eta_n(rm, n) \in E\},$$

where E is the set of even numbers, and the average number of hypercycles in $G_{r,n,T}$ can be written in the following form:

$$\mathbf{E}S(A_{r,n,T}) = \sum_{m=1}^{T} \binom{T}{m} P_E(rm, n), \tag{3.5.3}$$

where

$$P_E(rm, n) = \mathbf{P}\{\eta_1(rm, n) \in E, \dots, \eta_n(rm, n) \in E\}.$$

Thus, to estimate $\mathbf{E}S(A_{r,n,T})$, we need to know the asymptotic behavior of $P_E(rm, n)$.

We consider a more general case and obtain the asymptotic behavior of the probabilities

$$P_R(s, n) = \mathbf{P}\{\eta_1(s, n) \in R, \dots, \eta_n(s, n) \in R\},$$

where R is a subset of the set of all nonnegative integers.

The joint distribution of the random variables $\eta_1(s, n), \dots, \eta_n(s, n)$ can be expressed as a conditional distribution of independent random variables $\xi_1, \dots, \xi_n$, identically distributed by the Poisson law with an arbitrary parameter λ, in the following way (see, e.g., [90]). For any nonnegative integers $s_1, \ldots, s_n$ such that $s_1 + \cdots + s_n = s$,

$$\mathbf{P}\{\eta_1(s, n) = s_1, \ldots, \eta_n(s, n) = s_n\}$$
$$= \mathbf{P}\{\xi_1 = s_1, \ldots, \xi_n = s_n \mid \xi_1 + \cdots + \xi_n = s\}.$$

Therefore

$$P_R(s, n) = \mathbf{P}\{\eta_1(s, n) \in R, \ldots, \eta_n(s, n) \in R\}$$
$$= \frac{\mathbf{P}\{\xi_1 \in R, \ldots, \xi_n \in R, \ \xi_1 + \cdots + \xi_n = s\}}{\mathbf{P}\{\xi_1 + \cdots + \xi_n = s\}}$$
$$= (\mathbf{P}\{\xi_1 \in R\})^n \frac{\mathbf{P}\{\xi_1 + \cdots + \xi_n = s \mid \xi_1 \in R, \ldots, \xi_n \in R\}}{\mathbf{P}\{\xi_1 + \cdots + \xi_n = s\}}.$$

We now introduce independent identically distributed random variables $\xi_1^{(R)}, \ldots, \xi_n^{(R)}$ with the distribution

$$\mathbf{P}\{\xi_1^{(R)} = k\} = \mathbf{P}\{\xi_1 = k \mid \xi_1 \in R\}, \quad k = 0, 1, \ldots.$$

It is not difficult to see that

$$\mathbf{P}\{\xi_1 + \cdots + \xi_n = s \mid \xi_1 \in R, \ldots, \xi_n \in R\} = \mathbf{P}\{\xi_1^{(R)} + \cdots + \xi_n^{(R)} = s\},$$

and therefore

$$P_R(s, n) = (\mathbf{P}\{\xi_1 \in R\})^n \frac{\mathbf{P}\{\xi_1^{(R)} + \cdots + \xi_n^{(R)} = s\}}{\mathbf{P}\{\xi_1 + \cdots + \xi_n = s\}}. \tag{3.5.4}$$

Let $x = s/n$ and choose the parameter λ of the Poisson distribution in such a way that

$$x = \mathbf{E}\xi_1^{(R)} = \sum_{k \in R} k \frac{\lambda^k e^{-\lambda}}{k!} / \mathbf{P}\{\xi_1 \in R\}.$$

Let d be the maximum span of the lattice on which the set R is situated and denote the lattice by Γ_R.

Theorem 3.5.2. *If $s, n \to \infty$ such that $n \in \Gamma_R$, then in any interval of the form $0 < x_0 \leq x \leq x_1 < \infty$,*

$$P_R(s, n) = (\mathbf{P}\{\xi_1 \in R\})^n \left(\frac{x^x e^\lambda}{\lambda^x e^x}\right)^n \frac{d\sqrt{x}}{\sigma} (1 + o(1))$$

uniformly in $x = s/n$, where the parameter λ of the Poisson distribution of the random variable ξ_1 is the root of the equation $x = \mathbf{E}\xi_1^{(R)}$, and $\sigma^2 = \mathbf{D}\xi_1^{(R)}$ (the variance).

Proof. The local limit theorem holds for the sum $\xi_1^{(R)} + \cdots + \xi_n^{(R)}$. Following the classical proof of the local limit theorem of Gnedenko [49], we prove that if $s, n \to \infty$ such that $n \in \Gamma_R$, then

$$\mathbf{P}\{\xi_1^{(R)} + \cdots + \xi_n^{(R)} = n\} = \frac{d}{\sigma\sqrt{2\pi n}}(1 + o(1))$$

uniformly in $x = s/n$ in any interval of the form $0 < x_0 \leq x \leq x_1 < \infty$, where $\sigma^2 = \mathbf{D}\xi_1^{(R)}$, and d is the span of the lattice Γ_R.

When we substitute the expression into (3.5.4) and take into account that the sum $\xi_1 + \cdots + \xi_n$ is distributed by the Poisson law with parameter λn, we obtain the assertion of the theorem. ∎

Note that (3.5.4) implies the estimate

$$P_R(s, n) \leq (\mathbf{P}\{\xi_1 \in R\})^n \frac{s! \, e^{\lambda n}}{(\lambda n)^s},$$

where $P_R(s, n)$ does not depend on λ, and on the right-hand side any positive value can be assigned to this parameter. Let $E = \{0, 2, \ldots\}$. In this case,

$$\mathbf{P}\{\xi_1 \in E\} = e^{-\lambda} \cosh \lambda,$$

and the estimate takes the form

$$P_E(s, n) \leq (\cosh \lambda)^n \frac{s!}{\lambda^s n^s}, \tag{3.5.5}$$

where $\lambda > 0$ can be chosen arbitrarily.

We now estimate

$$\mathbf{E}S(A_{r,n,T}) = \sum_{m=1}^{T} \binom{T}{m} P_E(rm, n).$$

Lemma 3.5.1. *If $r \geq 3$ is fixed, and $T, n \to \infty$ such that $T/n \to \alpha$, then for any $\varepsilon > 0$, there exists $\delta > 0$ such that*

$$\sum_{1 \leq m \leq \delta T} \binom{T}{m} P_E(rm, n) \leq \varepsilon.$$

Proof. First we point out that

$$\mathbf{P}\{\xi_1^{(E)} = 2k\} = \mathbf{P}\{\xi_1 = 2k \mid \xi_1 \in E\} = \frac{\lambda^{2k}}{(2k)! \cosh \lambda}, \quad k = 0, 1, \ldots,$$

$$\mathbf{E}\xi_1^{(E)} = \lambda \tanh \lambda.$$

Put $x = rm/n$ and choose the parameter λ of the Poisson distribution in such a way that $x = \lambda \tanh \lambda$. From (3.5.5), it follows that

$$P_E(rm, n) \leq (\cosh \lambda)^n \frac{(rm)!}{\lambda^{rm} n^{rm}}.$$

Since the value of x becomes small for sufficiently small δ, we can assume that $\lambda \leq 1$ in the domain of summation. For such λ,

$$\lambda^2/4 \leq x = \lambda \tanh \lambda \leq \lambda^2,$$

and therefore

$$\cosh \lambda \leq e^{\lambda^2} \leq e^{4x}.$$

We now estimate the sum. It is easy to see that

$$\sum_{1 \leq m \leq \delta T} \binom{T}{m} P_E(rm, n) \leq \sum_{1 \leq m \leq \delta T} \frac{T^m}{m!} (\cosh \lambda)^n \frac{(rm)!}{\lambda^{rm} n^{rm}}$$

$$\leq \sum_{1 \leq m \leq \delta T} T^m e^{4xn} \frac{(rm)^{rm-m}}{\lambda^{rm} n^{rm}}$$

$$\leq \sum_{1 \leq m \leq \delta T} \left(\frac{T}{n}\right)^m e^{4rm} \frac{1}{x^{rm/2}} \left(\frac{rm}{n}\right)^{m(r-1)}$$

$$= \sum_{1 \leq m \leq \delta T} \left(\left(\frac{T}{n}\right)^{r/2} e^{4r} \left(\frac{m}{T}\right)^{r/2-1} r^{r/2-1}\right)^m$$

$$= \sum_{1 \leq m \leq \delta T} \left(\left(\frac{T}{n}\right)^{r/2} r^{r/2-1} e^{4r} \delta^{r/2-1}\right)^m.$$

Since T/n tends to a constant, the last sum can be made arbitrarily small by choosing a sufficiently small δ. ∎

Lemma 3.5.2. *If r is fixed, and $T, n \to \infty$ such that $T/n \to \alpha$, $0 < \alpha < 1$, then for any $\varepsilon > 0$, there exists $\delta > 0$ such that*

$$\sum_{(1-\delta)T \leq m \leq T} \binom{T}{m} P_E(rm, n) \leq \varepsilon.$$

Proof. Put $\lambda = rm/n$ and let an integer m_0 be chosen such that $m_0/T \leq \delta$. With such a choice of λ, by (3.5.5),

$$\sum_{m=T-m_0}^{T} \binom{T}{m} P_E(rm, n) \leq \sum_{m=T-m_0}^{T} \binom{T}{m} (\cosh \lambda)^n \frac{(rm)!}{(rm)^{rm}}.$$

Since in the domain of summation, λ is greater than some positive constant, there exists $q < 1$ such that

$$e^{-\lambda} \cosh \lambda = (1 + e^{-2\lambda})/2 \le q.$$

By using the inequality

$$(rm)! \le c(rm)^{rm} e^{-rm} (rT)^{1/2},$$

where c is a constant, we obtain

$$\sum_{m=T-m_0}^{T} \binom{T}{m} P_E(rm, n) \le c(rT)^{1/2} \sum_{m=T-m_0}^{T} \binom{T}{m} (e^{-\lambda} \cosh \lambda)^n$$

$$\le c(rT)^{1/2} q^n \sum_{m=T-m_0}^{T} \binom{T}{m} \frac{q^m (1-q)^{T-m}}{q^T (1-q)^{m_0}}$$

$$\le c(rT)^{1/2} \left(\frac{q}{(1-q)^{m_0/(n-T)}} \right)^{n-T}.$$

Since $q, \alpha < 1$, the value $m_0/(n-T)$ can be made arbitrarily small by choosing a sufficiently small δ, and therefore the value $q/(1-q)^{m_0/(n-T)}$ can be made smaller than some $Q < 1$. Thus, for a sufficiently small δ, the right-hand side tends to zero under the conditions of the lemma. ∎

Proof of Theorem 3.5.1. We now estimate the middle part of the sum. As $T/n \to \alpha$ and $\delta \le m/T \le 1 - \delta$, the values $x = rm/n$ lie in an interval of the form $0 < x_0 \le x \le x_1 < \infty$. When we apply Theorem 3.5.2, we obtain for even rm,

$$P_E(rm, n) = (\mathbf{P}\{\xi_1 \in E\})^n \left(\frac{x}{\lambda e} \right)^{rm} e^{\lambda n} \frac{2\sqrt{x}}{\sigma} (1 + o(1))$$

uniformly in x, $x_0 \le x \le x_1$, where $x = \mathbf{E}\xi_1^{(E)} = \lambda \tanh \lambda$, $\sigma^2 = \mathbf{D}\xi_1^{(E)} = \lambda^2 + x - x^2$.

From $\mathbf{P}\{\xi_1 \in E\} = e^{-\lambda} \cosh \lambda$, we obtain the final estimate: As $T, n \to \infty$, $T/n \to \alpha$,

$$P_E(rm, n) = (\cosh \lambda)^n \left(\frac{x}{\lambda e} \right)^{xn} \frac{2\sqrt{x}}{\sigma} (1 + o(1)) \tag{3.5.6}$$

uniformly in m, $\delta \le m/T \le 1 - \delta$.

Setting $p = m/T$, $q = 1 - p$, and using the normal approximation to the binomial distribution show that, as $T \to \infty$,

$$\binom{T}{m} = \binom{T}{m} p^m q^{T-m} (p^m q^{T-m})^{-1} = \frac{1}{p^m q^{T-m} \sqrt{2\pi T pq}} (1 + o(1))$$

uniformly in m, $\delta \le m/T \le 1 - \delta$.

Let $a = T/n$ and write $p = m/T$ in terms of $x = rm/n$ and a. Then

$$p = \frac{m}{T} = \frac{x}{ar}, \qquad q = 1 - \frac{m}{T} = \frac{ar - x}{ar},$$

and the estimate of $\binom{T}{m}$ takes the following form. As $T \to \infty$, $\delta \le m/T \le 1 - \delta$,

$$\binom{T}{m} = \frac{ar}{\sqrt{2\pi x(ar - x)an}} \left(\frac{(ar)^{ar}(ar - x)^x}{x^x(ar - x)^{ar}} \right)^{n/r} (1 + o(1)) \qquad (3.5.7)$$

uniformly in m.

We combine the estimates (3.5.6) and (3.5.7) and obtain

$$\binom{T}{m} P_E(rm, n) = (f(a, x))^n \frac{2ar\sqrt{x}}{\sigma\sqrt{2\pi x(ar - x)an}} (1 + o(1)),$$

where

$$f(a, x) = \cosh \lambda \left(\frac{x}{\lambda e} \right)^x \left(\frac{ar - x}{x} \right)^{x/r} \left(\frac{ar}{ar - x} \right)^a,$$

$$x = \lambda \tanh \lambda.$$

The function $f(a, x)$ increases as a increases,

$$f'_x(a, x) = f(a, x) \log \left(\frac{x}{\lambda} \left(\frac{ar - x}{x} \right)^{1/r} \right) \to -\infty, \quad x \to 0,$$

and the derivative $f'_x(a, x)$ has no more than two zeros. Therefore the system of equations

$$f(a, x) = 1,$$

$$f'_x(a, x) = 0, \qquad (3.5.8)$$

$$\lambda \tanh \lambda = x$$

has the unique solution $(\alpha_r, x_r, \lambda_r)$; at this point, the function $f(\alpha_r, x)$ as a function of x attains its maximum, which is equal to 1. Therefore, for all x, $0 < x < ar$,

$$f(\alpha_r, x) \le f(\alpha_r, x_r) = 1.$$

In addition,

$$f(a, x) < f(\alpha_r, x) \le 1, \quad a < \alpha_r,$$

$$f(a, x_r) > f(\alpha_r, x_r) = 1, \quad a > \alpha_r.$$

This implies that the middle part of the sum tends to zero for $\alpha < \alpha_r$ and tends to infinity for $\alpha > \alpha_r$.

If we consider the estimates for the tails of the sum in Lemmas 3.5.1 and 3.5.2, we obtain the assertion of Theorem 3.5.1 because system (3.5.8) can be easily transformed to the form mentioned in the statement of the theorem. ∎

It would be interesting to find the limit distribution of the number of hypercycles. Up to now, no one has succeeded even in proving that $S(A_{r,n,T})$ tends in probability to infinity as $T, n \to \infty$, $T/n \to \alpha > \alpha_r$.

3.6. Reconstructing the true solution

We consider the system of equations in GF(2):

$$x_{i(t)} + x_{j(t)} = b_t, \quad t = 1, \ldots, T, \tag{3.6.1}$$

where the pairs $(i(t), j(t))$, $t = 1, \ldots, T$, are independent identically distributed two-dimensional random vectors that take values (i, j), $i < j$, $i, j = 1, \ldots, n$, with equal probabilities $\binom{n}{2}^{-1}$.

In Section 3.1, we interpreted a system similar to (3.6.1) as the result of T trials performed with the aim of classifying n objects by random pairwise comparisons, and we set $b_t = 0$ if the comparison of $x_{i(t)}$ and $x_{j(t)}$ showed that these objects were from the same class, and $b_t = 1$ otherwise, for $t = 1, \ldots, T$. If the comparisons are not absolutely right, then the result of a comparison may deviate from the true value. Suppose that $X^* = (x_1^*, \ldots, x_n^*)$ is the vector of true values of the unknowns, and the column-vector $B^* = (b_1^*, \ldots, b_T^*)$ is obtained by substituting X^* into the left-hand side of system (3.6.1):

$$AX^* = B^*, \tag{3.6.2}$$

where A is the matrix of system (3.6.1).

If the measurements are not precise, then it is natural to suppose that

$$b_t = b_t^* + \varepsilon_t, \quad t = 1, \ldots, T,$$

where $\varepsilon_1, \ldots, \varepsilon_T$ are independent identically distributed random variables that do not depend on A and take the values 0 and 1. These random variables can be interpreted as errors. Let

$$p = \frac{1 - \Delta}{2} = \mathbf{P}\{\varepsilon_1 = 1\}, \qquad q = \frac{1 + \Delta}{2} = \mathbf{P}\{\varepsilon_1 = 0\}, \tag{3.6.3}$$

where Δ is called the excess.

The problem is to estimate or reconstruct the vector $X^* = (X_1^*, \ldots, x_n^*)$ on the basis of the matrix A and the right-hand side $B = (b_1, \ldots, b_T)$ of system (3.6.1).

In a similar situation over the field of real numbers, an estimate of the true solution of a system of linear equations with perturbed right-hand sides can be found by the least-square method. Under some conditions on the matrix and the errors in the right-hand sides, the least-square method provides an estimate that converges to the true solution as the number of equations tends to infinity. In contrast to the field of real numbers, in GF(2) a good estimate $\hat{X} = (\hat{x}_1, \ldots, \hat{x}_n)$ coincides with the true solution $X^* = (x_1^*, \ldots, X_n^*)$ with probability tending to 1 as $T \to \infty$.

As usual, we associate the graph $\Gamma_{n,T}$ with the left-hand side of system (3.6.1). The graph $\Gamma_{n,T}$ has n labeled vertices corresponding to the unknowns $x_1, \ldots, x_n$ and T edges $e_t = (i(t), j(t)), t = 1, \ldots, T$. The edges $e_1, \ldots, e_T$ are independent and assume the $n(n-1)/2$ possible values with equal probability. Therefore, the graph $\Gamma_{n,T}$ may have multiple edges.

It is clear that along with the vector X^*, the vector $\bar{X}^* = (\bar{x}_1^*, \ldots, \bar{x}_n^*)$ with elements $\bar{x}_i^* = x_i^* + 1, t = 1, \ldots, n$, satisfies the system (3.6.2). The pair X^*, $\bar{X}^*$ is uniquely determined by the system (3.6.2) if the graph $\Gamma_{n,T}$ is connected, in other words, if the system (3.6.2) contains all the unknowns and is not decomposed into subsystems with disjoint sets of unknowns.

Denote by $p_{n,T}$ the probability that the graph $\Gamma_{n,T}$ is connected. It follows from Theorem 2.3.8 that if $n, T \to \infty$ such that $T = n \log n + an + o(n)$, where a is a constant, then

$$p_{n,T} \to e^{-e^{-a}}.$$

Thus, if $n, T \to \infty$ and the pair X^*, $\bar{X}^*$ is determined by the system (3.6.2) with probability tending to 1, then

$$\frac{T}{n \log n} = 1 + \frac{w_n}{\log n},$$

where $w_n \to \infty$.

In this section, we present three algorithms for reconstructing the true solution of system (3.6.1) with perturbed right-hand sides. We first describe the reconstruction method that can be called the voting algorithm. This algorithm consists of correcting the right-hand sides $b_1, \ldots, b_T$ of the system (3.6.1) by the majority rule. Let the system (3.6.1) contain the subsystem with $m_{ij}, i < j$, equations:

$$x_i + x_j = a_{ij}^{(1)},$$
$$\cdots \tag{3.6.4}$$
$$x_i + x_j = a_{ij}^{(m_{ij})}.$$

The true value of $a_{ij}^{(1)}, \ldots, a_{ij}^{(m_{ij})}$ equals $a_{ij}^* = x_i^* + x_j^*$.

We set $a_{ij} = 1$ if

$$a_{ij}^{(1)} + \cdots + a_{ij}^{(m_{ij})} > m_{ij}/2,$$

and $a_{ij} = 0$ otherwise.

Under some conditions, system (3.6.1) is indecomposable and $a_{ij} = a_{ij}^*$ for all $i, j = 1, \ldots, n$; thus the true solution is reconstructed.

Denote by $P(n, T)$ the probability of reconstructing the true solution of system (3.6.1) by the voting algorithm, that is,

$$P(n, T) = \mathbf{P}\{a_{ij} = a_{ij}^*, \ i, j = 1, \ldots, n\}.$$

Theorem 3.6.1. *If $n, T \to \infty$ and $\Delta \to 0$ such that*

$$\frac{\Delta^2 T}{n^2 \log n} \to \infty,$$

then $P(n, T) \to 1$.

Proof. Let

$$\mu(n, T) = \min_{1 \le i, j \le n} m_{ij},$$

where the minimum is over all subsystems of the form (3.6.4). It is clear that

$$P(n, T) = \mathbf{P}\{\mu(n, T) > m\} P_m(n, T) + \mathbf{P}\{\mu(n, T) \le m\} \bar{P}_m(n, T), \quad (3.6.5)$$

where $P_m(n, T)$ and $\bar{P}_m(n, T)$ are the conditional probabilities of reconstructing the true solution under the conditions $\{\mu(n, T) > m\}$ and $\{\mu(n, T) \le m\}$, respectively.

We obtain a rough estimate for the probability $\mathbf{P}\{\mu(n, T) > m\}$. It is clear that $\mathbf{P}\{\mu(n, T) > m\}$ is the probability that each cell contains more than m particles in the classical scheme of allocating T particles into $\binom{n}{2}$ cells. Denote by η_i the number of particles in the ith cell and put $\xi_i = 1$ if $\eta_i \le m$, and $\xi_i = 0$ if $\eta_i > m$, $i = 1, \ldots, \binom{n}{2}$. By (1.1.1),

$$\mathbf{P}\{\mu(n, T) \le m\} = \mathbf{P}\{\xi_1 + \cdots + \xi_{\binom{n}{2}} > 0\}$$

$$\le \binom{n}{2} \mathbf{E}\xi_1 = \binom{n}{2} \mathbf{P}\{\eta_1 \le m\}.$$

The random variable η_1 has the binomial distribution with T trials and the probability of success $\binom{n}{2}^{-1}$. Since $\alpha = \mathbf{E}\eta_1 = T/\binom{n}{2} \to \infty$, the normal approximation is valid for this distribution. We choose $m = \alpha(1 - \Delta)$, assume that $(\Delta\sqrt{\alpha})^3/T \to 0$, and estimate the probability $\mathbf{P}\{\eta_1 \le m\}$. By taking into account the choice of m and the equality $\mathbf{D}\eta_1 = \alpha(1 + o(1))$, we obtain

$$\mathbf{P}\{\eta_1 \le n\} = \mathbf{P}\{(\eta_1 - \alpha)/\sqrt{\mathbf{D}\eta_1} \le (m - \alpha)/\sqrt{\mathbf{D}\eta_1}\}$$

$$= \mathbf{P}\{(\eta_1 - \alpha)/\sqrt{\mathbf{D}\eta_1} \le \Delta\sqrt{\alpha}(1 + o(1))\}$$

$$= \frac{1}{\sqrt{2\pi}} \int_{-\infty}^{-\Delta\sqrt{\alpha}} e^{-u^2/2} \, du(1 + o(1)).$$

Hence, there exists a constant c such that

$$\mathbf{P}\{\eta_1 \leq m\} \leq c e^{-\Delta^2 \alpha/2}.$$

Thus, for $m = \alpha(1 - \Delta)$,

$$\mathbf{P}\{\mu(n, T) \leq m\} \to 0 \tag{3.6.6}$$

because $\Delta^2 \alpha / \log n \to \infty$ and $n^2 e^{-\Delta\alpha/2} \to 0$.

Now we have to show that under the conditions of the theorem, $P_m(n, T) \to 1$. In other words, we have to prove that $a_{ij} = a_{ij}^*$ for all $i, j = 1, \ldots, n$ with probability tending to 1. The additional requirement of the indecomposability of the system (3.6.1) or of the connectedness of the graph $\Gamma_{n,T}$ is obviously fulfilled.

Recall that $b_t = b_t^* + \varepsilon_t$. We may assume that in the subsystem (3.6.4),

$$a_{ij}^{(k)} = a_{ij}^* + \varepsilon_{ij}^{(k)}, \quad k = 1, \ldots, m_{ij},$$

where the random variables $\varepsilon_{ij}^{(1)}, \ldots, \varepsilon_{ij}^{(m_{ij})}$ are independent and have the same distribution as $\varepsilon_1, \ldots, \varepsilon_T$ from the right-hand side of (3.6.1). Denote by $\xi(n, T)$ the number of wrong decisions, that is, the number of realized events $\{a_{ij} \neq a_{ij}^*\}$, $i, j = 1, \ldots, n$. Now let $\xi_{ij} = 1$ if $\varepsilon_{ij}^{(1)} + \cdots + \varepsilon_{ij}^{(m_{ij})} > m_{ij}/2$, and $\xi_{ij} = 0$ otherwise. It is clear that the number of wrong decisions can be represented in the form

$$\xi(n, T) = \sum_{i<j} \xi_{ij},$$

and

$$1 - P_m(n, T) = \mathbf{P}\{\xi(n, T) > 0 \mid \mu(n, T) > m\}$$

$$\leq \binom{n}{2} \mathbf{E}(\xi_{12} \mid \mu(n, T) > m\}$$

$$= \binom{n}{2} \mathbf{P}\{\xi_{12} = 1 \mid \mu(n, T) > m\}. \tag{3.6.7}$$

Now we derive estimates for

$$\mathbf{P}\{\xi_{12} = 1 \mid \mu(n, T) > m\} = \mathbf{P}\{\varepsilon_{12}^{(1)} + \cdots + \varepsilon_{12}^{(m_{12})} > m_{12}/2 \mid \mu(n, T) > m\}.$$

The random variables $\varepsilon_{12}^{(1)}, \ldots, \varepsilon_{12}^{(m_{12})}$ are independent and have the same distribution as the random variables $\varepsilon_1, \ldots, \varepsilon_T$ from the right-hand side of system (3.6.1). We set $S_k = \varepsilon_1 + \cdots + \varepsilon_k$ and estimate

$$\mathbf{P}\{S_k > k/2\} = \mathbf{P}\{S_k - \mathbf{E}S_k > k\Delta/2\}.$$

Here, and later in this section, we use the following inequality of exponential type for the sum S_k that was proposed by Hoeffding [59] and can be found in [122] (see

Theorem 1.1.16). For any positive Δ,

$$\mathbf{P}\{S_k - \mathbf{E}S_k > k\Delta/2\} \le e^{-k\Delta^2/2}. \qquad (3.6.8)$$

Therefore

$$\mathbf{P}\{\varepsilon_{ij}^{(1)} + \cdots + \varepsilon_{ij}^{(m_{ij})} > m_{ij}/2 \mid \mu(n, T) > m\} \le e^{-m\Delta^2/2},$$

and from (3.6.7), we obtain

$$1 - P_m(n, T) \le \binom{n}{2} e^{-m\Delta^2/2}. \qquad (3.6.9)$$

For $m = \alpha(1 - \Delta)$, $\alpha = T/\binom{n}{2}$, under the conditions of Theorem 3.6.1, the right-hand side of (3.6.9) tends to zero. Thus, the assertion of the theorem now follows from (3.6.5), (3.6.6), and (3.6.9).￭

We now describe the second algorithm for reconstructing the true solution of system (3.6.1), which can be called the method of coordinate testing.

We choose a vector $X^{(0)} = (x_1^{(0)}, \ldots, x_n^{(0)})$ by random sampling from the set of all n-dimensional vectors over GF(2). Denote by $B^{(0)} = (b_1^{(0)}, \ldots, b_T^{(0)})$ the column-vector obtained by substituting $X^{(0)}$ for X in the left-hand side of (3.6.1). Let $\beta(X^{(0)})$ be the number of the coordinates of $B^{(0)}$ that coincide with the corresponding coordinates of the vector $B = (b_1, \ldots, b_T)$ of the right-hand sides of system (3.6.1). We construct a vector $X^{(1)} = (x_1^{(1)}, \ldots, x_n^{(1)})$ from $X^{(0)}$ and system (3.6.1) and show that, with probability tending to 1, the vector $X^{(1)}$ coincides with the true solution X^*.

Therefore we consider the vectors

$$X_{i,0} = \left(x_1^{(0)}, \ldots, x_{i-1}^{(0)}, 0, x_{i+1}^{(0)}, \ldots, x_n^{(0)}\right),$$

$$X_{i,0} = \left(x_1^{(0)}, \ldots, x_{i-1}^{(0)}, 1, x_{i+1}^{(0)}, \ldots, x_n^{(0)}\right)$$

and calculate the values $\beta(X_{i,0})$ and $\beta(X_{i,1})$, defined for the vectors $X_{i,0}$ and $X_{i,1}$ in the same way $\beta(X^{(0)})$ was defined for $X^{(0)}$.

For $i = 1, \ldots, n$, let

$$x_i^{(1)} = \begin{cases} 0 & \text{if } \beta(X_{i,0}) \ge \beta(X_{i,1}), \\ 1 & \text{if } \beta(X_{i,0}) < \beta(X_{i,1}). \end{cases}$$

Denote by $\xi(X)$ the number of coordinates of the vectors X and X^* that coincide. The value

$$\eta(X) = \max\left(\xi(X), \xi(\bar{X})\right),$$

where $\bar{X} = (\bar{x}_1, \ldots, \bar{x}_n) = (x_1 + 1, \ldots, x_n + 1)$, is called the number of coincidences.

Lemma 3.6.1. *If* $n \to \infty$, *then the distribution of the random variable* $(2\eta(X^{(0)}) - n)/\sqrt{n}$ *converges weakly to the distribution of the modulus of the random variable that has the normal distribution with parameters* $(0, 1)$.

Proof. Since the vector $X^{(0)}$ is chosen from the set of all n-dimensional vectors by random sampling with equal probabilities, the random variable $S_n = \xi(X^{(0)})$ has the binomial distribution with parameters $(n, 1/2)$. From the obvious equality $\xi(X) + \xi(\bar{X}) = n$, the random variable $\eta(X^{(0)})$ is represented in the form

$$\eta(X^{(0)}) = \max(S_n, n - S_n).$$

It is clear that

$$\frac{1}{\sqrt{n}} (2\eta(X^{(0)}) - n) = \max\left(\frac{2S_n - n}{\sqrt{n}}, \frac{n - 2S_n}{\sqrt{n}}\right)$$

$$= \frac{1}{\sqrt{n}} |2S_n - n|,$$

and the assertion of Lemma 3.6.1 follows from the convergence of the distribution of $(2S_n - n)/\sqrt{n}$ to the normal distribution with parameter $(0, 1)$. ∎

We can now prove the following assertion concerning the algorithm of coordinate testing.

Theorem 3.6.2. *If* $n, T \to \infty$ *and* $\Delta \to 0$ *such that*

$$\frac{\Delta^2 T}{n^2 \log n} \to \infty,$$

then

$$\mathbf{P}\{X^{(1)} = X^*\} \to 1.$$

Proof. For definiteness, assume

$$\xi(X^{(0)}) \geq \xi(\bar{X}^{(0)}).$$

The coordinates of $X^{(0)}$ that coincide with the corresponding coordinates of X^* are called true, whereas those that do not coincide are called wrong.

For the algorithm of coordinate testing to lead to the true solution, the following obvious conditions must be fulfilled. For each coordinate of the vector $X^{(0)}$, the value of $\beta(X^{(0)})$ must increase if we replace the wrong value of the coordinate by the true value, and the value of $\beta(X^{(0)})$ must strictly decrease if we replace the true value by the wrong one.

We separate all the equations of the system (3.6.1) that contain x_i, and denote the number of such equations by n_i. Replacing $x_i^{(0)}$ by $\bar{x}_i^{(0)}$ changes the contribution in $\beta(X^{(0)})$ of these equations only, and each equation containing x_i contributes

1 or -1. If $x_i^{(0)}$ is wrong, then the increment of $\beta(X^{(0)})$ due to replacing $x_i^{(0)}$ by $\bar{x}_i^{(0)}$ is equal to the random variable $\beta_i(X^{(0)})$ such that $(\beta_i(X^{(0)}) + n_i)/2$ has the binomial distribution with parameters (n_i, p_i), where p_i is the probability that the coincidence in a fixed equation containing x_i appears after substituting $\bar{x}_i^{(0)}$ for $x_i^{(0)}$, provided $\bar{x}_i^{(0)}$ is wrong. It is not difficult to see that

$$p_i = vq + (1 - v)p, \qquad (3.6.10)$$

where $q = \mathbf{P}\{b_i = b_i^*\}$, $p = 1 - q$, and v is the probability that the second variable in the equation has the true value. The second variable takes values from the set

$$\left\{x_1^{(0)}, \ldots, x_n^{(0)}\right\} \setminus x_i^{(0)}$$

with equal probabilities. Therefore $v = (k - 1)/(n - 1)$, where k is the number of true coordinates of $X^{(0)}$, which equals $\xi(X^{(0)})$ under the assumption that $\xi(X^{(0)}) \geq \xi(\bar{X}^{(0)})$. It follows from Lemma 3.6.1 and equality (3.6.10) that

$$
\begin{aligned}
p_i &= \frac{(1 + \Delta)k}{2(n - 1)} + \left(1 - \frac{k}{n - 1}\right)\frac{1 - \Delta}{2} \\
&= \frac{1}{2} + \frac{\left(2\xi(X^{(0)}) - n + 1\right)\Delta}{2(n - 1)},
\end{aligned}
$$

which we write as

$$p_i = \frac{1}{2} + \frac{|\xi_n|\Delta}{2\sqrt{n}}, \qquad (3.6.11)$$

where

$$\xi_n = \frac{\left(2\xi(X^{(0)}) - n + 1\right)\sqrt{n}}{n - 1}.$$

By assumption, ξ_n is asymptotically normal with parameters $(0, 1)$. Therefore

$$\mathbf{P}\{|\xi_n| > \left(\Delta^2 T/(n^2 \log n)\right)^{-1/4}\} \to 1 \qquad (3.6.12)$$

because $\Delta^2 T/(n^2 \log n) \to \infty$.

Next, we find a lower bound for n_i, $i = 1, \ldots, n$. To this end, we take into account only the first variable in each equation. Then we obtain the classical scheme of equiprobable allocation of T particles into n cells, and by applying the corresponding results on the distribution of the minimum of contents of cells [90], we find that

$$\mathbf{P}\left\{\min_{1 \leq i \leq n} n_i \geq T/(2n)\right\} \to 1. \qquad (3.6.13)$$

For the increments $\xi_i(X^{(0)})$, we have

$$P\{\xi_i(X^{(0)}) < 0 \mid x_i^{(0)} \text{ is wrong}\}$$
$$= P\{(\xi(X^{(0)}) + n_i)/2 < n_i/2 \mid x_i^{(0)} \text{ is wrong}\}$$
$$= P\{S_{n_i} < n_i/2\},$$

where S_{n_i} has the binomial distribution with parameters (n_i, p_i). From (3.6.11), we find that

$$P\{S_{n_i} < n_i/2\} = P\{S_{n_i} - \mathbf{E} S_{n_i} < \Delta |\xi_n| n_i/(2\sqrt{n})\}.$$

When we use estimate (3.6.8) of the exponential type for the binomial distribution and take into account (3.6.12) and (3.6.13), we obtain

$$P\{S_{n_i} < n_i/2\} \leq \exp\left\{-\frac{\Delta^2 T}{n^2}\left(\frac{\Delta^2 T}{n^2 \log n}\right)^{-1/2}\right\}.$$

In a similar way, we obtain the bound

$$P\{\xi_i(X^{(0)}) > 0 \mid x_i^{(0)} \text{ is true}\} \leq \exp\left\{-\frac{\Delta^2 T}{n^2}\left(\frac{\Delta^2 T}{n^2 \log n}\right)^{-1/2}\right\}.$$

Therefore an upper estimate for the probability of at least one wrong decision while testing all the coordinates of the vector $X^{(0)}$ is

$$\sum_{i=1}^{n}\left(P\{\xi_i(X^{(0)}) < 0 \mid x_i^{(0)} \text{ is wrong}\} + P\{\xi_i(X^{(0)}) > 0 \mid x_i^{(0)} \text{ is true}\}\right)$$

$$\leq 2n \exp\left\{-\frac{\Delta^2 T}{n^2}\left(\frac{\Delta^2 T}{n^2 \log n}\right)^{-1/2}\right\}$$

and tends to zero under the conditions of the theorem because

$$\Delta^2 T/(n^2 \log n) \to \infty. \qquad \blacksquare$$

With the help of a preliminary search of the n-dimensional vectors, it is possible to select an initial vector $X^{(0)}$ with a great number $\eta(X^{(0)})$ of coordinates coinciding with the corresponding coordinates of the true solution X^*. If the algorithm for coordinate testing begins with this initial vector, then a much smaller number of equations is needed to reconstruct the true solution. This number is comparable to the number of edges needed for the graph $\Gamma_{n,T}$ to be connected.

Theorem 3.6.3. *If $n, T \to \infty$ and $\Delta \to 0$ such that*

$$\frac{\Delta^2 T}{n \log n} \to \infty,$$

then there exists an algorithm that reconstructs the true solution of system (3.6.1) *with probability tending to 1.*

Proof. The algorithm, which gives the true solution under the conditions of the theorem, begins with a preliminary search of an initial vector $X^{(0)}$ with a large number of coincidences with the true vector X^*. The choice of $X^{(0)}$ is determined by a search of all n-dimensional vectors. To this end, we choose the level $l = Tq - u_T \sqrt{T}$, where $q = \mathbf{P}\{b_t = b_t^*\} = (1+\Delta)/2$ and $u_T = \Delta\sqrt{T}/18$, and select the vectors X for which $\beta(X) \geq l$. Recall that $\beta(X)$ is the number of coincident coordinates of the vector $B = (b_1, \ldots, b_T)$ and the vector of the right-hand sides of system (3.6.1) that are obtained when X is substituted into the left-hand side of the system.

The vector X^* will be selected with probability tending to 1. Indeed,

$$\mathbf{P}\{\beta(X^*) < Tq - u_T\sqrt{T}\} = \mathbf{P}\{S_T - \mathbf{E}S_T < -u_T\sqrt{T}\},$$

where S_T is the number of successes in T independent trials with the probability of success equal to $q = (1 + \Delta)/2$. By using estimate (3.6.8), we find that

$$\mathbf{P}\{\beta(X^*) < l\} \leq e^{-u_T^2/2},$$

and the complementary probability $\mathbf{P}\{\beta(X^*) \geq l\} \to 1$ because $u_T \to \infty$.

If $\xi(X) = s$, then the probability of the coincidence of a fixed component of the right-hand sides is

$$p(s) = \frac{qs(s-1)}{n(n-1)} + \frac{q(n-s)(n-s-1)}{n(n-1)} + \frac{2s(1-q)(n-s)}{n(n-1)},$$

and, since $q = (1+\Delta)/2$, we find

$$p(s) = \frac{1}{2} + \frac{\Delta(2s-n)(2s-n+1)}{2n(n-1)}.$$

For example, let $s < 2n/3$. Then $p(s) < 1/2 + \Delta/9$, beginning with some n, and for any fixed X with $\xi(X) = s < 2n/3$,

$$\mathbf{P}\{\beta(X) \geq l\} = \mathbf{P}\{S_T \geq Tq - u_T\sqrt{T}\}$$
$$\leq \mathbf{P}\{S_T - \mathbf{E}S_T \geq 7\Delta T/18 - u_T\sqrt{T}\}$$
$$= \mathbf{P}\{S_T - \mathbf{E}S_T \geq \Delta T/3\},$$

where S_T is the number of successes in T independent trials with probability $p(s)$ of success.

By using the inequality (3.6.8) of exponential type, we find

$$\mathbf{P}\{\beta(X) \geq l\} \leq e^{-\Delta^2 T/18}.$$

The probability that none of the vectors X with $\xi(X) < 2n/3$ will be selected does not exceed $2^n e^{-\Delta^2 T/18}$, and under the conditions of the theorem this probability tends to zero. Thus, with the help of the exhaustive search, it is possible to select, with probability tending to 1, a vector $X^{(0)}$ such that $\xi(X^{(0)}) \geq 2n/3$. Beginning the algorithm for coordinate testing with this vector $X^{(0)}$, we find, using the notations introduced in the proof of Theorem 3.6.2, that

$$\mathbf{P}\{\xi_i(X^{(0)}) < 0 \mid x_i^{(0)} \text{ is wrong}\} = \mathbf{P}\{S_{n_i} - \mathbf{E}S_{n_i} < -\Delta|\xi_n|n_i/(2\sqrt{n})\}.$$

Using estimate (3.6.8) and taking into account that with probability tending to 1, $|\xi_n| > \sqrt{n}/3$ for the selected vector and $n_i \geq T/(2n)$, we find the estimate

$$\mathbf{P}\{\xi_i(X^{(0)}) < 0 \mid x_i^{(0)} \text{ is wrong}\} \leq \mathbf{P}\{S_{n_i} - \mathbf{E}S_{n_i} < -\Delta n_i/6\}$$
$$\leq e^{-\Delta^2 T/(36n)}.$$

Similarly we obtain

$$\mathbf{P}\{\xi_i(X^{(0)}) > 0 \mid x_i^{(0)} \text{ is true}\} \leq e^{-\Delta^2 T/(36n)}.$$

As in the proof of Theorem 3.6.2, an upper bound for the probability of at least one wrong decision, while all n coordinates of $X^{(0)}$ are tested, is $2^n e^{-\Delta^2 T/(36n)}$ and tends to zero under the conditions of the theorem. ∎

Thus, if we use the exhaustive search, then the true solution can be reconstructed under the condition $\Delta^2 T/(n \log n) \to \infty$. If the number of equations T is such that $\Delta^2 T/(n^2 \log n) \to \infty$, then the reconstruction can be realized by the voting algorithm, which is more economical with respect to the number of operations. Clearly, there is considerable interest in the algorithms that lead to the true solution with probability tending to 1 under intermediate conditions on the number of equations T and do not require the exhaustive search of all 2^n vectors.

Let us describe an algorithm that will be referred to as A_2. Consider all $\binom{T}{2}$ equations obtained as the pairwise unions of the equations of the system (3.6.1). Among the equations obtained by this operation, there are equations that contain either four, or two, or zero unknowns each. Denote by S_2 the subsystem that includes all the equations with two unknowns each. The algorithm A_2 ends with the application of the voting algorithm to the subsystem S_2. The following theorem gives the conditions under which the algorithm A_2 reconstructs the true solution.

Theorem 3.6.4. *If $n, T \to \infty$ and $\Delta \to 0$ such that*

$$\frac{\Delta^4 T^2}{n^3 \log n} \to \infty,$$

then the algorithm A_2 reconstructs the true solution with probability tending to 1.

Proof. Let i and j be arbitrary, assume $i < j$, and consider all equations of the system S_2 of the form

$$x_i + x_j = b_{ij}^{(1)},$$

$$\cdots \tag{3.6.14}$$

$$x_i + x_j = b_{ij}^{(m_{ij})}.$$

The equality $m_{ij} = m$ means that the graph $\Gamma_{n,T}$ corresponding to system (3.6.1) contains exactly m vertices, say $v_1, \ldots, v_m$, such that the graph $\Gamma_{n,T}$ contains the edges $(v_1, i), (v_1, j), \ldots, (v_m, i), (v_m, j)$. The right-hand sides $b_{ij}^{(1)}, \ldots, b_{ij}^{(m_{ij})}$ are the pairwise sums of $2m_{ij}$ independent random variables, and therefore they are independent and, according to Lemma 3.2.1, take the true value $b_{ij}^* = x_i^* + x_j^*$ with probability $(1 + \Delta^2)/2$ and the wrong value with probability $(1 - \Delta^2)/2$.

Let $b_{ij} = 1$ if $b_{ij}^{(1)} + \cdots + b_{ij}^{(m_{ij})} > m_{ij}/2$, and $b_{ij} = 0$ otherwise. As in the proof of Theorem 3.6.1, we denote by $\mu(n, T)$ the minimum value of m_{ij} over all subsystems of the form (3.6.14). As in (3.6.5), the probability $P(n, T)$ of reconstructing the true solution can be represented in the form

$$P(n, T) = \mathbf{P}\{\mu(n, T) > m\} P_m(n, T) + \mathbf{P}\{\mu(n, T) \le m\} \bar{P}_m(n, T), \tag{3.6.15}$$

where $P_m(n, T)$ and $\bar{P}_m(n, T)$ are the conditional probabilities of reconstructing the true solution by the majority method under the condition that $\{\mu(n, T) > m\}$ and $\{\mu(n, T) \le m\}$, respectively.

As in the proof of Theorem 3.6.1, we need to estimate $\mathbf{P}\{\mu(n, T) > m\}$, but here this estimation is more laborious.

Let $\xi_{ij} = 1$ if $m_{ij} \le m$, and $\xi_{ij} = 0$ if $m_{ij} > m$, $i < j$, $i, j = 1, \ldots, n$. It is clear that

$$\mathbf{P}\{\mu(n, T) \le m\} = \mathbf{P}\left\{ \sum_{i<j} \xi_{ij} > 0 \right\} \le \binom{n}{2} \mathbf{P}\{m_{12} \le m\}. \tag{3.6.16}$$

Let $\mu_i = 1$ if the edges $(i + 2, 1)$ and $(i + 2, 2)$ occur in $\Gamma_{n,T}$, and $\mu_i = 0$ otherwise; and $\nu_i = 1$ if exactly one of the edges $(i + 2, 1)$, $(i + 2, 2)$ occurs in $\Gamma_{n,T}$, and $\nu_i = 0$ if the edges $(i + 2, 1)$ and $(i + 2, 2)$ do not occur in $\Gamma_{n,T}$. The random variable m_{12} can be represented as the following sum of indicators:

$$m_{12} = \mu_1 + \cdots + \mu_{n-2},$$

and

$$P\{m_{12} \le m\} = \sum_{k=0}^{m} \sum_{\{i_1,...,i_k\}} p_{i_1...i_k}, \qquad (3.6.17)$$

where $p_{i_1...i_k}$ is the probability that $\mu_{i_1}, \ldots, \mu_{i_k}$ take the value 1 and all the other random variables take the value 0.

It is not difficult to see that (M_t, N_t), where

$$M_t = \mu_1 + \cdots + \mu_t, \qquad N_t = \nu_1 + \cdots + \nu_t,$$

is a Markov chain because (μ_{t+1}, ν_{t+1}) depends only on the number of edges used to construct the random variables $\mu_1, \ldots, \mu_t, \nu_1, \ldots, \nu_t, t = 1, \ldots, n-2$. More precisely, let

$$p(t \mid Y_{t-1}, Z_{t-1}) = P\{\mu_t = 1 \mid M_{t-1} = Y_{t-1}, N_{t-1} = Z_{t-1}\},$$

$$q(t \mid Y_{t-1}, Z_{t-1}) = P\{\mu_t = 0 \mid M_{t-1} = Y_{t-1}, N_{t-1} = Z_{t-1}\}.$$

By using this notation, we can write the probability $p_{i_1...i_k}$ that $\mu_{i_1}, \ldots, \mu_{i_k}$ take the value 1 and all the other random variables take the value 0 in the form

$$p_{i_1...i_k} = P\{\mu_{i_1} = 1, \ldots, \mu_{i_k} = 1,$$

$$\mu_i = 0, \ i \ne i_1, \ldots, i_k \mid \nu_1 = z_1, \ldots, \nu_{n-2} = z_{n-2}\}$$

$$= q(1 \mid Y_0, Z_0) \cdots q(i_1 - 1 \mid Y_{i_1-2}, Z_{i_1-2}) p(i_1 \mid Y_{i_1-1}, Z_{i_1-1})$$

$$\times q(i_1 + 1 \mid Y_{i_1}, Z_{i_1}) \cdots q(n - 2 \mid Y_{n-3}, Z_{n-3}),$$

where $Z_0 = Y_0 = 0$, $Z_t = z_1 + \cdots + z_t$, and Y_t is the number of $i_1, \ldots, i_k$ that do not exceed t.

We now estimate the probabilities $p(t \mid Y, Z)$ and $q(t \mid Y, Z)$. It is clear that $p(t \mid Y, Z) + q(t \mid Y, Z) = 1$, and the probability $p(t \mid Y, Z)$ does not depend on t and equals the probability $p_2(s, N)$ that two fixed places corresponding to the edge $(1, t)$, $(2, t)$ will be occupied after allocating $s = T - 2Y - Z$ edges into $N = \binom{n}{2} - Z$ places in the classical scheme of allocation of particles. Therefore

$$p_2(s, N) = \frac{s!}{1! \, 1! \, (s-2)! \, N^2} \left(1 - \frac{2}{N}\right)^{s-2}$$

$$+ \sum_{k+l>2, k, l \le 1} \frac{s!}{k! \, l! \, (s-k-l)! \, N^{k+l}} \left(1 - \frac{2}{N}\right)^{s-k-l},$$

and we have the following estimates:

$$p_2(s, N) \geq \frac{s(s-1)}{N^2}\left(1-\frac{2}{N}\right)^{s-2},$$

$$p_2(s, N) \leq \frac{s(s-1)}{N^2}\left(1-\frac{2}{N}\right)^{s-2}$$

$$+ \frac{s(s-1)}{N^2}\sum_{k+l>0}\frac{(s-2)!}{k!\,l!\,(s-k-l-2)!\,N^{k+l}}\left(1-\frac{2}{N}\right)^{s-k-l-2}$$

$$= \frac{s(s-1)}{N^2}\left(1-\frac{2}{N}\right)^{s-2} + \frac{s(s-1)}{N^2}\left(1-\left(1-\frac{2}{N}\right)^{s-2}\right).$$

Since

$$T - 3n \leq s = T - Y - Z \leq T,$$

$$n(n-3)/2 \leq N = \binom{n}{2} - Z \leq n(n-1)/2,$$

we obtain for all $k = 0, 1, \ldots, n-2$,

$$p_{i_1 \ldots i_k} \leq P^k Q^{n-k-2},$$

where

$$P = \max p(t \mid Y_{t-1}, Z_{t-1}),$$

$$Q = 1 - \min p(t \mid Y_{t-1}, Z_{t-1}).$$

Therefore it follows from (3.6.16) and (3.6.17) that

$$\mathbf{P}\{m_{12} \leq m\} \leq (P+Q)^{n-2}\mathbf{P}\{\xi_1 + \cdots + \xi_{n-2} \leq m\},$$

where $\xi_1, \ldots, \xi_{n-2}$ are independent identically distributed random variables,

$$\mathbf{P}\{\xi_1 = 1\} = P/(P+Q), \qquad \mathbf{P}\{\xi_1 = 0\} = Q/(P+Q).$$

As $n, T \to \infty$ and $T/\binom{n}{2} \to 0$,

$$\frac{P}{P+Q} = \frac{4T^2}{n^4}\left(1 + O(T/n^2)\right),$$

and under the conditions of the theorem,

$$(P+Q)^{n-2} = 1 + o(1). \tag{3.6.18}$$

The random variable $\zeta_{n-2} = \xi_1 + \cdots + \xi_{n-2}$ has the binomial distribution with parameters $(n-2, P/(P+Q))$.

Let $\alpha = \mathbf{E}\zeta_{n-2} = (n-2)P/(P+Q)$ and $m = \alpha(1-\Delta^2)$. We assume that T is not too large, so that $\Delta^4 T^2/n^{10/3} \to 0$. Then, for sufficiently large n,

$$\mathbf{P}\{\zeta_{n-2} \le m\} \le \mathbf{P}\{(\zeta_{n-2} - \mathbf{E}\zeta_{n-2})/\sqrt{\mathbf{D}\zeta_{n-2}} \le -\Delta^2\sqrt{\alpha}/2(1+o(1))\}$$

$$= \frac{1}{\sqrt{2\pi}} \int_{-\infty}^{-\Delta^2\sqrt{\alpha}/2} e^{-u^2/2}\, du(1+o(1)),$$

and there exists a constant c such that

$$\mathbf{P}\{\zeta_{n-2} \le m\} \le ce^{-\Delta^4\alpha/8}. \tag{3.6.19}$$

Thus, by virtue of (3.6.16), (3.6.18), and (3.6.19),

$$\mathbf{P}\{\mu(n,T) \le m\} \le cn^2 e^{-\Delta^4\alpha/8} \to 0 \tag{3.6.20}$$

because, under the conditions of the theorem, $\Delta^4 T^2/(n^3 \log n) \to \infty$, and consequently, $n^2 e^{-\Delta^4\alpha/8} \to 0$.

As in the proof of Theorem 3.6.1, we have to show that under the conditions of Theorem 3.6.4, the system S_2 is indecomposable and $P(n,T) \to 1$. In other words, we have to show that $b_{ij} = b_{ij}^*$ for all $i, j = 1, \ldots, n$ with probability tending to 1.

By the same reasoning as in the proof of Theorem 3.6.1, for $m = \alpha(1-\Delta^2)$, we obtain

$$1 - P_m(n,T) \le \binom{n}{2} e^{-m\Delta^4/4}, \tag{3.6.21}$$

and under the conditions of Theorem 3.6.4, the right-hand side of (3.6.21) tends to zero.

The assertion of the theorem follows from (3.6.15), (3.6.20), and (3.6.21). ∎

3.7. Notes and references

The theory of systems of random equations in finite fields was developed by the Russian mathematicians V. E. Stepanov, G. V. Balakin, I. N. Kovalenko, A. A. Levitskaya, and others. The connection between systems of equations in GF(2) and graphs was first pointed out and used by Stepanov. The notion of a critical set was introduced in [79] (see also [13] and [85]).

The theory of recurring sequences and shift registers mentioned in Section 3.1 can be found in [50] and [156].

Theorems 3.2.1 and 3.2.2 were proved by Kovalenko in [92]. This brilliant result initiated a series of investigations of similar problems that were carried out by Kovalenko and his school. These investigations developed in two directions. The first direction concerns extensions of Theorems 3.2.1 and 3.2.2 to matrices over more general algebraic structures. It is not difficult to see that by virtue of the Markovian character of the process $\rho_n(t)$, a recurrence relation for $p_{n,T}(k) = \mathbf{P}\{\rho_n(T) = k\}$ can be derived and used for the proof of Theorem 3.2.1. In this way, the extension of the result to a finite field with q elements can be easily obtained [93]. Let the elements of $T \times n$ matrix $A = \|\alpha_{t,j}\|$ in GF(q) take the values $0, 1, \ldots, q - 1$ with equal probabilities, then the $p_{n,T}(k)$ for any $k = 0, 1, \ldots$ satisfy the equation

$$p_{n,T}(k) = z^n p_{n,T-1}(k) + (1 - z^n) p_{n-1,T-1}(k), \qquad (3.7.1)$$

where $z = 1/q$. Indeed, if the first row of A is a zero vector, then $\rho_n(T) = \rho_n(T - 1)$, and if the row contains at least one nonzero element, then $\rho_n(T) = \rho_{n-1}(T - 1) + 1$. It follows from (3.7.1) that if $s \geq 0$ and m are fixed integers, $m + s \geq 0$, $n \to \infty$, and $T = n + m$, then

$$\mathbf{P}\{\rho_n(T) = n - s\} \to q^{-s(m+s)} \prod_{i=s+1}^{\infty} \left(1 - \frac{1}{q^i}\right) \prod_{i=1}^{m+s} \left(1 - \frac{1}{q^i}\right)^{-1}. \qquad (3.7.2)$$

The investigations in the second direction concern the bounds of invariance of the results of Theorems 3.2.1 and 3.2.2 with respect to the deviations of the distribution of elements of the matrix A from the equiprobable distribution. The problem of the invariance and a proof of Theorem 3.2.3 are given in [91, 92]. A modified proof of Theorem 3.2.3 is contained in [93].

Theorem 3.2.4 can be easily extended to any moment of a fixed order of the number of solutions, but that is not sufficient for the proof of the invariance property, since the limit distribution (3.7.2) does not satisfy the sufficient conditions of the unique reconstruction by its moments; hence, Theorem 1.1.3 cannot be applied.

Levitskaya [96, 97] presents results on the number of solutions of linear random systems over arbitrary rings and the corresponding results on the invariance of the moment and the limit distributions. These results are summarized in [93], where, in particular, the exact bounds for the invariance are given for random linear systems in arbitrary finite rings. For the system considered in Theorem 3.2.3, the exact bounds for $p_{ij}^{(n)}$ have the form

$$\delta_n \leq p_{ij}^{(n)} \leq 1 - \delta_n,$$

where $\delta_n = (\log n + x_n)/n$ and $x_n \to \infty$ arbitrarily slowly as $n \to \infty$.

Matrices that satisfy condition (3.3.1) were considered by Balakin [12], who also proved Theorems 3.3.1 and 3.3.2. Closer investigation of the estimates used in our proof of Theorem 3.3.1 allows us to obtain the following assertions.

Theorem 3.7.1. *If* $n \to \infty$,

$$T = n + \beta_n \log n,$$

$\beta_n \to -\infty$, $\beta_n = o(n/\log n)$, and condition (3.3.1) holds, then the distribution of $s(A)$ converges to the Poisson distribution with parameter e^{-x}.

Theorem 3.7.2. *If $n \to \infty$,*

$$T = n + \beta \log n + o(\log n),$$

β *is a constant, and condition* (3.3.1) *holds, then the distribution of $s(A)$ converges to the Poisson distribution with parameter e^{-x} if $\beta < 0$, and with parameter $e^{-x-\beta}$ if $\beta > 0$.*

Theorems 3.3.1, 3.3.2, 3.7.1, and 3.7.2 give a complete description of the behavior of the rank of such matrices, except for the case $\beta = 0$, where the behavior is unknown. Note that in [12], the analogues of Theorems 3.3.1, 3.7.1, and 3.7.2 are proved for the systems over GF(q), $q \geq 2$ (see also [86]), and the connection between the rank of a matrix in GF(q) and other characteristics such as the permanent rank and rank of lines is considered. The initial results on the ranks of random matrices are presented in [38] and [11].

Stepanov began investigating systems of linear equations of the form (3.4.1) with the help of their relations to random graphs. In particular, he proved Theorems 3.4.1 and 3.4.2. Now the theory of random graphs provides a basis for obtaining the results on the systems of random equations with coefficients taking their values with equal probabilities. If the coefficients of a system are essentially nonequiprobable, then there are no standard approaches to investigating its properties. Only a few results are known for such systems. We remark that at this time, graph theory is not sufficiently developed to answer questions about nonequiprobable cases. Only the method of moments (see Theorem 1.1.3) and the so-called direct methods are used to solve these problems. Theorem 3.4.3 is a corollary to Theorem 2.4.1 proved in [88] by the method of moments.

Theorems 3.4.4 and 3.4.5 are proved in [83]. The asymptotics of the probability of consistency of a system of linear equations in GF(2) (and in more general algebraic structures) with independent random coefficients that take the values 0 and 1 with equal probabilities have been obtained by Levitskaya [98] (see also [93]). This probability takes only two values and is the same for all possible right-hand sides of the system that are not the zero vector. It follows from Theorems 3.4.4 and 3.4.5 that the probability of consistency of the system (3.4.1) depends on the number of 1's in the vector of the right-hand sides of the system (see also [83]).

The results of Section 3.5 on the behavior of the probability of consistency of the system (3.5.1) can be found in [13] (see also [85]). Theorem 3.5.1 is proved by the author, but the critical values α_r were first obtained by Balakin under slightly different assumptions on the matrix $A_{r,n,T}$. These results are extended to GF(q) in [89]. The proof of Theorem 3.5.2 is given in [87].

We can consider the probability of the consistency of a system from the point of view of mathematical statistics. Consider, for example, the system (3.4.1) and assume the following two hypotheses on the distribution of the right-hand sides of the system. Let the hypothesis H_0 be the existence of a vector $X^* = (x_1^*, \ldots, x_n^*)$, which is interpreted as the true solution of the system, and $b_t = x_{i(t)}^* + x_{j(t)}^*$, $t = 1, \ldots, T$. Under hypothesis H_0, system (3.4.1) is always consistent. Under the alternative hypothesis H_1, the right-hand sides $b_1, \ldots, b_T$ are independent random variables that are independent of the left-hand side of the system and take the values 0 and 1 with equal probabilities. To distinguish between the hypotheses H_0 and H_1, we can use the consistency of the system as a test: If the system is consistent, we accept the hypothesis H_0, and we accept H_1 otherwise. Therefore the hypothesis H_0 is never rejected if it is true, and the error of the first kind, the probability of rejecting H_0 if it is true, is zero. The error of the second kind, the probability of accepting H_0 if it is wrong, is equal to the probability of consistency of the system (3.4.1). Thus, the probability of consistency is the main characteristic in the statistical problem of testing the hypotheses H_0 and H_1.

Section 3.6 is devoted to the other statistical problems that consist of reconstructing the true solution on the basis of a system of random equations with distorted right-hand sides. These results can be found in the paper [84].

4

Random permutations

4.1. Random permutations and the generalized scheme of allocation

Denote by S_n the set of all one-to-one mappings of the set $X_n = \{1, 2, \ldots, n\}$ into itself. This set contains $n!$ elements. We consider a random permutation σ that equals any element of S_n with probability $(n!)^{-1}$.

A permutation $s \in S_n$ can be written as

$$
s = \begin{pmatrix} 1, & 2, & \ldots, & n \\ s_1, & s_2, & \ldots, & s_n \end{pmatrix},
$$

where s_k is the image of k under the mapping s, $k = 1, \ldots, n$. The mapping s can be represented also by the graph $\Gamma_n^{(s)} = \Gamma(X_n, W_n)$ whose vertex set is X_n, and the edge set W_n consists of the arcs (k, s_k) directed from k to s_k, $k = 1, \ldots, n$. Since exactly one arc enters each vertex and exactly one arc emanates from each vertex, the graph $\Gamma_n^{(s)}$ consists of the connected components that are cycles, which are called the cycles of the permutation s.

Denote by Γ_n the random graph corresponding to the random permutation σ, which takes the values s with equal probabilities. It is obvious that $\mathbf{P}\{\Gamma_n = \Gamma_n^{(s)}\} = (n!)^{-1}$.

In Section 1.3, we showed that the generalized scheme of allocation introduced in Section 1.2 can be applied to a wide class of problems related to the behavior of the connected components of random graphs. In Example 1.3.1, we showed that the generalized scheme can be used in the study of random permutations. Recall that in the generalized scheme, we separate the subset of graphs with exactly N components, assign one of the $N!$ possible orders to the set of these components, and denote by $\eta_1, \ldots, \eta_N$ the sizes of the components. If there exist nonnegative identically distributed random variables $\xi_1, \ldots, \xi_N$

such that for any integers $k_1, \ldots, k_N$,

$$\mathbf{P}\{\eta_1 = k_1, \ldots, \eta_N = k_N\} = \mathbf{P}\{\xi_1 = k_1, \ldots, \xi_N = k_N \mid \xi_1 + \cdots \xi_N = n\},$$

(4.1.1)

we say that the generalized scheme determined by the random variables $\xi_1, \ldots, \xi_N$ is applied to the random graph.

As was shown in Example 1.3.1, the generalized scheme that corresponds to the random graph Γ_n of a random permutation from S_n is determined by the random variables $\xi_1, \ldots, \xi_N$ with the distribution

$$\mathbf{P}\{\xi_1 = k\} = \frac{x^k}{k \log(1 - x)}, \quad k = 1, 2 \ldots, \quad 0 < x < 1, \qquad (4.1.2)$$

since the number of elements in S_n is $a_n = n!$ and the number of connected realizations of the random graph Γ_n is $b_n = (n - 1)!$.

For the random permutations, the corresponding generating functions have the form

$$A(x) = \sum_{n=0}^{\infty} \frac{a_n x^n}{n!} = \frac{1}{1 - x},$$

$$B(x) = \sum_{n=0}^{\infty} \frac{b_n x^n}{n!} = -\log(1 - x).$$

Thus the study of various characteristics of random permutations can be accomplished with the help of the generalized scheme. This is demonstrated for the most part in [78].

Recall some combinatorial identities that follow from the general results of Section 1.3.

Let ν_n be the number of cycles in a random permutation from S_n. Lemma 1.3.3 gives the equality

$$\mathbf{P}\{\nu_n = N\} = \frac{(B(x))^N}{N! x^n} \mathbf{P}\{\xi_1 + \cdots + \xi_N = n\}. \qquad (4.1.3)$$

Denote by α_r the number of cycles of length r in a random permutation from S_n, $r = 1, \ldots, n$. According to Lemma 1.3.7, for any nonnegative integers $m_1, \ldots, m_n$,

$$\mathbf{P}\{\alpha_1 = m_1, \ldots, \alpha_n = m_n\} = \prod_{r=1}^{n} \frac{1}{r^{m_r} m_r!} \qquad (4.1.4)$$

if $m_1 + 2m_2 + \cdots + nm_n = n$, and the probability is zero otherwise.

Let us introduce the generating function

$$\varphi_n(t_1, \ldots, t_n) = \sum_{m_1, \ldots, m_n} \mathbf{P}\{\alpha_1 = m_1, \ldots, \alpha_n = m_n\} t_1^{m_1} \cdots t_n^{m_n}$$

$$= \sum_{M_n} \frac{1}{m_1! \cdots m_n!} \left(\frac{t_1}{1}\right)^{m_1} \left(\frac{t_2}{2}\right)^{m_2} \cdots \left(\frac{t_n}{n}\right)^{m_n},$$

where the summation is over the set of integers

$$M_n = \{m_i \geq 0, \ i = 1, \ldots, n, \ m_1 + 2m_2 + \cdots + nm_n = n\}.$$

Put $\varphi_0 = 0$. It is not difficult to see that $\varphi_n(t_1, \ldots, t_n)$ is the coefficient of u^n in the expansion of $\exp\{ut_1 + u^2 t_2/2 + \cdots\}$:

$$\varphi(u, t_1, t_2, \ldots) = \sum_{n=0}^{\infty} \varphi_n(t_1, \ldots, t_n) u^n = \exp\left\{\sum_{n=1}^{\infty} \frac{u^n t_n}{n}\right\}. \tag{4.1.5}$$

The generating function (4.1.5) was obtained by Goncharov and was the basis of his pioneering investigations of random permutations [53]. In [78], the approach based on the generalized scheme of allocations was used in such investigations. In the next sections, we will present some examples of how the generalized scheme of allocation can be applied to random permutations. This will supplement the investigations presented in [78].

4.2. The number of cycles

It is well known that the number of cycles ν_n in a random permutation from S_n is asymptotically normal with parameters $(\log n, \log n)$ as $n \to \infty$. More precisely, as $n \to \infty$,

$$\mathbf{P}\{\nu_n = N\} = \frac{1}{\sqrt{2\pi \log n}} e^{-u^2/2}(1 + o(1)) \tag{4.2.1}$$

uniformly in the integers N such that $u = (N - \log n)/\sqrt{\log n}$ lies in any fixed finite interval.

The approach based on the generalized scheme of allocation makes it possible to obtain the asymptotics of the probability $\mathbf{P}\{\nu_n = N\}$ for all possible values of $N = N(n)$ as $n \to \infty$. According to (4.1.3), for any integer N,

$$\mathbf{P}\{\nu_n = N\} = \frac{(-\log(1-x))^N}{N! \, x^n} \mathbf{P}\{\xi_1 + \cdots + \xi_N = n\}, \tag{4.2.2}$$

where the parameter x can be taken arbitrarily from the interval $(0, 1)$, and ξ_1, $\ldots, \xi_N$ are independent identically distributed random variables with distribution (4.1.2).

Thus, to study the asymptotic behavior of the distribution of ν_n, it is sufficient to obtain the corresponding local limit theorems for the sum

$$\zeta_N = \xi_1 + \cdots + \xi_N,$$

where the parameter x in the distribution of the summands can be chosen so that obtaining the local theorems becomes simple.

We begin with $x = 1 - 1/n$ and prove a series of limit theorems that make it possible to describe the behavior of the probability $\mathbf{P}\{\nu_n = N\}$ for the values of N not too far from $\log n$.

Theorem 4.2.1. *If $n \to \infty$, $N = \gamma \log n + o(\log n)$, where γ is a constant, $0 < \gamma < \infty$, then*

$$\mathbf{P}\{\zeta_N = k\} = \frac{1}{n\Gamma(\gamma)} z^{\gamma-1} e^{-z}(1 + o(1))$$

uniformly in the integers k such that $z = k/n$ lies in any interval of the form $0 < z_0 \le z \le z_1$ and z_0 and z_1 are constants.

Before proving the theorem, we obtain some auxiliary results. We have chosen $x = 1 - 1/n$. For such x,

$$\mathbf{P}\{\xi_1 = k\} = \frac{(1 - 1/n)^k}{k \log n}, \quad k = 1, 2, \ldots, \tag{4.2.3}$$

and the characteristic function of the random variable ξ_1 equals

$$\varphi_n(t) = -\frac{1}{\log n} \log\left(1 - e^{it} + \frac{1}{n}e^{it}\right).$$

Represent $\varphi_n(t)$ in the form

$$\varphi_n(t) = -\frac{1}{\log n}\left(\log\left(\frac{1}{n} - it\right) + \log(1 + \psi_1(t) + \psi_2(t))\right), \tag{4.2.4}$$

where

$$\psi_1(t) = \frac{1 + it - e^{it}}{1/n - it}, \qquad \psi_2(t) = \frac{e^{it} - 1}{n(1/n - it)}.$$

For $\psi_1(t)$ and $\psi_2(t)$, the following estimates are valid:

$$|\psi_1(t)| \le \frac{|e^{it} - 1 - it|}{|t|} \le \frac{|t|}{2}, \tag{4.2.5}$$

$$|\psi_2(t)| \le \frac{|e^{it} - 1|}{n|t|} \le \frac{1}{n}. \tag{4.2.6}$$

By using the explicit form of $\varphi_n(t)$, the representation (4.2.4) and the bounds (4.2.5) and (4.2.6), we obtain the following estimates of $\varphi_n(t)$.

Lemma 4.2.1. *If $n \to \infty$, $N = \gamma \log n + o(\log n)$, where γ is a constant, $0 < \gamma < \infty$, then for any fixed t,*

$$\varphi_n^N \left(\frac{t}{n}\right) \to \frac{1}{(1-it)^\gamma}.$$

Lemma 4.2.2. *If $n \to \infty$, $N = \gamma \log n + o(\log n)$, where γ is a constant, $0 < \gamma < \infty$, then there exist positive constants ε and c such that for $|t/n| \le \varepsilon$,*

$$\left| \varphi_n^N(t/n) \right| \le c|t|^{-\gamma}.$$

Lemma 4.2.3. *If $n \to \infty$, then for $0 < \varepsilon \le |t| \le \pi$, where ε is an arbitrary constant, there exists a constant c such that for sufficiently large n,*

$$|\varphi_n(t)| \le c/\log n.$$

Lemma 4.2.4. *If $n \to \infty$, then there exists a positive constant ε such that for $|t/n| \le \varepsilon$,*

$$\left| \frac{1}{n} \varphi_n' \left(\frac{t}{n}\right) \right| \le \frac{2}{(1+|t|)\log n}.$$

As follows from Lemma 4.2.1, as $n \to \infty$ and $N = \gamma \log n + o(\log n)$, where γ is a constant, $0 < \gamma < \infty$, the distributions of the normalized sums ζ_N/n converge to the gamma distribution with characteristic function $(1 - it)^{-\gamma}$ and density $z^{\gamma-1}e^{-z}/\Gamma(\gamma)$, $z > 0$. Actually, as stated in Theorem 4.2.1, these distributions become close locally.

Proof of Theorem 4.2.1. By the inversion formula, the probability

$$P\{\zeta_N = k\} = P\{\zeta_N/n = z\}$$

can be represented in the form

$$P\{\zeta_N/n = z\} = \frac{1}{2\pi n} \int_{-\pi n}^{\pi n} e^{-itz} \varphi_n^N(t/n)\, dt,$$

and

$$\frac{1}{\Gamma(\gamma)} z^{\gamma-1} e^{-z} = \int_{-\infty}^{\infty} \frac{e^{-itz}}{(1-it)^\gamma}\, dt.$$

Hence,

$$2\pi n P\{\zeta_N/n = z\} - 2\pi e^{-z} = I_1 + I_2 + I_3 + I_4,$$

where

$$I_1 = \int_{-A}^{A} e^{-itz} \left(\varphi_n^N(t/n) - (1-it)^{-\gamma} \right) dt,$$

$$I_2 = \int_{A<|t|<\varepsilon n} e^{-itz} \varphi_n^N(t/n)\, dt,$$

$$I_3 = \int_{\varepsilon n \le |t| \le \pi n} e^{-itz} \varphi_n^N(t/n)\, dt,$$

$$I_4 = -\int_{A<|t|} e^{-itz}(1-it)^{-\gamma}\, dt,$$

with the constants ε and A to be chosen later.

By Lemma 4.2.1, $\varphi_n^N(t/n) \to (1-it)^{-\gamma}$ for any fixed t. By Theorem 1.1.9, this means that the convergence is uniform with respect to t in any finite interval. Therefore $I_1 \to 0$ for any fixed A as $n \to \infty$.

By Lemma 4.2.3, for sufficiently large n,

$$|I_3| \le 2\pi n (c/\log n)^N \le 2\pi n e^{-2N/\gamma},$$

and, for $N = \gamma \log n + o(\log n)$, the right-hand side tends to zero as $n \to \infty$.

To estimate I_2 and I_4, we integrate by parts. For I_4, this leads to

$$\int_{A}^{\infty} e^{-itz}(1-it)^{-\gamma}\, dt = -\frac{e^{-itz}}{iz(1-it)^{\gamma}}\Big|_{A}^{\infty} + \frac{\gamma}{z}\int_{A}^{\infty} e^{-itz}(1-it)^{-\gamma-1}\, dt.$$

Therefore

$$|I_4| \le \frac{2}{z(1+A^2)^{\gamma/2}} + \frac{2\gamma}{z}\int_{A}^{\infty} \frac{dt}{(1+t^2)^{(\gamma+1)/2}}$$

$$\le \frac{2}{zA^\gamma} + \frac{2\gamma}{z}\int_{A}^{\infty} \frac{dt}{t^{\gamma+1}} \le \frac{c_4}{A^\gamma},$$

where c_4 is a constant, and I_4 can be made arbitrarily small by the choice of sufficiently large A.

Similarly,

$$\int_{A}^{\infty} e^{-itz} \varphi_n^N\left(\frac{t}{n}\right) dt$$

$$= -\frac{e^{-itz}}{iz}\varphi_n^N\left(\frac{t}{n}\right)\Big|_{A}^{\varepsilon n} + \frac{N}{izn}\int_{A}^{\varepsilon n} e^{-itz}\varphi_n^{N-1}\left(\frac{t}{n}\right)\varphi_n'\left(\frac{t}{n}\right) dt.$$

By using the estimates of Lemmas 4.2.2, 4.2.3, and 4.2.4, we obtain

$$|I_2| \leq \frac{2}{z}\left|\varphi_n^N\left(\frac{A}{n}\right)\right| + \frac{2}{z}\left|\varphi_n^N(\varepsilon)\right| + \frac{2N}{zn}\int_A^{\varepsilon n}\left|\varphi_n^{N-1}\left(\frac{t}{n}\right)\varphi_n'\left(\frac{t}{n}\right)\right|dt$$

$$\leq c_2\left(\frac{1}{A^\gamma} + \left(\frac{c}{\log n}\right)^N\right) + c_3\int_A^\infty\frac{dt}{t^{\gamma+1}},$$

where c, c_2, and c_3 are constants.

If we choose sufficiently large A and n, we can make $|I_2|$ arbitrarily small. ∎

Now we can prove the following theorem on the behavior of the probability $P\{\nu_n = N\}$.

Theorem 4.2.2. *If $n \to \infty$ and $N = \gamma \log n + o(\log n)$, where γ is a constant, $0 < \gamma < \infty$, then*

$$P\{\nu_n = N\} = \frac{(\log n)^N}{N!\, n\Gamma(\gamma)}(1 + o(1)).$$

Proof. For $x = 1 - 1/n$, the representation (4.2.2) takes the form

$$P\{\nu_n = N\} = \frac{(\log n)^N}{N!\,(1 - 1/n)^n}P\{\zeta_N = n\}, \qquad (4.2.7)$$

where $\zeta_N = \xi_1 + \cdots + \xi_N$ is the sum of independent identically distributed random variables with distribution (4.2.3). By Theorem 4.2.1,

$$P\{\zeta_N = n\} = \frac{1}{n\Gamma(\gamma)}e^{-1}(1 + o(1)).$$

By substituting this expression into (4.2.7), we obtain the assertion of Theorem 4.2.2. ∎

The case where $\gamma = N/\log n \to 0$ is described by the following theorem.

Theorem 4.2.3. *If $n \to \infty$ and $\gamma = N/\log n \to 0$, then*

$$P\{\zeta_N = n\} = NP\{\xi_1 = n\}(1 + o(1)) = \frac{e^{-1}}{n\Gamma(\gamma)}(1 + o(1)).$$

Proof. Taking into account that $\gamma < 1/2$ beginning with some n, we choose the level $n(1 - \gamma)$ and represent the probability $P\{\zeta_N = n\}$ as follows:

$$P\{\zeta_N = n\} = P\{\zeta_N = n,\ \xi_i \leq n(1 - \gamma),\ i = 1, \ldots, n\}$$

$$+ NP\{\zeta_N = n,\ \xi_N > n(1 - \gamma)\}. \qquad (4.2.8)$$

Since

$$P\{\xi_1 = m\} = \frac{e^{-1}}{n \log n}(1 + o(1))$$

uniformly in m, $n \geq m \geq n(1 - \gamma)$, we see that

$$P\{\zeta_N = n, \ \xi_N > n(1 - \gamma)\} = \sum_{m > n(1-\gamma)} P\{\xi_N = m, \ \zeta_{N-1} = n - m\}$$

$$= \frac{e^{-1}}{n \log n} P\{\zeta_{N-1} \leq \gamma n\}(1 + o(1)). \quad (4.2.9)$$

We now prove

$$P\{\zeta_{N-1} \leq \gamma n\} \to 1. \qquad (4.2.10)$$

Show that the random variable $\zeta_N/(\gamma n)$ converges in probability to zero. By the representation (4.2.4) and the estimates (4.2.5) and (4.2.6),

$$\varphi_n\left(\frac{t}{\gamma n}\right) = -\frac{1}{\log n}\left(\log\left(\frac{1}{n} - \frac{it}{\gamma n}\right) + O\left(\frac{1}{\gamma n}\right)\right)$$

$$= 1 - \frac{\log(\gamma - it) - \log \gamma}{\log n} + O\left(\frac{1}{\gamma n \log n}\right),$$

and if $\gamma = N/\log n \to 0$, then

$$\varphi_n^N\left(\frac{t}{\gamma n}\right) = \left(1 - \frac{\log(\gamma - it) - \log \gamma}{\log n} + O\left(\frac{1}{\gamma n \log n}\right)\right)^N \to 1.$$

Thus, the characteristic function of $\zeta_N/(\gamma n)$ converges to the characteristic function of the random variable that assumes the value 0 with probability 1, and we obtain (4.2.10).

With some technical difficulties, it can be proved that under the conditions of the theorem,

$$P\{\zeta_n = n, \ \xi_i \leq n(1 - \gamma), \ i = 1, \ldots, n\} = o(1/(n \log n)).$$

The assertion of the theorem follows from this relation and the relations (4.2.8), (4.2.9), and (4.2.10). ∎

Theorem 4.2.4. *If $n \to \infty$ and $\gamma = N/\log n \to 0$, then*

$$P\{\nu_n = N\} = \frac{\gamma (\log n)^N}{N! \, n}(1 + o(1)).$$

Proof. The assertion of the theorem follows immediately from Theorem 4.2.3 and representation (4.2.7) if we take into account that the gamma function $\Gamma(\gamma) = 1/\gamma(1 + o(1))$ as $\gamma \to 0$. ∎

Now consider the case where $N/\log n \to \infty$. We distinguish four subcases: $\alpha = n/N \to \infty$, $\alpha \to c > 1$, $\alpha \to 1$ with $m = n - N \to \infty$, and $\alpha \to 1$ with m fixed.

Let $\alpha \to \infty$. We must select the value of the parameter x so that $\mathbf{E}\zeta_N$ is close to n. Since for ξ_1 with distribution (4.1.2),

$$\mathbf{E}\xi_1 = -\frac{x}{(1-x)\log(1-x)},$$

we choose x, $0 < x < 1$, such that

$$-\frac{x}{(1-x)\log(1-x)} = \alpha, \qquad (4.2.11)$$

where $\alpha = n/N$. This equation is approximately satisfied if we take

$$x = 1 - \frac{1}{\alpha \log \alpha}.$$

If $N/\log n \to \infty$, then $x = 1 - 1/(\alpha \log \alpha)$ is farther from the singular point $x = 1$ than $x = 1 - 1/n$, and therefore the normal approximation is valid for the sum ζ_N.

Theorem 4.2.5. *If $n, N \to \infty$ such that $N/\log n \to \infty$, $\alpha = n/N \to \infty$ and the parameter $x = 1 - 1/(\alpha \log \alpha)$ and $\sigma_\alpha = \alpha\sqrt{\log \alpha}$, then*

$$\mathbf{P}\{\zeta_N = k\} = \frac{1}{\sigma_\alpha\sqrt{2\pi N}}e^{-z^2/2}(1 + o(1))$$

uniformly in the integers k such that $z = (k - n)/(\sigma_\alpha\sqrt{N})$ lies in any fixed finite interval.

Proof. The characteristic function of the random variable ξ_1 is

$$\varphi_n(t) = \frac{\log\left(1 - xe^{it}\right)}{\log(1-x)}.$$

It is easy to see that for any fixed t, as $N/\log n \to \infty$ and $\alpha = n/N \to \infty$,

$$e^{-it\alpha}\varphi_n\left(\frac{t}{\sigma_\alpha\sqrt{N}}\right) = 1 - \frac{t^2}{2N} + o\left(\frac{1}{N}\right).$$

Denote by $\psi_n(t)$ the characteristic function of $\zeta_N^* = (\zeta_N - n)/(\sigma_\alpha\sqrt{N})$, then under the conditions of the theorem for any fixed t,

$$\psi_n(t) = \left(e^{-it\alpha}\varphi_n\left(\frac{t}{\sigma_\alpha\sqrt{N}}\right)\right)^N \to e^{-t^2/2},$$

and the distribution of ζ_N^* converges weakly to the normal distribution with parameters $(0, 1)$.

The local convergence can be proved by the standard reasoning and we omit this technical part of the proof of Theorem 4.2.5. ∎

From Theorem 4.2.5 and representation (4.2.2), we obtain the following assertion.

Theorem 4.2.6. *If $n, N \to \infty$ such that $N/\log n \to \infty$, $\alpha = n/N \to \infty$, then*

$$\mathbf{P}\{\nu_n = N\} = \frac{(-\log(1-x))^N}{N! \, x^n \sigma_\alpha \sqrt{2\pi N}} (1 + o(1)),$$

where $x = 1 - 1/(\alpha \log \alpha)$ and $\sigma_\alpha = \alpha \sqrt{\log \alpha}$.

The following theorem for the case where α tends to a constant greater than 1 can be proved in the same way as Theorem 4.2.5.

Theorem 4.2.7. *If $n, N \to \infty$ and there exist constants α_0 and α_1 such that $1 < \alpha_0 \le \alpha \le \alpha_1$, the parameter $x = x_\alpha$, where x_α is the unique solution of equation (4.2.11) in the interval $(0, 1)$, and*

$$\sigma_x^2 = -\frac{x_\alpha \log(1 - x_\alpha) + x_\alpha^2}{(1 - x_\alpha)^2 \log^2(1 - x_\alpha)},$$

then

$$\mathbf{P}\{\zeta_N = k\} = \frac{1}{\sigma_x \sqrt{2\pi N}} e^{-z^2/2} (1 + o(1))$$

uniformly in the integers k such that $z = (k - n)/(\sigma_x \sqrt{2\pi N})$ lies in any fixed finite interval.

Proof. The proof is similar to the proof of Theorem 4.2.5 and we omit the details. Note only that $\alpha = \mathbf{E}\xi_1$ and $\sigma_x^2 = \mathbf{D}\xi_1$ for $x = x_\alpha$. ∎

Using Theorem 4.2.7 and representation (4.2.2), we obtain the following assertion on the distribution of ν_n.

Theorem 4.2.8. *If $n, N \to \infty$ and there exist constants α_0 and α_1 such that $1 < \alpha_0 \le \alpha \le \alpha_1$, then*

$$\mathbf{P}\{\nu_n = N\} = \frac{(-\log(1 - x_\alpha))^N}{N! \, x_\alpha^n \sigma_x \sqrt{2\pi N}} (1 + o(1)),$$

where x_α is the unique solution of equation (4.2.11) in the interval $(0, 1)$, and

$$\sigma_x^2 = -\frac{x_\alpha \log(1 - x_\alpha) + x_\alpha^2}{(1 - x_\alpha)^2 \log^2(1 - x_\alpha)}.$$

The asymptotic normality of ζ_N is preserved if $\alpha = n/N \to 1$ slowly, as specified below.

Theorem 4.2.9. *If $n, N \to \infty$ such that $\alpha = n/N \to 1$ and $m = n - N \to \infty$, and the parameter $x = x_\alpha$, where x_α is the unique solution of equation (4.2.11) in the interval $(0, 1)$, then*

$$\mathbf{P}\{\zeta_N = k\} = \frac{1}{\sqrt{2\pi m}} e^{-z^2/2}(1 + o(1))$$

uniformly in the integers k such that $z = (k - n)/\sqrt{m}$ lies in any fixed finite interval.

The proof is similar to the proof of Theorem 4.2.5 and we omit the details.

From Theorem 4.2.9 and representation (4.2.2), we obtain the following assertion on the behavior of $\mathbf{P}\{\nu_n = N\}$.

Theorem 4.2.10. *If $n, N \to \infty$ such that $\alpha = n/N \to 1$ and $m = n - N \to \infty$, then*

$$\mathbf{P}\{\nu_n = N\} = \frac{(-\log(1 - x_\alpha))^N}{N! \, x_\alpha^n \sqrt{2\pi m}}(1 + o(1)),$$

where x_α is the unique solution of equation (4.2.11).

It is not difficult to see that if $m^2/N \to 0$, then

$$x_\alpha = \frac{2m}{N}(1 + O(m/N)),$$

and consequently

$$(-\log(1 - x_\alpha))^N = x_\alpha^N \left(1 + x_\alpha/2 + O(x_\alpha^2)\right)^N = x_\alpha^N e^m \left(1 + O(m^2/N)\right),$$

$$x_\alpha^m = \frac{2^m m^m}{N^m}\left(1 + O(m^2/N)\right).$$

Therefore it follows from Theorem 4.2.9 that if $n, N \to \infty$, $\alpha = n/N \to 1$, $m \to \infty$ and $m^2/N \to 0$, then

$$\mathbf{P}\{\nu_n = N\} = \frac{N^m}{N! \, 2^m m!}(1 + o(1)).$$

Finally we consider the case where m is bounded.

Theorem 4.2.11. *If $N \to \infty$ and the parameter $x = 1/N$, then for any fixed $k = 0, 1, \ldots,$*

$$\mathbf{P}\{\zeta_N - N = k\} \to \frac{1}{2^k k!} e^{-1/2}.$$

Proof. By expanding the characteristic function $\varphi(t)$ of the random variable ξ_1 with parameter $x = 1/N$, we obtain for any fixed t,

$$\varphi(t) = \frac{\log\left(1 - xe^{it}\right)}{\log(1 - x)} = e^{it}\left(1 + \frac{x}{2}(e^{it} - 1) + O(x^2)\right).$$

If $x = 1/N$ and $N \to \infty$, then the characteristic function of $\zeta_N - N$ is equal to $(1 + (e^{it} - 1)/(2N) + O(N^{-2}))^N$ and tends to $e^{(e^{it}-1)/2}$. This means that the distribution of $\zeta_N - N$ converges to the Poisson distribution with parameter $1/2$.

∎

From this theorem and representation (4.2.2), we obtain the following assertion, which completes the description of the asymptotic behavior of the distribution of ν_n.

Theorem 4.2.12. *If* $n \to \infty$, $n/N \to 1$, *and* $m = n - N$ *is fixed, then*

$$\mathbf{P}\{\nu_n = N\} = \frac{N^m}{N! \, 2^m m!}(1 + o(1)).$$

It is not difficult to see that Theorems 4.2.2, 4.2.4, 4.2.6, 4.2.8, 4.2.10, and 4.2.12 give a complete description of the asymptotic behavior of the distribution of the number of cycles in a random permutation of degree n as $n \to \infty$.

4.3. Permutations with restrictions on cycle lengths

In this section, we present some results on permutations with restrictions on their cycle lengths. Let R be a subset of the set of natural numbers. We consider the set $S_{n,R}$ of all permutations of degree n with cycle lengths from the set R. One of the first questions that arises in this situation concerns the asymptotic behavior of the number $a_{n,R}$ of elements in $S_{n,R}$. This problem is far from being completely solved. Here we describe some of the solutions provided by an approach based on the generalized scheme of allocation.

Let the uniform distribution be defined on $S_{n,R}$ and let $\nu_{n,R}$ be the total number of cycles in a random permutation from this set. Put $b_{n,R} = (n-1)!$ if $n \in R$, and $b_{n,R} = 0$ otherwise. It is easy to see that

$$\mathbf{P}\{\nu_{n,R} = N\} = \frac{n!}{N! \, a_{n,R}} \sum_{n_1 + \cdots + n_N = n} \frac{b_{n_1,R} \cdots b_{n_N,R}}{n_1! \cdots n_N!}. \qquad (4.3.1)$$

We introduce independent identically distributed random variables $\xi_1^{(R)}, \ldots,$ $\xi_N^{(R)}$ with distribution

$$\mathbf{P}\{\xi_1^{(R)} = k\} = \frac{b_{k,R} x^k}{k! \, B_R(x)} = \frac{x^k}{k B_R(x)}, \qquad k \in R, \qquad (4.3.2)$$

where

$$B_R(x) = \sum_{k=1}^{\infty} \frac{b_{k,R} x^k}{k!} = \sum_{k \in R} \frac{x^k}{k}, \qquad x > 0.$$

By using these random variables, we can rewrite (4.3.1) in the form

$$P\{\nu_{n,R} = N\} = \frac{n! \, (B_R(x))^N}{N! \, x^n a_{n,R}} P\{\xi_1^{(R)} + \cdots + \xi_N^{(R)} = n\}. \tag{4.3.3}$$

Hence, summing over N, we obtain

$$a_{n,R} = \frac{n!}{x^n} e^{B_R(x)} \sum_{N=1}^{\infty} \frac{(B_R(x))^N}{N!} e^{-B_R(x)} P\{\xi_1^{(R)} + \cdots + \xi_N^{(R)} = n\}. \tag{4.3.4}$$

It is clear that above we have repeated the general approach of Section 1.3 for the case of the set $S_{n,R}$, and relations (4.3.1), (4.3.3), and (4.3.4) are the realizations of the general relations (1.3.1), (1.3.10), and (1.3.11), respectively.

To find the asymptotics of the numbers $a_{n,R}$, it is sufficient to choose an appropriate value of the parameter x, substitute it into the expression of the distribution (4.3.2), and then prove a local limit theorem for the sum of independent random variables with this distribution.

We succeed in obtaining results on $a_{n,R}$ only if the structure of R has some regularity. In the general case, the asymptotics of $a_{n,R}$ is unknown.

To demonstrate the approach, we consider first a simple case where R is the set E of even numbers.

Theorem 4.3.1. *If* $n \to \infty$, *then*

$$a_{n,E} = 2 \left(\frac{n}{e}\right)^n (1 + o(1)) \tag{4.3.5}$$

for even n, *and* $a_{n,E} = 0$ *for odd* n.

Proof. To prove the theorem, we use the representation (4.3.4). We consider the random variables $\xi_1^{(E)}, \ldots, \xi_N^{(E)}$ with distribution (4.3.2), where $R = E = \{2, 4, \ldots\}$, and

$$B_R(x) = B_E(x) = \sum_{k \in R} \frac{x^{2k}}{2k} = -\frac{1}{2} \log \left(1 - x^2\right).$$

The random variables $\xi_i = \xi_i^{(E)}/2$, $i = 1, \ldots, N$, are independent identically distributed, and

$$P\{\xi_1 = k\} = -\frac{x^{2k}}{2k \log(1 - x^2)}, \quad k = 1, 2, \ldots. \tag{4.3.6}$$

If we choose $x = \sqrt{1 - 1/n}$, then this distribution coincides with distribution (4.2.3) from the previous section, and according to Theorem 4.2.1, if $n \to \infty$, $N = \gamma \log n + o(\log n)$, where γ is a constant, $0 < \gamma < \infty$, then

$$P\{\xi_1 + \cdots + \xi_N = k\} = \frac{1}{n\Gamma(\gamma)} z^{\gamma-1} e^{-z} (1 + o(1))$$

uniformly in the integers k such that $z = k/n$ lies in any fixed interval of the form $0 < z_0 \leq z \leq z_1$, where z_0 and z_1 are constants.

Since

$$\varsigma_N^{(E)} = \xi_1^{(E)} + \cdots + \xi_N^{(E)} = 2(\xi_1 + \cdots + \xi_N),$$

we obtain that if $n \to \infty$, $N = (\log n)/2 + o(\log n)$, and n is even, then

$$\mathbf{P}\{\varsigma_N^{(E)} = n\} = \mathbf{P}\{\xi_1 + \cdots + \xi_N = n/2\} = \frac{\sqrt{2}}{n\sqrt{\pi}}e^{-1/2}(1 + o(1)). \quad (4.3.7)$$

For odd n, this probability equals zero.

To obtain $a_{n,R}$ with the help of relation (4.3.4), we have to sum the probabilities $\mathbf{P}\{\varsigma_N^{(E)} = n\}$ with the Poisson coefficients. To this end, we need to estimate these probabilities for all N. We show that for all N,

$$\mathbf{P}\{\varsigma_N^{(E)} = n\} \leq \frac{2N}{n \log n}. \quad (4.3.8)$$

This bound is a consequence of the following chain of estimates. It follows from (4.3.2) that

$$\mathbf{P}\{\varsigma_N^{(E)} = n\} = \frac{x^n}{(B_E(x))^N} \sum_{K(n,N)} \frac{1}{k_1 \ldots k_N},$$

where

$$K(n, N) = \{k_1, \ldots, k_N : k_1 + \cdots + k_N = n, \ k_1, \ldots, k_N \in R\}.$$

Hence,

$$\mathbf{P}\{\varsigma_N^{(E)} = n\} = \frac{x^n}{n(B_E(x))^N} \sum_{K(n,N)} \left(\frac{1}{k_1 \cdots k_{N-1}} + \cdots + \frac{1}{k_2 \cdots k_N} \right)$$

$$\leq \frac{N}{n(B_E(x))^N} \sum_{K(n,N)} \frac{x^{k_1}}{k_1} \cdots \frac{x^{k_{N-1}}}{k_{N-1}}$$

$$\leq \frac{N}{n(B_E(x))^N} \left(\sum_{1 \leq k \leq n, \, k \in R} \frac{x^k}{k} \right)^{N-1} \leq \frac{N}{n B_E(x)}.$$

We obtain relation (4.3.8) because $B = B_E(x) = (\log n)/2$.

We split the sum

$$S = \sum_{N=1}^{n/2} \frac{B^N e^{-B}}{N!} \mathbf{P}\{\varsigma_N^{(E)} = n\}$$

into four summands, dividing the domain of summation into four parts:

$$A_1 = \{N: 1 \leq N < B - B^{3/4}\},$$

$$A_2 = \{N: B - B^{3/4} \leq N \leq B + B^{3/4}\},$$

$$A_3 = \{N: B + B^{3/4} < N \leq B + B^2\},$$

$$A_4 = \{N: B + B^2 < N \leq n/2\}.$$

It is not difficult to see that relation (4.3.7) is satisfied uniformly in $N \in A_2$. Therefore

$$S_2 = \sum_{N \in A_2} \frac{B^N e^{-B}}{N!} P\{\zeta_N^{(E)} = n\}$$

$$= \sqrt{\frac{2}{\pi}} e^{-1/2} \frac{1}{n} \sum_{N \in A_2} \frac{B^N e^{-B}}{N!} (1 + o(1)) = \frac{\sqrt{2}}{n\sqrt{\pi e}} (1 + o(1)),$$

since $B = (\log n)/2$, and as $N \to \infty$,

$$\sum_{N \in A_2} \frac{B^N e^{-B}}{N!} \to 1.$$

The remaining part of the sum is $o(1/n)$. Indeed, by applying estimate (4.3.8), we obtain

$$S_1 \leq \frac{2B}{n \log n} \sum_{N \in A_1} \frac{B^N e^{-B}}{N!} = \frac{1}{n} \sum_{N \in A_1} \frac{B^N e^{-B}}{N!},$$

and $S_1 = o(1/n)$ because

$$\sum_{N \in A_1} \frac{B^N e^{-B}}{N!} \to 0$$

as $n \to \infty$.

It follows from (4.3.8) that

$$S_3 \leq \frac{\log n}{2n} \sum_{N \in A_3} \frac{B^N e^{-B}}{N!}.$$

If we use the normal approximation for the Poisson distribution, we find that

$$\sum_{N \in A_3} \frac{B^N e^{-B}}{N!} \leq \sum_{N \geq B + B^{3/4}} \frac{B^N e^{-B}}{N!} \leq c_1 \int_{B^{1/4}}^{\infty} e^{-u^2/2} \, du \leq c_2 e^{-\sqrt{B}/2},$$

where c_1 and c_2 are constants. Hence, $S_3 = o(1/n)$.

Similarly, by using (4.3.8), we obtain

$$S_4 \leq \frac{1}{\log n} \sum_{N \in A_4} \frac{B^N e^{-B}}{N!} \leq \frac{1}{\log n} \sum_{N \geq B^2} \left(\frac{Be}{N}\right)^N e^{-B}$$

$$\leq \frac{1}{\log n} \sum_{N \geq B^2} \left(\frac{e}{B}\right)^N e^{-B}.$$

Hence, $S_4 = o(1/n)$ because $e/B < e^{-1}$ for n sufficiently large.

If we combine the estimates of S_1, S_2, S_3, and S_4, we obtain

$$S = \sqrt{\frac{2}{\pi e} \frac{1}{n}} (1 + o(1)).$$

Substituting this expression into (4.3.4) and expanding $n!$ by the Stirling formula give the assertion of the theorem.   ∎

The analogous result is valid for the number of permutations for which R is the set of odd numbers.

We turn now to the case where the set R is not as regular as E. Let $R(k)$ be the number of elements of R that are not greater than k. Set $R(0) = 0$. In the sequel, we assume that

$$\lim_{k \to \infty} R(k)/k = \rho, \quad 0 \leq \rho \leq 1.$$

In this case, ρ is called the density of R in the set of natural numbers.

We will find the asymptotics of $a_{n,R}$ under the following additional conditions on the set R.

(1) There exists a positive integer r such that, for any nonnegative integer s, the set $R \cap \{s + 1, \ldots, s + r\}$ cannot be embedded in any integer lattice with a step not equal to 1.

(2) The generating function $F(z)$ of the set R has a finite number m of poles at the points $z_l = e^{2\pi i l/m}$, $l = 0, 1, \ldots, m - 1$, on the unit circle $|z| = 1$; in other words, it is of the form

$$F(z) = \sum_{k \in R} z^k = P(z)/(1 - z^m), \qquad (4.3.9)$$

where $P(z)$ is a polynomial.

Note that, since the coefficients of the series $F(z)$ take a finite number of values, by Szegő's theorem (see, for example, [19]), there are only two possibilities for $F(z)$: Either $F(z)$ has the form (4.3.9), or the set of singular points of $F(z)$ is dense everywhere on the unit circle, and therefore $F(z)$ cannot be extended outside the unit circle. We consider here only the first case. In this case, the coefficients of $F(z)$, with exception of some initial numbers, form a periodic sequence with period m, and, therefore, the set R has density $\rho = l/m$, where l is the number of units in the period.

Consider independent identically distributed random variables $\xi_1, \ldots, \xi_N$ with distribution

$$\mathbf{P}\{\xi_1 = k\} = \frac{x^k}{kB(x)}, \quad k \in R, \tag{4.3.10}$$

where

$$x = 1 - \frac{1}{n}, \qquad B(x) = \sum_{k \in R} \frac{x^k}{k}.$$

Theorem 4.3.2. *Suppose that R has the density $\rho > 0$ and satisfies conditions (1) and (2), $n \to \infty$, $N = \rho \log n + o(\log n)$. Then*

$$n\mathbf{P}\{\xi_1 + \cdots + \xi_N = k\} = y^{\rho-1}e^{-y}/\Gamma(\rho)(1 + o(1))$$

uniformly in the integers k such that $y = k/n$ lies in any fixed interval of the form $0 < y_0 \le y \le y_1 < \infty$.

With the aid of Theorem 4.3.2 and relation (4.3.9), we prove the following assertions.

Theorem 4.3.3. *Suppose that R has the density $\rho > 0$ and satisfies conditions (1) and (2). Then, as $n \to \infty$,*

$$a_{n,R} = (n-1)! \, e^{B_{n,R}} / \Gamma(\rho)(1 + o(1)), \tag{4.3.11}$$

where

$$B_{n,R} = \sum_{k \in R} \frac{1}{k}\left(1 - \frac{1}{n}\right)^k.$$

Since $\sum_{k=1}^{\infty}(1 - 1/n)^k/k = \log n$, the assertion (4.3.11) can be written in the form

$$a_{n,R} = n! \, e^{-L_{n,R}} / \Gamma(\rho)(1 + o(1)),$$

where

$$L_{n,R} = \sum_{k \notin R} \frac{1}{k}\left(1 - \frac{1}{n}\right)^k.$$

Theorem 4.3.4. *Suppose that R has the density $\rho > 0$ and satisfies conditions (1) and (2). Then, as $n \to \infty$,*

$$\mathbf{P}\{\nu_{n,R} = N\} = \frac{1}{\sqrt{2\pi B_{n,R}}} \exp\left\{-\frac{(N - B_{n,R})^2}{2B_{n,R}}\right\}(1 + o(1))$$

uniformly in the integers N such that $(N - B_{n,R})/\sqrt{B_{n,R}}$ lies in any fixed finite interval.

To prove Theorem 4.3.2, we establish some auxiliary results. The characteristic function of distribution (4.3.10) is

$$\varphi(t) = \sum_{k \in R} \frac{x^k e^{itk}}{k B(x)} = \frac{B(xe^{it})}{B(x)}.$$

Lemma 4.3.1. *If R has the density $\rho > 0$, then, as $n \to \infty$,*

$$\varphi\left(\frac{t}{n}\right) = 1 - \frac{\log(1 - it)}{\log n} + o\left(\frac{1}{\log n}\right)$$

for any fixed t.

Proof. We first derive some auxiliary estimates. It is easy to see that

$$\sum_{k \in R} x^k = \sum_{k=1}^{\infty} x^k (R(k) - R(k-1))$$

$$= \sum_{k=1}^{\infty} x^k R(k) - \sum_{k=1}^{\infty} x^k R(k-1)$$

$$= (1 - x) \sum_{k=1}^{\infty} x^k R(k).$$

Set $\varepsilon = \log n$. For such ε,

$$\sum_{1 \le k \le \varepsilon} x^k R(k) \le \sum_{1 \le k \le \varepsilon} k \le \log^2 n,$$

and, since R has positive density,

$$\sum_{k > \varepsilon} x^k R(k) = \sum_{k > \varepsilon} \frac{k x^k R(k)}{k} = \sum_{k > \varepsilon} k x^k \rho (1 + o(1)).$$

Thus, as $n \to \infty$,

$$\sum_{k \in R} x^k = \rho(1 - x) \sum_{k=1}^{\infty} k x^k (1 + o(1)) + O\left(\frac{\log^2 n}{n}\right)$$

$$= \rho(1 - x) \left(\frac{x}{(1 - x)^2}(1 + o(1))\right) + O\left(\frac{\log^2 n}{n}\right)$$

$$= \rho n + o(n). \tag{4.3.12}$$

Similarly we obtain the estimate

$$B(x) = \sum_{k \in R} \frac{x^k}{k} = \rho \log n + o(n). \tag{4.3.13}$$

We now write the characteristic function in the form

$$\varphi\left(\frac{t}{n}\right) = \frac{B(xe^{it/n})}{B(x)} = 1 + \frac{1}{B(x)}\left(B(xe^{it/n}) - B(x)\right).$$

It is easy to see that

$$B(xe^{it/n}) - B(x)$$

$$= \sum_{k \in R} \frac{1}{k} x^k (e^{itk/n} - 1)$$

$$= \sum_{k=1}^{\infty} \frac{1}{k} x^k (e^{itk/n} - 1)(R(k) - R(k-1))$$

$$= \sum_{k=1}^{\infty} \frac{1}{k} x^k R(k) \left(e^{itk/n} - 1 - \frac{xk}{k+1}(e^{it(k+1)/n} - 1) \right)$$

$$= \sum_{k=1}^{\infty} \frac{1}{k} x^k R(k) \left(e^{itk/n}(1 - e^{it/n}) + \frac{1}{n}(e^{it(k+1)/n} - 1) \right.$$

$$\left. + \frac{1}{k+1}(e^{it(k+1)/n} - 1) + \frac{1}{n(k+1)}(e^{it(k+1)/n} - 1) \right).$$

First of all, we estimate the part that does not contribute essentially to the sum. If t is fixed and $n \to \infty$, then

$$\sum_{k=1}^{\infty} \frac{x^k R(k)}{k(k+1)n}(e^{it(k+1)/n} - 1) = O\left(\sum_{k=1}^{\infty} \frac{x^k}{n^2}\right) = O\left(\frac{1}{n}\right).$$

We transform the other parts of the sum as follows:

$$\sum_{k=1}^{\infty} \frac{x^k R(k)}{k} e^{itk/n}(1 - e^{it/n})$$

$$= -it \sum_{k=1}^{\infty} \frac{x^k R(k)}{kn} e^{itk/n} + \sum_{k=1}^{\infty} \frac{x^k R(k)}{k} e^{itk/n}\left(1 + \frac{it}{n} - e^{it/n}\right)$$

$$= -it \sum_{k=1}^{\infty} \frac{x^k R(k)}{kn} e^{itk/n} + O\left(\frac{1}{n^2}\sum_{k=1}^{\infty} x^k\right)$$

$$= -it \sum_{k=1}^{\infty} \frac{x^k R(k)}{kn} e^{itk/n} + O\left(\frac{1}{n}\right) \tag{4.3.14}$$

and

$$\sum_{k=1}^{\infty} \frac{x^k R(k)}{kn} \left(e^{it(k+1)/n} - 1\right)$$

$$= \sum_{k=1}^{\infty} \frac{x^k R(k)}{kn} \left(e^{itk/n} - 1\right) + \sum_{k=1}^{\infty} \frac{x^k R(k)}{kn} e^{itk/n} \left(e^{it/n} - 1\right)$$

$$= \sum_{k=1}^{\infty} \frac{x^k R(k)}{kn} \left(e^{itk/n} - 1\right) + O\left(\frac{1}{n}\right). \qquad (4.3.15)$$

Similarly,

$$\sum_{k=1}^{\infty} \frac{x^k R(k)}{k(k+1)} \left(e^{it(k+1)/n} - 1\right) = \sum_{k=1}^{\infty} \frac{x^k R(k)}{k^2} \left(e^{itk/n} - 1\right) + O\left(\frac{\log n}{n}\right).$$

Set $\varepsilon = \log n$ and $E = n \log n$. Then

$$\left| \sum_{k \leq \varepsilon} \frac{x^k R(k)}{kn} e^{itk/n} \right| \leq \frac{1}{n} \sum_{k \leq \varepsilon} x^k = O\left(\frac{\log n}{n}\right),$$

$$\left| \sum_{k \leq \varepsilon} \frac{x^k R(k)}{kn} \left(e^{itk/n} - 1\right) \right| \leq \frac{1}{n^2} \sum_{k \leq \varepsilon} kx^k = O\left(\frac{\log^2 n}{n^2}\right),$$

$$\left| \sum_{k \leq \varepsilon} \frac{x^k R(k)}{k^2} \left(e^{itk/n} - 1\right) \right| \leq \frac{1}{n} \sum_{k \leq \varepsilon} x^k = O\left(\frac{\log n}{n}\right).$$

In exactly the same way,

$$\left| \sum_{k \geq E} \frac{x^k R(k)}{kn} e^{itk/n} \right| \leq \frac{1}{n} \sum_{k \geq E} x^k \leq x^E \leq e^{-\log n} = \frac{1}{n},$$

$$\left| \sum_{k \geq E} \frac{x^k R(k)}{kn} \left(e^{itk/n} - 1\right) \right| \leq \frac{1}{n^2} \sum_{k \geq E} kx^k \leq \frac{\log n}{n} + \frac{1}{n},$$

$$\left| \sum_{k \geq E} \frac{x^k R(k)}{k^2} \left(e^{itk/n} - 1\right) \right| \leq \frac{1}{n} \sum_{k \geq E} x^k \leq \frac{1}{n}.$$

It is clear that

$$R(k)/k = \rho + o(1), \qquad x^k = e^{-k/n}(1 + o(1))$$

uniformly in k, $\varepsilon \le k \le E$. Hence,

$$\sum_{\varepsilon \le k \le E} \frac{x^k R(k)}{kn} e^{itk/n} = \rho \sum_{\varepsilon \le k \le E} \frac{1}{n} e^{-k(1-it)/n} + o\left(\frac{1}{n} \sum_{\varepsilon \le k \le E} e^{-k/n} \right)$$

$$= \rho \sum_{\varepsilon \le k \le E} \frac{1}{n} e^{-k(1-it)/n} + o(1).$$

Similarly,

$$\sum_{\varepsilon \le k \le E} \frac{x^k R(k)}{kn} \left(e^{itk/n} - 1 \right) = \rho \sum_{\varepsilon \le k \le E} \frac{1}{n} e^{-k/n} \left(e^{itk/n} - 1 \right) + o(1),$$

$$\sum_{\varepsilon \le k \le E} \frac{x^k R(k)}{k^2} \left(e^{itk/n} - 1 \right) = \rho \sum_{\varepsilon \le k \le E} \frac{n}{k} \frac{1}{n} e^{-k/n} \left(e^{itk/n} - 1 \right) + o(1).$$

The sums in the right-hand sides of these relations are integral sums of integrable functions. Therefore, as $n \to \infty$, their limits exist and equal

$$\int_0^\infty e^{-(1-it)z} dz = \frac{1}{1-it},$$

$$\int_0^\infty e^{-z} \left(e^{itz} - 1 \right) dz = \frac{1}{1-it} - 1,$$

$$\int_0^\infty \frac{1}{z} e^{-z} \left(e^{itz} - 1 \right) dz = -\log(1-it),$$

respectively. Thus, as $n \to \infty$, for any fixed t,

$$B\left(xe^{it/n} \right) - B(x) = -\rho \log(1-it) + o(1),$$

and hence,

$$\varphi\left(\frac{t}{n} \right) = 1 - \frac{\log(1-it)}{\log n} + o\left(\frac{1}{\log n} \right).$$

∎

Lemma 4.3.1 implies that for any fixed t, as $n \to \infty$ and $N = \rho \log n + o(\log n)$,

$$\varphi^N\left(\frac{t}{n} \right) = \frac{1}{(1-it)^\rho} + o(1), \qquad (4.3.16)$$

and for the normalized sum $(\xi_1 + \cdots + \xi_N)/n$ the limit distribution is the distribution with the characteristic function $(1-it)^{-\rho}$ that has the density $y^{\rho-1} e^{-y} / \Gamma(\rho)$.

To prove the local convergence of the distributions, we have to estimate $\varphi(t/n)$ outside a neighborhood of zero.

Lemma 4.3.2. *Suppose that R has the density $\rho > 0$ and satisfies conditions (1) and (2). Then, for any $\varepsilon > 0$, there exists $q < 1$ such that for $\varepsilon \leq |t| \leq \pi$,*

$$|\varphi(t)| \leq q.$$

Proof. Let k_1, k_2, and k_3 be integers and a_{k_1}, a_{k_2}, $a_{k_3} > 0$. It is easy to verify that

$$
\left| a_{k_1} e^{itk_1} + a_{k_2} e^{itk_2} + a_{k_3} e^{itk_3} \right|^2 = (a_{k_1} + a_{k_2} + a_{k_3})^2
$$
$$
- 2 a_{k_1} a_{k_2} (1 - \cos t (k_2 - k_1))
$$
$$
- 2 a_{k_1} a_{k_3} (1 - \cos t (k_3 - k_1))
$$
$$
- 2 a_{k_2} a_{k_3} (1 - \cos t (k_3 - k_2)).
$$

For $a > 0$ and $\delta \geq 0$,

$$a - \sqrt{a^2 - \delta} \geq \delta / (2a).$$

Therefore

$$
a_{k_1} + a_{k_2} + a_{k_3} - \left| a_{k_1} e^{itk_1} + a_{k_2} e^{itk_2} + a_{k_3} e^{itk_3} \right|
$$
$$
\geq \frac{a_{k_1} a_{k_2} (1 - \cos t (k_2 - k_1))}{a_{k_1} + a_{k_2} + a_{k_3}}
$$
$$
+ \frac{a_{k_1} a_{k_3} (1 - \cos t (k_3 - k_1))}{a_{k_1} + a_{k_2} + a_{k_3}}
$$
$$
+ \frac{a_{k_2} a_{k_3} (1 - \cos t (k_3 - k_2))}{a_{k_1} + a_{k_2} + a_{k_3}}. \tag{4.3.17}
$$

Suppose now that, as in condition (1), the integers k_1, k_2, and k_3 do not lie on any lattice with a step greater than 1 and are contained in an interval of length r. Then, for $\varepsilon \leq |t| \leq \pi$, the three cosines from the right-hand side of (4.3.17) do not simultaneously take the value 1. Moreover, since k_1, k_2, and k_3 are contained in an interval of length r, their differences can take only a finite number of values. Therefore, there exists $\alpha > 0$ such that for $\varepsilon \leq |t| \leq \pi$,

$$
(1 - \cos t (k_2 - k_1)) + (1 - \cos t (k_3 - k_1)) + (1 - \cos t (k_3 - k_2)) \geq 3\alpha \tag{4.3.18}
$$

uniformly in all such k_1, k_2, and k_3.

We now let $a_k = x^k / k$, $k = 1, 2, \ldots$, and suppose condition (1) holds for $k_1 > k_2 > k_3$. It follows from (4.3.17) and (4.3.18) that

$$
a_{k_1} + a_{k_2} + a_{k_3} - \left| a_{k_1} e^{itk_1} + a_{k_2} e^{itk_2} + a_{k_3} e^{itk_3} \right| \geq \alpha a_r a_{k_1}. \tag{4.3.19}
$$

Write the characteristic function $\varphi(t)$ in the form

$$\varphi(t) = \sum_{l=0}^{\infty} \sum_{k=rl+1}^{rl+r} \frac{a_k}{B(x)} (R(k) - R(k-1)) e^{itk}.$$

From every set $\{rl+1, \ldots, rl+r\}$, select, according to condition (1), three integers k_{1l}, k_{2l}, and k_{3l} from R that do not lie on any integer lattice with a step not equal to 1. Using estimate (4.3.19) gives

$$a_{k_{1l}} + a_{k_{2l}} + a_{k_{3l}} - \left| a_{k_{1l}} e^{itk_{1l}} + a_{k_{2l}} e^{itk_{2l}} + a_{k_{3l}} e^{itk_{3l}} \right| \geq \alpha a_r a_{k_1 l} \geq \alpha a_r a_{rl+r}.$$

Therefore, taking into account that $R(k_{il}) - R(k_{il} - 1) = 1$ for $i = 1, 2, 3$ and $l = 0, 1, \ldots$ yields

$$B(x)|\varphi(t)| \leq \sum_{l=0}^{\infty} \sum_{k=rl+1}^{rl+r} a_k (R(k) - R(k-1))$$

$$- \sum_{l=0}^{\infty} (a_{k_{1l}} + a_{k_{2l}} + a_{k_{3l}})$$

$$+ \sum_{l=0}^{\infty} \left| a_{k_{1l}} e^{itk_{1l}} + a_{k_{2l}} e^{itk_{2l}} + a_{k_{3l}} e^{itk_{3l}} \right|$$

$$\leq B(x) - \frac{\alpha a_r}{r} \sum_{l=0}^{\infty} \frac{x^{r(l+1)}}{l+1}. \tag{4.3.20}$$

Inequalities (4.3.20) imply the assertion of Lemma 4.3.2 because r is fixed, $x = 1 - 1/n$, and, as $n \to \infty$,

$$B(x) = \rho \log n + o(\log n), \qquad \sum_{l=0}^{\infty} \frac{x^{r(l+1)}}{l+1} = -\log\left(1 - x^r\right) = \log n + o(\log n).$$

∎

Lemma 4.3.3. *Suppose that R has the density $\rho > 0$ and satisfies conditions (1) and ((2). Then there exist c_1 and $\varepsilon > 0$ such that for every $l = 0, 1, \ldots, m-1$ and $|t/n - 2\pi l/m| \leq \varepsilon$, for sufficiently large n,*

$$\frac{1}{n} \left| \varphi'\left(\frac{t}{n}\right) \right| \leq \frac{c_1}{\sqrt{1 + (t - 2\pi l n/m)^2} \log n}.$$

Proof. We start by estimating

$$\frac{1}{n} \varphi'\left(\frac{t}{n}\right) = \frac{i}{n B(x)} \sum_{k \in R} x^k e^{itk/n}.$$

By condition (2), there exist c, $\delta > 0$ such that for $|z| < 1$, $|z_l - z| \leq \delta$, $l = 0, 1, \ldots, m - 1$,

$$\left| \sum_{k \in R} z^k \right| \leq \frac{c}{|z_l - z|}, \qquad l = 0, 1, \ldots, m - 1. \tag{4.3.21}$$

Set $z = x e^{it/n}$, where $x = 1 - 1/n$. It is clear that $|z| < 1$ and there exists $\varepsilon > 0$ such that if $|t/n - 2\pi l/m| \leq \varepsilon$, then $|z_l - z| \leq \delta$ for sufficiently large n. Therefore, (4.3.21) implies that for $|t/n - 2\pi l/m| \leq \varepsilon$, $l = 0, 1, \ldots, m - 1$,

$$\left| \varphi' \left(\frac{t}{n} \right) \right| \leq \frac{1}{B(x)} \left| \sum_{k \in R} z^k \right| \leq \frac{cn}{B(x)|1 - xe^{i(t/n - 2\pi l/m)}|}$$

$$\leq \frac{c_2 n}{B(x)\sqrt{1 + (t - 2\pi l n/m)^2}}.$$

Since $B(x) = \rho \log n + o(\log n)$, there exists c_1 such that for every $l = 0, 1, \ldots, m - 1$, if $|t/n - 2\pi l/m| \leq \varepsilon$, then

$$\left| \varphi' \left(\frac{t}{n} \right) \right| \leq \frac{c_1 n}{\log n \sqrt{1 + (t - 2\pi l n/m)^2}}$$

for sufficiently large n.   ∎

We now proceed to estimate the characteristic function $\varphi(t)$ in the intermediate range of t. Obtaining the estimate involves some technical difficulties. So, for the sake of greater clarity, we first treat the case $R = \mathbf{N}$. In this case,

$$p_k = \mathbf{P}\{\xi_1 = k\} = \frac{x^k}{k B(x)}, \qquad k = 1, 2, \ldots,$$

$$B(x) = -\log(1 - x) = \log n.$$

Consider the random variable $\tilde{\xi} = \xi_1 - \xi_2$. Its distribution is symmetric, and for $m \geq 0$,

$$\tilde{p}_m = \mathbf{P}\{\tilde{\xi} = m\} = \sum_{k=1}^{\infty} p_k \, p_{k+m}.$$

Let

$$\tilde{\varphi}(t) = \sum_m \tilde{p}_m e^{itm} = \tilde{p}_0 + 2 \sum_{m=1}^{\infty} \tilde{p}_m \cos tm.$$

It is clear that the characteristic function $\varphi(t)$ of the random variable ξ_1 is related to $\tilde{\varphi}(t)$ by the equality $\tilde{\varphi}(t) = |\varphi(t)|^2$. To estimate $\tilde{\varphi}(t)$, we use a standard inequality (see, e.g., [49]): For $t > 0$,

$$1 - \tilde{\varphi}(t) = 2 \sum_{m=1}^{\infty} \tilde{p}_m (1 - \cos tm) \geq 2 \sum_{s=0}^{\infty} \sum_{m \in M_s} \tilde{p}_m, \tag{4.3.22}$$

where

$$M_s = \left\{ m: \frac{\pi}{2t} + \frac{2\pi s}{t} \leq m \leq \frac{3\pi}{2t} + \frac{2\pi s}{t} \right\}.$$

Lemma 4.3.4. For $m_0 > 0$,

$$2 \sum_{m \geq m_0} \tilde{p}_m \geq \sum_{l \geq 2m_0} p_l \sum_{k=1}^{\infty} p_k.$$

Proof. By using $l = m + k$ as the variable of summation, we obtain

$$2 \sum_{m \geq m_0} \tilde{p}_m = 2 \sum_{m \geq m_0} \sum_{k=1}^{\infty} p_k p_{k+m} = \sum_{l \geq m_0+1} \sum_{k=1}^{l-m_0} p_l p_k + \sum_{l=1}^{\infty} \sum_{k=l+m_0}^{\infty} p_k p_l. \tag{4.3.23}$$

The right-hand side of (4.3.23) is estimated from below by the quantity

$$\sum_{l \geq 2m_0} p_l \sum_{k=1}^{\infty} p_k = \sum_{l \geq 2m_0} p_l.$$

To see this, it is sufficient to delete the first terms in the first sum from (4.3.23), retaining

$$\sum_{l \geq 2m_0} p_l \sum_{k=1}^{l-m_0} p_k,$$

and, in the second sum from (4.3.23), to shift the domain of summation to $2m_0$, giving

$$\sum_{l \geq 2m_0} p_l \sum_{k=l-m_0+1}^{\infty} p_k,$$

which does not exceed the second sum from (4.3.23) by the monotonicity of the probabilities. ∎

Lemma 4.3.5. For $0 < t \leq \pi$,

$$1 - \tilde{\varphi}(t) \geq \frac{1}{3} \sum_{k \geq \pi/t} p_k.$$

Proof. Note that the summation on the right-hand side of (4.3.22) occurs over integers m from an interval of length π/t. If we enumerate intervals of such a length on the positive semi-axis starting at the point $\pi/(2t)$, the domain of summation will consist of the intervals labeled by odd numbers. Notice that the sequence of probabilities p_k, $k = 1, 2, \ldots$, is monotone, and the numbers of integer points in any two intervals of length π/t differ by at most 1. Therefore each interval of length π/t for $0 < t \leq \pi$ contained in the right-hand side of the sum (4.3.22) contributes not less than one-third of the total sum of the two following intervals: the interval itself and the interval adjoined to it on the right side, which does not belong to the initial domain of summation. (Note that, as $t \to \infty$, the number of integer points in one interval increases and its contribution to the sum tends to $1/2$.) Therefore, (4.3.22) implies

$$1 - \tilde{\varphi}(t) \geq 2 \sum_{s=0}^{\infty} \sum_{m \in M_s} \tilde{p}_m \geq \frac{2}{3} \sum_{m \geq \pi/(2t)} \tilde{p}_m.$$

By applying Lemma 4.3.4, we obtain the assertion of Lemma 4.3.5.   ∎

It remains to estimate the sum of the form $\sum_{k \geq a} p_k$ from below. If we use the inequality

$$1 - 1/n \geq e^{-1/(n-1)},$$

we obtain

$$\sum_{k \geq a} \frac{x^k}{k} \geq \sum_{k \geq a} \frac{e^{-k/(n-1)}}{k/(n-1)} \frac{1}{n-1} \geq \int_{a/(n-1)}^{\infty} \frac{e^{-y}}{y} dy \geq c_3 - \log \frac{a}{n-1},$$

(4.3.24)

where c_3 is a constant.

We use Lemma 4.3.5, set $a = \pi n/|t|$ in (4.3.24), and obtain for $|t|/n \leq \pi$,

$$1 - \tilde{\varphi}\left(\frac{t}{n}\right) \geq \frac{1}{3} \sum_{l \geq \pi n/|t|} p_l \geq \frac{1}{3B(x)} \left(c_3 - \log \frac{\pi n}{|t|(n-1)}\right)$$

$$\geq \frac{1}{3B(x)} (\log |t| + c_4),$$

where c_4 is a constant. Hence, we go on to estimate $\varphi(t/n)$ and find that

$$\left|\varphi\left(\frac{t}{n}\right)\right| \leq \left(1 - \frac{\log |t| + c_4}{3 \log n}\right)^{1/2} \leq \exp\left\{-\frac{\log |t| + c_4}{6 \log n}\right\}.$$

If $N > \frac{1}{2} \log n$, then

$$\left|\varphi\left(\frac{t}{n}\right)\right|^N \leq \exp\left\{-\frac{1}{12} \log |t| + \frac{c_4}{12 \log n}\right\} \leq c_5 |t|^{-1/12}.$$

(4.3.25)

We now return to the case $R \subset \mathbf{N}$. We retain the notation $\varphi(t)$ and $\tilde{\varphi}(t)$ for the characteristic functions and set

$$p_k = \mathbf{P}\{\xi_1 = k\} = \frac{a_k x^k}{B(x)}, \quad k \in R, \qquad B(x) = \sum_{k \in R} a_k x^k, \quad a_k = \frac{1}{k};$$

$$\delta_R(k) = 0 \text{ for } k \notin R, \text{ and } \delta_R(k) = 1 \text{ for } k \in R.$$

Lemma 4.3.6. *Suppose that R has the density $\rho > 0$ and satisfies conditions (1) and (2). Then, for $|t|/n \leq \pi$ and $N \geq \frac{1}{2}\rho \log n$,*

$$\left| \varphi\left(\frac{t}{n}\right) \right|^N \leq c_6 |t|^{-1/(12r^2\rho)},$$

where r is defined in condition (1) and c_6 is a constant.

Proof. We revise the arguments leading to estimate (4.3.25). Inequality (4.3.22) now takes the following form: For $t > 0$,

$$1 - |\varphi(t)|^2 = \frac{2}{B^2(x)} \sum_{s=0}^{\infty} \sum_{m \in M_s} \sum_{k=1}^{\infty} a_k x^k \delta_R(k) a_{k+m} x^{k+m} \delta_R(k+m),$$

where

$$M_s = \left\{ m: \frac{\pi}{2t} + \frac{2\pi s}{t} \leq m \leq \frac{3\pi}{2t} + \frac{2\pi s}{t} \right\}.$$

We retain only one summand in each interval of length r, replace this summand by the minimum value over the interval, and use the transition from the sum over one interval of length r to one-third of the sum over the interval of twice the length. Then we obtain for $t > 0$,

$$\sum_{s=0}^{\infty} \sum_{m \in M_s} a_{k+m} x^{k+m} \delta_R(k+m) \geq \frac{1}{3} \sum_{rl \geq \pi/(2t)} a_{k+rl} x^{k+rl}.$$

Once again, we preserve only one summand in each interval of length r and get

$$1 - |\varphi(t)|^2 = \frac{2}{3B^2(x)} \sum_{k=1}^{\infty} a_k x^k \delta_R(k) \sum_{l \geq \pi/(2tr)} a_{k+rl} x^{k+rl}$$

$$\geq \frac{2}{3B^2(x)} \sum_{m=1}^{\infty} a_{rm} x^{rm} \sum_{l \geq \pi/(2tr)} a_{rm+rl} x^{rm+rl}$$

$$= \frac{2}{3B^2(x)r^2} \sum_{l \geq \pi/(2tr)} \sum_{m=1}^{\infty} \frac{x^{rm}}{m} \frac{x^{r(m+l)}}{m+l}.$$

The assertion of Lemma 4.3.4 is based on the monotonicity of the probabilities p_k, $k = 1, 2, \ldots$. The summands of the last double sum are similar to the summands of the sum in Lemma 4.3.4, and the values x^{rk}/k, $k = 1, 2, \ldots$, are also monotonic. Therefore we may use Lemma 4.3.4 and obtain

$$1 - |\varphi(t)|^2 \geq \frac{2}{3B^2(x)r^2} \sum_{l \geq \pi/(2tr)} \frac{x^{rl}}{l} \sum_{m=1}^{\infty} \frac{x^{rm}}{m} = -\frac{\log(1 - x^r)}{3B^2(x)r^2} \sum_{l \geq \pi/(tr)} \frac{x^{rl}}{l}.$$

For a fixed r, the estimate (4.3.24) remains true. Therefore, by taking into account the asymptotics $B(x) = \rho \log n + o(\log n)$ and $-\log(1 - x^r) = \log n + o(\log n)$, we find

$$1 - \left|\varphi\left(\frac{t}{n}\right)\right|^2 \geq \frac{1}{3r^2\rho^2 \log n}(\log |t| + c).$$

Hence,

$$\left|\varphi\left(\frac{t}{n}\right)\right| \leq \left(1 - \frac{\log |t| + c}{3r^2\rho^2 \log n}\right)^{1/2} \leq \exp\left\{-\frac{\log |t| + c}{6r^2\rho^2 \log n}\right\},$$

and for $N \geq \frac{1}{2}\rho \log n$,

$$\left|\varphi\left(\frac{t}{n}\right)\right|^N \leq c_6 |t|^{-1/(12r^2\rho)},$$

where c_6 is a constant.   ∎

Proof of Theorem 4.3.2. Consider the sum $\zeta_N = \xi_1 + \cdots + \xi_N$ of independent identically distributed random variables with distribution (4.3.10). As we have seen, Lemma 4.3.1 implies that, as $n \to \infty$ and $N = \rho \log n + o(\log n)$, the distribution of ζ_N/n converges weakly to the distribution with density

$$u^{\rho-1}e^{-u}/\Gamma(\rho), \quad u > 0.$$

We now prove the local convergence of these distributions. For an integer k, let $y = k/n$. By the inversion formula,

$$P\{\zeta_N = k\} = P\left\{\frac{1}{n}\zeta_N = y\right\} = \frac{1}{2\pi n} \int_{-\pi n}^{\pi n} e^{-ity}\varphi^N\left(\frac{t}{n}\right) dt,$$

where $\varphi(t)$ is the characteristic function of the distribution (4.3.10). The density of the limit distribution at a point $u > 0$ can be represented by the integral

$$\frac{u^{\rho-1}e^{-u}}{\Gamma(\rho)} = \frac{1}{2\pi} \int_{-\infty}^{\infty} \frac{1}{(1 - it)^\rho} e^{-itu} du.$$

Hence,

$$2\pi n P\{\zeta_N = k\} - 2\pi y^{\rho-1}e^{-y}/\Gamma(\rho) = I_1 + I_2 + I_3,$$

where

$$I_1 = \int_{-A}^{A} e^{-ity} \left(\varphi^N \left(\frac{t}{n} \right) - \frac{1}{(1-it)^\rho} \right) dt,$$

$$I_2 = -\int_{A \le |t|} e^{-ity} \frac{1}{(1-it)^\rho} dt,$$

$$I_3 = \int_{A \le |t| \le \pi n} e^{-ity} \varphi^N \left(\frac{t}{n} \right) dt,$$

and the constant A in the integrals is to be chosen later.

By (4.3.16), for any fixed A, the integral I_1 tends to zero as $n \to \infty$ and $N = \rho \log n + o(\log n)$.

To estimate the integrals I_2 and I_3, we integrate by parts. For I_2, this yields

$$\int_A^\infty \frac{e^{-ity}}{(1-it)^\rho} dt = -\frac{e^{-ity}}{iy(1-it)^\rho} \Big|_A^\infty + \frac{\rho}{y} \int_A^\infty \frac{e^{-ity}}{(1-it)^{\rho+1}} dt.$$

Hence,

$$|I_2| \le \frac{2}{y(1+A^2)^{\rho/2}} + \frac{2}{y} \int_A^\infty \frac{dt}{(1+t^2)^{(\rho+1)/2}},$$

and $|I_2|$ can be made arbitrarily small by the choice of A.

Similarly,

$$\int_A^{\pi n} e^{-ity} \varphi^N \left(\frac{t}{n} \right) dt = \frac{e^{-ity}}{iy} \varphi^N \left(\frac{t}{n} \right) \Big|_a^{\pi n} + I,$$

where

$$I = \frac{N}{iy} \int_A^{\pi n} e^{-ity} \varphi^{N-1} \left(\frac{t}{n} \right) \frac{1}{n} \varphi' \left(\frac{t}{n} \right) dt.$$

Therefore

$$|I_3| \le \frac{2}{y} |\varphi(\pi)|^N + \frac{2}{y} \left| \varphi \left(\frac{A}{n} \right) \right|^N + |I|.$$

When we use the estimates of Lemmas 4.3.2 and 4.3.6, we obtain

$$|\varphi(\pi)|^N \le q^N, \quad q < 1; \qquad |\varphi(A/n)|^N \le c_5 A^{-1/(12r^2\rho)}.$$

Hence these summands can be made arbitrarily small. It remains to estimate the integral I. Choose ε such that Lemma 4.3.3 is valid, and represent I as the sum of three integrals:

$$I = I_1(\varepsilon) + I_2(\varepsilon) + I_3(\varepsilon),$$

where

$$I_1(\varepsilon) = \frac{N}{iy} \int_{A \leq |t| \leq \varepsilon n} e^{-ity} \varphi^{N-1}\left(\frac{t}{n}\right) \frac{1}{n} \varphi'\left(\frac{t}{n}\right) dt,$$

$I_2(\varepsilon)$ is the integral over the sum of ε-neighborhoods of the poles of $F(z)$, that is, over the sum that equals

$$\bigcup_{l=1}^{m-1}\left[-\varepsilon n + \frac{2\pi l n}{m}, \varepsilon n + \frac{2\pi l n}{m}\right],$$

and $I_3(\varepsilon)$ is the integral over the remaining set

$$A_\varepsilon = [-\pi n, -\varepsilon n] \cup [\varepsilon n, \pi n] \setminus \bigcup_{l=1}^{m-1}\left[-\varepsilon n + \frac{2\pi l n}{m}, \varepsilon n + \frac{2\pi l n}{m}\right].$$

By using Lemmas 4.3.3 and 4.3.6, we find

$$|I_1(\varepsilon)| \leq \frac{2 N c_1 c_6}{y \log n} \int_A^\infty \frac{1}{(1+t^2)^{1/2}} t^{-1/(12 r^2 \rho)} dt,$$

and for $y \geq y_0 > 0$, the value $|I_1(\varepsilon)|$ can be made arbitrarily small by the choice of a sufficiently large A.

By using Lemma 4.3.3, we find

$$\frac{1}{n}\int_{-\varepsilon n + 2\pi l n/m}^{\varepsilon n + 2\pi l n/m} \left|\varphi'\left(\frac{t}{n}\right)\right| dt \leq \frac{c_1}{\log n}\int_{-\varepsilon n + 2\pi l n/m}^{\varepsilon n + 2\pi l n/m} \frac{dt}{\sqrt{1 + (t - 2\pi l n/m)^2}}$$

$$= \frac{c_1}{\log n}\int_{-\varepsilon n}^{\varepsilon n} \frac{dt}{\sqrt{1 + t^2}},$$

and there exists a constant c_7 such that for a fixed ε,

$$\int_{-\varepsilon n}^{\varepsilon n} \frac{dt}{\sqrt{1+t^2}} \leq c_7 \log n.$$

Therefore, we use the estimate of Lemma 4.3.2 and find that for $y \geq y_0 > 0$,

$$|I_2(\varepsilon)| \leq \frac{m c_1 c_4 N q^{N-1}}{y \log n} \int_{-\varepsilon n}^{\varepsilon n} \frac{dt}{\sqrt{1+t^2}} \leq m c_1 c_4 c_7 y_0^{-1} N q^{N-1},$$

and under the conditions of Theorem 4.3.2, the right-hand side tends to zero.

For $t \in A_\varepsilon$,

$$|\varphi'(t/n)| \leq c_8 / B(x),$$

where c_8 is a constant that is the upper bound of $|F(z)|$ for $|z| = x$ not in the neighborhoods of the poles. By using this estimate and the estimate of Lemma 4.3.2, we find

$$|I_3(\varepsilon)| \le \frac{N}{y} \int_{A_\varepsilon} \left|\varphi^{N-1}\left(\frac{t}{n}\right)\right| \frac{1}{n} \left|\varphi'\left(\frac{t}{n}\right)\right| dt$$

$$\le \frac{c_8 N}{yn B(x)} q^{N-1} \int_{A_\varepsilon} dt$$

$$\le \frac{c_8 N}{yB(x)} q^{N-1}.$$

Under the conditions of the theorem, the last term of this chain of inequalities tends to zero for $y \ge y_0 > 0$. ∎

It is easy to see that by first choosing a sufficiently large A and then a sufficiently large n, we can make the difference being estimated arbitrarily small. Note that the difference is bounded uniformly with respect to N, and hence, there exists a constant c_9 such that for $y \ge y_0 > 0$ and for all N,

$$\mathbf{P}\{\zeta_N = k\} \le c_9/n. \tag{4.3.26}$$

Proof of Theorem 4.3.3. In (4.3.8), divide the domain of summation into two parts: $N_1 = \{N: |N - B(x)| \le N^{2/3}\}$ and $N_2 = \{N: |N - B(x)| > N^{2/3}\}$.

It is not difficult to see that the assertion of Theorem 4.3.2 is fulfilled uniformly in $N \in N_1$. Therefore

$$n\mathbf{P}\{\zeta_N = n\} = e^{-1}/\Gamma(\rho)(1 + o(1))$$

uniformly in $N \in N_1$, so

$$\sum_{N \in N_1} \frac{(B(x))^N}{N!} e^{-B(x)} \mathbf{P}\{\zeta_N = n\}$$

$$= \frac{e^{-1}}{n\Gamma(\rho)} \sum_{N \in N_1} \frac{(B(x))^N}{N!} e^{-B(x)}(1 + o(1)) = \frac{e^{-1}}{n\Gamma(\rho)}(1 + o(1)).$$

We use the estimate (4.3.26) and obtain

$$\sum_{N \in N_2} \frac{(B(x))^N}{N!} e^{-B(x)} \mathbf{P}\{\zeta_N = n\} \le \frac{c_9}{n} \sum_{N \in N_2} \frac{(B(x))^N}{N!} e^{-B(x)}.$$

Since the sum on the right-hand side of this inequality tends to zero, the total sum in (4.3.4) equals $(en\Gamma(\rho))^{-1}(1 + o(1))$. It remains to note that

$$x^n = e^{-1}(1 + o(1)), \qquad B(x) = B_{n,R} = \sum_{k \in R} \frac{1}{k}\left(1 - \frac{1}{n}\right)^k.$$

∎

Proof of Theorem 4.3.4. According to (4.3.3),

$$P\{\nu_{n,R} = N\} = \frac{n!(B(x))^N}{N!x^n a_{n,R}} P\{\zeta_N = n\}.$$

If we substitute the corresponding expressions for $a_{n,R}$ and $P\{\zeta_N = n\}$, we obtain

$$P\{\nu_{n,R} = N\} = \frac{(B(x))^N}{N!} e^{-B(x)}(1 + o(1))$$

for $N = B(x) + o(B(x))$. We note that $B(x) = B_{n,R}$ and that the expression obtained above holds uniformly in N such that $(N - B(x))/\sqrt{B(x)}$ lies in any fixed finite interval; thus, we obtain the assertion of Theorem 4.3.4. ∎

4.4. Notes and references

The probabilistic approach that is now commonly used in combinatorics was first formulated in an explicit form and applied in the investigations of the symmetric group S_n by V. L. Goncharov [51, 52, 53]. For the random variables $\alpha_1, \ldots, \alpha_n$, he found the joint distribution (4.1.4) and the generating function (4.1.5). For the total number of cycles $\nu_n = \alpha_1 + \cdots + \alpha_n$, he proved that, as $n \to \infty$,

$$E\nu_n = \log n + \gamma + o(1),$$

$$\sqrt{D\nu_n} = \sqrt{\log n} - (\pi^2/2 - \gamma/2)/\sqrt{\log n} + o(\sqrt{\log n}).$$

Goncharov also proved that the distribution of $(\nu_n - \log n)/\sqrt{\log n}$ converges to the standard normal distribution, and the distribution of α_r converges to the Poisson distribution with parameter $1/r$.

Let β_{ν_n} be the length of the maximum cycle in a random permutation from S_n. Goncharov [51, 53] showed that

$$P\{\beta_{\nu_n} < m\} = \sum_{h=0}^{\infty} \frac{(-1)^h}{h!} S_h(m, n),$$

where

$$S_0(m, n) = 1, \quad S_h(m, n) = \sum_{\substack{k_1 + \cdots + k_h \leq n, \\ k_1, \ldots, k_h \geq m}} \frac{1}{k_1 \cdots k_h}, \quad h \geq 1.$$

Let

$$I_0(x, 1 - x) = 1, \quad I_h(x, 1 - x) = \int\limits_{\substack{x_1 + \cdots + x_h < 1 - x, \\ x_1, \ldots, x_h > x}} \frac{dx_1 \cdots dx_h}{x_1 \cdots x_h}, \quad 0 < x < 1.$$

Goncharov proved that, as $n \to \infty$, the random variable β_{ν_n}/n has the distribution with the density

$$\phi(x) = \frac{1}{x} \sum_{h=0}^{\lambda-1} \frac{(-1)^h}{h!} I(x, 1-x), \quad \frac{1}{1+\lambda} \le x \le \frac{1}{\lambda},$$

which, as is clear from the preceding formula, is defined by different analytic expressions on the sequential intervals of the form $[1/(1+\lambda), 1/\lambda]$, where λ is an integer. For example,

$$\phi(x) = \frac{1}{x}, \quad \frac{1}{2} \le x \le 1;$$

$$\phi(x) = \frac{1}{x}\left(1 - \log\frac{1-x}{x}\right), \quad \frac{1}{3} \le x \le \frac{1}{2}.$$

Although Goncharov investigated the cycle structure of random permutations in great detail, these problems continue to be of significant interest to mathematicians. V. F. Kolchin [71] proposed an approach based on the generalized scheme of allocation. The results on the asymptotic properties of random permutations obtained with the help of this approach are presented in [78]. Note that, among the others, the asymptotic logarithmic normality of the middle terms of the series of order statistics composed of the lengths of cycles, and the local limit theorem on the convergence of the distribution of the total number of cycles ν_n to the normal distribution were first proved by this method. It is clear that this approach makes it possible to investigate the asymptotic behavior of the local probabilities $P\{\nu_n = N\}$ for all possible values of $N = N(n)$ as $n \to \infty$. These investigations were carried out in [109, 115, 117, 146, 147, 148]. In Section 4.2, the results of these investigations are presented. Theorems 4.2.1 and 4.2.2 were proved by Yu. Pavlov in [115, 117]; and Theorems 4.2.5, 4.2.6, 4.2.9, 4.2.10, and 4.2.12 were proved by L. M. Volynets in [146, 147, 148].

Methods of estimating the rate of convergence in limit theorems for sums of independent random variables are well developed in the theory of probability. Therefore the approach that reduces the study of characteristics of random permutations to problems concerning the sums of independent summands provides an obvious way to obtain the limit theorems containing estimates of the rate of convergence. The estimates under the conditions of Theorem 4.2.1 were obtained by Yu. Pavlov [117] and for $\gamma = 1$ by A. Pavlov [109]. The following result of Volynets [146] provides a better bound than the one given in [109].

Theorem 4.4.1. *If $n \to \infty$, $N = \log n + x\sqrt{\log n}$, $x/\sqrt{\log n} \to 0$, then*

$$P\{\nu_n = N\} = \frac{(\log n)^N}{N!n}\left(1 + O\left(\frac{|x|}{\sqrt{\log n}} + \frac{1}{\log n}\right)\right).$$

Volynets [146] proved this theorem by using the approach based on the general-ized scheme of allocation.

Let Σ_n be the set of all single-valued mappings of the set $\{1, \ldots, n\}$ into itself. In particular, $S_n \subset \Sigma_n$. The random mappings from Σ_n were first studied by J. B. Kruskal [94] and B. Harris [57], and many studies have considered subsets of Σ_n, which are distinguished from Σ_n by various constraints on the mappings. We mention only the articles by V. N. Sachkov [128, 129, 130], in which the mappings have the height of less than a fixed number, and cycle lengths are from a fixed set; the articles by A. A. Grusho [54, 55], which treat the subset $\Sigma_{n,r}$ that consists of the mappings from Σ_n whose vertex degrees are not greater than r; the articles by Yu. Pavlov [114, 115] considering the characteristics of the mappings with exactly m components (the case $m = 1$ is considered by G. N. Bagaev in [8, 9]); and the article by J. Arney and E. A. Bender [5], which treats mappings with constraints on degrees of the vertices. The research in these directions began in the early seventies and is still ongoing. In our opinion, the most surprising results concerning mappings with constraints were obtained by I. B. Kalugin [64], which we summarize.

Let $\Sigma_{n,R}$ be the subset of mappings from Σ_n such that the degrees of the vertices take values only from a set R that contains zero and does not coincide with the set $\{0, 1\}$.

Let $\xi(\lambda)$ be a random variable with the distribution

$$\mathbf{P}\{\xi(\lambda) = k\} = \frac{\lambda^k e^{-\lambda}}{k! \, P(R, \lambda)}, \quad k \in R,$$

where λ is a positive constant and

$$P(R, \lambda) = \sum_{k \in R} \frac{\lambda^k e^{-\lambda}}{k!}.$$

There exists α_R such that $\mathbf{E}\xi(\alpha_R) = 1$. Denote by B_R the variance of the random variable $\xi(\alpha_R)$. For the number of cyclic vertices $\lambda_R^{(n)}$ and the height $\tau_{n,R}$ of the random mapping from $\Sigma_{n,R}$, the following assertions are well known [64, 78].

Theorem 4.4.2. *If $n \to \infty$, then*

$$\sqrt{n/B_R} \, \mathbf{P}\{\lambda_R^{(n)} = k\} = z e^{-z^2/2}(1 + o(1))$$

uniformly in the integers k such that $z = k\sqrt{B_R/n}$ lies in any interval of the form $0 < z_0 \leq z \leq z_1 < \infty$.

Theorem 4.4.3. *If $n \to \infty$, then for any fixed $x > 0$,*

$$\mathbf{P}\{\sqrt{B_R/n} \, \tau_{n,R} \leq x\} \to \sum_{k=-\infty}^{\infty} (-1)^k e^{-k^2 x^2/2}.$$

An unexpected result appears if we consider the set $\Sigma^*_{n,R}$ of mappings from Σ_n defined as follows. If in the graph of a mapping from Σ_n we delete the edges that connect the cyclic vertices, we obtain a graph consisting of trees. The set $\Sigma^*_{n,R}$ contains the mappings from Σ_n such that the degree of any vertex of the trees takes a value in R. Thus the difference in the restrictions on the degrees in $\Sigma^*_{n,R}$ and $\Sigma_{n,R}$ seems to be insignificant because only the restrictions on the degrees of cyclic vertices differ by 1. But the sets $\Sigma_{n,R}$ and $\Sigma^*_{n,R}$ have a substantial difference in the structure of their corresponding random graphs.

Let $\Lambda_R^{(n)}$ and $\tau^*_{n,R}$, respectively, be the number of cyclic vertices and the height of a random mapping from the set $\Sigma^*_{n,R}$ with uniform distribution. For the random variable $\xi(\lambda)$, set

$$a_R = \mathbf{E}\xi(1), \qquad b_R^2 = \mathbf{D}\xi(1).$$

If R does not coincide with the set of all nonnegative integers, then $a_R < 1$.

Theorem 4.4.4. *If $n \to \infty$, then*

$$\mathbf{P}\{\Lambda_R^{(n)} = k\} = \frac{1}{b_R\sqrt{2\pi n}}e^{-z^2/2}(1 + o(1))$$

uniformly in the integers k such that $z = (k - (1 - a_R)n)/(b_R\sqrt{n})$ lies in any fixed finite interval.

Theorem 4.4.5. *If $n \to \infty$ and $t = t(n)$ is such that $na_R^t \to \beta$, where β is a constant, then for any fixed integer m,*

$$\mathbf{P}\{\tau^*_{n,R} \le t + m\} = \exp\{-k_\beta \beta a_R^m\} + o(1),$$

where the constant k_β depends only on β and the set R.

Since $t = t(n)$ is of order $\log n$, the random mappings from $\Sigma^*_{n,R}$ have many cyclic vertices and, as a consequence, have the height of order $\log n$ rather than $\sqrt{n}$ as in the case for the mappings from $\Sigma_{n,R}$. A satisfactory explanation for this situation is not known.

In Section 4.3, we considered the set $S_{n,R}$ of all permutations of degree n with cycle lengths from a fixed set R. The interest in such sets may be partly explained by their connection with the equations involving permutations, which we will look at in the next section. Another reason for investigating the set $S_{n,R}$ and similar sets of mappings with various restrictions is the possibility (see [5]) of approximating more complicated sets of combinatorial objects by such sets with relatively simple constraints. Partly for these reasons, the asymptotic behavior of the number $a_{n,R}$ of elements in $S_{n,R}$ has been considered in some recent studies [25, 80, 102, 149, 153, 154].

The generating function $f(z)$ for the numbers $a_{n,R}$ of elements in $S_{n,R}$ is

$$f(z) = \sum_{n-0}^{\infty} \frac{a_{n,R} z^n}{n!} = \exp\left\{\sum_{r \in R} \frac{z^r}{r}\right\}.$$

Therefore it is convenient to apply the saddle-point method to obtain the asymptotics of $a_{n,R}$. By this method, the cases in which the elements of R form an arbitrary arithmetic progression are considered in [25, 107]; see also [130].

The application of the Tauberian-type theorems is another approach that has been used in the investigations of this problem [153, 154, 155]. Let $R(n)$ be the number of elements of R that are not greater than n and let $|A|$ be the number of elements in A.

Theorem 4.4.6.   *Let* $n \to \infty$,

$$R(n)/n \to \rho, \quad 0 < \rho \leq 1, \tag{4.4.1}$$

and for $m \geq n$, $m = O(n)$,

$$\frac{1}{n} |k: k \leq n, \ k \in R, \ m - k \in R| \to \rho^2. \tag{4.4.2}$$

Then

$$a_{n,R} = (n-1)! \exp\{l_{n,R} - \gamma\rho\} / \Gamma(\rho)(1 + o(1)), \tag{4.4.3}$$

where

$$l_{n,R} = \sum_{r \in R, r \leq n} \frac{1}{r},$$

γ *is the Euler constant, and* Γ *is the Euler gamma function.*

Conditions (4.4.1) and (4.4.2) indicate that the set R is similar to a typical realization of a random set containing each positive integer with probability ρ independent of the other integers.

As examples of the sets R that satisfy conditions (4.4.1) and (4.4.2), we may take sets of the form

$$R = \{k: \{g(k)\} \in \Delta\}, \tag{4.4.4}$$

where $g(t)$ is a real-valued function of $t \geq 0$, $\{x\}$ is the fractional part of x, and Δ is an interval or a finite union of intervals from $[0, 1]$ with the Lebesgue measure ρ.

A. L. Yakymiv [154, 155] proved that a set R of the form (4.4.4) satisfies conditions (4.4.1) and (4.4.2) if

$$g(t) = t^{\alpha} l(t),$$

where α is a noninteger positive number, $l(t)$ is a slowly varying function, and as $t \to \infty$,

$$\frac{d^n}{dt^n}l(t) = o\left(t^{-n}l(t)\right), \quad n = 1, \ldots, [\alpha] + 2.$$

Let $\alpha_{r,R}$ be the number of cycles of length r in a random permutation of $S_{n,R}$ and let $\nu_{n,R} = \alpha_{1,R} + \cdots + \alpha_{n,R}$ be its total number of cycles. Yakymiv [154, 155] proved the following assertions.

Theorem 4.4.7. *Suppose that conditions* (4.4.1) *and* (4.4.2) *are satisfied and* $n \to \infty$. *Then the distribution of the random variable* $(\nu_{n,R} - l_{n,R})/\sqrt{\rho \log n}$ *converges weakly to the standard normal distribution, and for any fixed* $r \in R$, *the distribution of* $\alpha_{r,R}$ *converges to the Poisson distribution with parameter* $1/r$.

A case of irregular behavior of $a_{n,R}$ is considered in [149].

Theorem 4.4.8. *If* $n \to \infty$ *and* $R = E \cup M$, *where* E *is the set of all even positive numbers and* M *is a set of odd numbers such that the series*

$$b = \sum_{m \in M} \frac{1}{m}$$

converges, then

$$a_{n,R} = \left(\frac{n}{e}\right)^n \left(e^b + e^{-b}\right)(1 + o(1))$$

for even n, *and*

$$a_{n,R} = \left(\frac{n}{e}\right)^n \left(e^b - e^{-b}\right)(1 + o(1))$$

for odd n.

Volynets [149] proved this theorem with the aid of relation (4.3.4), in which she uses the representation

$$\mathbf{P}\{\xi_1^{(R)} + \cdots + \xi_N^{(R)} = n\}$$
$$= \sum_{s,m} \mathbf{P}\{\nu = m, \eta = s\}\mathbf{P}\{\xi_1^{(E)} + \cdots + \xi_{N-m}^{(E)} = n - s\}.$$

Here the variables $\xi_1^{(R)}, \ldots, \xi_N^{(R)}$ have the parameter x equal to $\sqrt{1 - 1/n}$, ν is the number of these variables taking values in M, η is the sum of these variables, and $\xi_1^{(E)}, \ldots, \xi_N^{(E)}$ are independent identically distributed random variables with the distribution

$$\mathbf{P}\{\xi_1^{(E)} = k\} = \frac{1(1 - 1/n)^{k/2}}{k \log n}, \quad k \in E.$$

Note that if $b \to 0$, the result of Theorem 4.4.8 transfers continuously to (4.3.5).

Theorems 4.3.2, 4.3.3, and 4.3.4 are given in [80]. It can be easily shown that the asymptotics (4.3.11) and (4.4.3) are identical. Thus, quite different sets of conditions yield coinciding results. This coincidence shows that there exist weaker conditions sufficient for the validity of the asymptotics (4.3.11). We give the detailed and cumbersome proof of Theorem 4.3.3 because we conjecture that condition (1) from this theorem and the existence of a positive density of R are sufficient for the validity of (4.3.11) and that it may be possible to simplify the proof.

The research on the sets $S_{n,R}$ of permutations with restrictions on the cycle lengths provides an example of a fruitful competition of various analytical methods of asymptotic analysis such as the saddle-point method, the application of Tauberian-type theorems, and the approach based on the generalized scheme of allocation.

Note that it would also be interesting to consider the cases where the density $\rho = 0$.

5

Equations containing an unknown permutation

5.1. A quadratic equation

If g and f are permutations of degree n, then the result of their sequential action $h = fg$ is a permutation of degree n called the product of g and f. The set S_n of all permutations of degree n with this operation is the well-known symmetric group of degree n. Therefore we can consider equations of the form

$$X^d = a, \qquad\qquad (5.1.1)$$

where d is a positive integer, $a \in S_n$, and X is an unknown permutation from S_n. In the previous chapter, we considered the set $S_{n,R}$ of all permutations of degree n with cycle lengths from a fixed set R and found the asymptotics for the number of elements in $S_{n,R}$ for some regular sets R. The interest in the sets of permutations $S_{n,R}$ may be partly explained by their connection with some equations involving permutations. For example, the set of all solutions of the equation

$$X^p = e \qquad\qquad (5.1.2)$$

in the symmetric group S_n, where e is the identity permutation and p is a prime number, is exactly the set $S_{n,R}$ with $R = \{1, p\}$. Indeed, a permutation X satisfies equation (5.1.2) if and only if its cycles are of the length 1 or p. Denote by $T_n^{(p)}$ the number of solutions of equation (5.1.2).

Theorem 5.1.1. *If p is a prime number, then*

$$T_n^{(p)} = \sum_{0 \le k \le n/p} \frac{1}{(n - pk)!\, k!\, p^k}.$$

Proof. Let σ be a random permutation from S_n. It is clear that

$$\mathsf{P}\{\sigma^p = e\} = T_n^{(p)}/n!,$$

and the study of $T_n^{(p)}$ is equivalent to the study of the probability $\mathbf{P}\{\sigma^p = e\}$. Since $T_n^{(p)} = a_{n,R}$, where $R = \{1, p\}$,

$$\{\sigma^p = e\} = \{\alpha_r = 0, \ r \neq 1, \ r \neq p\} = \{\alpha_1 + p\alpha_p = n\},$$

where α_r is the number of cycles of length r in a random permutation from S_n. By (4.1.4),

$$\mathbf{P}\{\alpha_1 = n - pk, \ \alpha_p = k, \ \alpha_r = 0, \ r \neq 1, \ r \neq p\} = \frac{1}{(n - pk)! \, k! \, p^k}.$$

Summing these probabilities over admissible values of k yields the assertion of the theorem. ∎

Set $a_{0,R} = 1$ and consider the generating function of the sequence $a_{n,R}$,

$$f_R(z) = \sum_{k=0}^{\infty} \frac{a_{n,R} z^n}{n!}.$$

Theorem 5.1.2.

$$f_R(z) = \exp\left\{ \sum_{r \in R} \frac{z^r}{r} \right\}.$$

Proof. According to (4.1.5),

$$\varphi(u, t_1, t_2, \ldots) = \sum_{n=0}^{\infty} \varphi_n(t_1, \ldots, t_n) u^n$$

$$= \exp\left\{ \sum_{n=1}^{\infty} \frac{u^n t_n}{n} \right\}, \tag{5.1.3}$$

where

$$\varphi_n(t_1, \ldots, t_n) = \sum_{m_1, \ldots, m_n} \mathbf{P}\{\alpha_1 = m_1, \ldots, \alpha_n = m_n\} t_1^{m_1} \ldots t_n^{m_n},$$

and α_r is the number of cycles of length r in a random permutation from S_n.

If we put $t_r = 1$ for $r \in R$ and $t_r = 0$ for $r \notin R$, we find that the corresponding generating function $\varphi_n(t_1, \ldots, t_n)$ is

$$\sum_{M_R} \mathbf{P}\{\alpha_r = m_r, \ r \in R, \ \alpha_r = 0, \ r \notin R\},$$

where

$$M_R = \left\{ m_1, \ldots, m_n : \sum_{r \in R} r m_r = n \right\}.$$

It is easy to see that

$$\sum_{M_R} P\{\alpha_r = m_r, \ r \in R, \ \alpha_r = 0, \ r \notin R\} = P\left\{\sum_{r \in R} r\alpha_r = n\right\}.$$

Thus, substituting $t_r = 1$ if $r \in R$ and $t_r = 0$ if $r \notin R$ into (5.1.3) shows that the generating function for

$$P\left\{\sum_{r \in R} r\alpha_r = n\right\} = \frac{a_{n,R}}{n!}$$

equals

$$\sum_{n=0}^{\infty} \frac{a_{n,R} u^n}{n!} = \exp\left\{\sum_{r \in R} \frac{u^r}{r}\right\}. \qquad (5.1.4)$$

∎

In view of Theorem 5.1.2, it is convenient to apply the saddle-point method to obtain asymptotics of $T_n^{(p)}$. In the next section, we will use a different approach based on the generalized scheme of allocation; however, for comparison, we now present the derivation of the asymptotics of $T_n^{(2)}$ by applying the saddle-point method.

Theorem 5.1.3. *As $n \to \infty$,*

$$T_n^{(2)} = \frac{1}{e^{1/4}\sqrt{2}} \left(\frac{n}{e}\right)^{n/2} e^{\sqrt{n}} \left(1 + O\left(n^{-1/4}\right)\right).$$

Proof. Since

$$1 + \sum_{n=1}^{\infty} \frac{T_n^{(2)} z^n}{n!} = e^{z + z^2/2},$$

by Cauchy's formula

$$F(n) = \frac{T_n^{(2)}}{n!} = \frac{1}{2\pi i} \int \frac{e^{z + z^2/2}}{z^{n+1}} dz,$$

integrating over an arbitrary contour that goes around the point $z = 0$. We can write

$$F(n) = \frac{1}{2\pi i} \int e^{z + z^2/2 - n \log z} \frac{dz}{z}$$

and choose the contour of integration to be the circle passing through the saddle point ϱ, where the derivative of the function

$$f(z) = z + \frac{z^2}{2} - n \log z$$

is zero. From the equation

$$f'(z) = 1 + z - \frac{n}{z} = 0,$$

we find that

$$\varrho = \sqrt{n + \tfrac{1}{4}} - \tfrac{1}{2}.$$

Thus, setting $z = \varrho e^{i\varphi}$, $\pi \le \varphi \le \pi$ shows that

$$F(n) = \frac{1}{2\pi i} \int e^{f(z)} \frac{dz}{z} = \frac{1}{2\pi} \int_{-\pi}^{\pi} e^{\varrho e^{i\varphi} + \varrho^2 e^{2i\varphi}/2 - n \log \varrho - in\varphi} \, d\varphi$$

$$= \frac{1}{2\pi \varrho^n} e^{\varrho + \varrho^2/2} \int_{-\pi}^{\pi} e^{\varrho(\cos \varphi - 1) + (\cos 2\varphi - 1)/2 + i\varrho \sin \varphi + (\varrho^2 \sin 2\varphi)/2 - in\varphi} \, d\varphi.$$

For the sake of brevity, we let $\alpha = \varrho \sin \varphi + (\varrho^2 \sin 2\varphi)/2 - n\varphi$ and write the integral in the form

$$F(n) = \frac{e^{\varrho + \varrho^2/2}}{2\pi \varrho^n} \int_{-\pi}^{\pi} \cos \alpha \, e^{\varrho(1 - \cos \varphi) - \varrho^2(1 - \cos 2\varphi)/2} d\varphi$$

$$+ \frac{i e^{\varrho + \varrho^2/2}}{2\pi \varrho^n} \int_{-\pi}^{\pi} \sin \alpha \, e^{\varrho(1 - \cos \varphi) - \varrho^2(1 - \cos 2\varphi)/2} d\varphi.$$

Since $F(n)$ is real, we see that

$$F(n) = \frac{e^{\varrho + \varrho^2/2}}{2\pi \varrho^n} \int_{-\pi}^{\pi} \cos \alpha \, e^{\varrho(1 - \cos \varphi) - \varrho^2(1 - \cos 2\varphi)/2} d\varphi.$$

We choose $\varepsilon = \varrho^{-3/4}$ and estimate the integral outside the ε-neighborhood of zero, as $n \to \infty$, taking into account that $\varrho = \sqrt{n + 1/4} - 1/2 \to \infty$. The integrand is even, so we only estimate the integral over φ, $0 \le \varphi \le \pi$. It is convenient to consider the graphs of the functions $\cos \varphi$ and $\cos 2\varphi$ included in the exponent. With the help of the graphs presented in Figure 5.1.1, we can easily see that

$$\left| \int_{\varepsilon}^{\pi/2} \cos \alpha \, e^{-\varrho(1 - \cos \varphi) - \varrho^2(1 - \cos 2\varphi)/2} d\varphi \right|$$

$$\le \int_{\varepsilon}^{\pi/2} e^{-\varrho^2(1 - \cos 2\varphi)/2} d\varphi \le \int_{\varepsilon}^{\pi/2} e^{-\varepsilon^2 \varrho^2/2} d\varphi \le \frac{\pi}{2} e^{-\varepsilon^2 \varrho^2/2} = \frac{\pi}{2} e^{-\sqrt{\varrho}/2},$$

since $1 - \cos 2\varepsilon \ge \varepsilon^2$ for sufficiently small ε.

Similarly,

$$\left| \int_{\pi}^{} {}^{/2\pi} \cos \alpha \, e^{-\varrho(1 - \cos \varphi) - \varrho^2(1 - \cos 2\varphi)/2} d\varphi \right| \le \int_{\pi/2}^{\pi} e^{-\varrho(1 - \cos \varphi)} d\varphi$$

$$\le \int_{\pi/2}^{\pi} e^{-\varrho} d\varphi = \frac{\pi}{2} e^{-\varrho}.$$

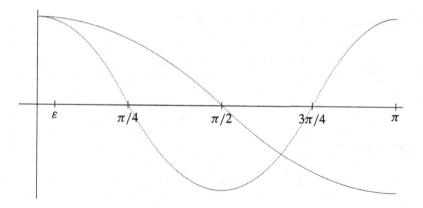

Figure 5.1.1. Graphs of $\cos\varphi$ and $\cos 2\varphi$

Thus

$$F(n) = \frac{e^{\varrho+\varrho^2/2}}{2\pi\varrho^n}\left(\int_{-\varepsilon}^{\varepsilon}\cos\alpha e^{-\varrho(1-\cos\varphi)-\varrho^2(1-\cos 2\varphi)/2}d\varphi + O(e^{-\sqrt{\varrho}/2})\right),$$

where $\varepsilon = \varrho^{-3/4}$. Since $\varrho + \varrho^2 - n = 0$, we find that, in a neighborhood of zero,

$$\alpha = \varrho\sin\varphi + \frac{\varrho^2}{2}\sin 2\varphi - n\varphi$$

$$= \varrho\varphi + \varrho^2\varphi - n\varphi + O(\varrho^2|\varphi|^3) = O(\varrho^2|\varphi|^3),$$

and therefore

$$\cos\alpha = 1 + O(\alpha^2) = 1 + O(\varrho^4\varphi^6).$$

The exponent of the integrand can be represented in the domain of integration as follows:

$$\varrho(1-\cos\varphi) + \frac{\varrho^2}{2}(1-\cos 2\varphi) = \frac{1}{2}(\varrho + 2\varrho^2)\varphi^2 + O(\varrho^2\varphi^4).$$

Thus, for $|\varphi| \le \varepsilon$,

$$\cos\alpha e^{\varrho(1-\cos\varphi)-\varrho^2(1-\cos 2\varphi)/2} = e^{-\varphi^2(\varrho+2\varrho^2)/2}(1 + O(\varrho^2\varphi^4 + \varrho^4\varphi^6))$$

$$= e^{-\varphi^2(\varrho+2\varrho^2)/2}(1 + O(\varrho^2\varepsilon^4 + \varrho^4\varepsilon^6))$$

$$= e^{-\varphi^2(\varrho+2\varrho^2)/2}(1 + O(\varrho^{-1/2})).$$

Therefore

$$\int_{-\varepsilon}^{\varepsilon}\cos\alpha e^{-\varrho(1-\cos\varphi)-\varrho^2(1-\cos 2\varphi)/2}d\varphi = \int_{-\varepsilon}^{\varepsilon}e^{-\varphi^2(\varrho+2\varrho^2)/2}d\varphi(1 + O(\varrho^{-1/2})).$$

The change of variables $\theta = \sqrt{\varrho + 2\varrho^2}\varphi$ gives

$$\frac{1}{\sqrt{2\pi}}\int_{-\varepsilon}^{\varepsilon}e^{-\varphi^2(\varrho+2\varrho^2)/2}d\varphi = \frac{1}{\sqrt{2\pi(\varrho+2\varrho^2)}}\int_{-\varepsilon\sqrt{\varrho+2\varrho^2}}^{\varepsilon\sqrt{\varrho+2\varrho^2}}e^{-\theta^2/2}d\theta$$

$$= \frac{1}{\sqrt{\varrho+2\varrho^2}}(1 + O(e^{-\sqrt{\varrho}})),$$

since as $x \to \infty$,

$$\int_x^\infty e^{-u^2/2}du = \frac{1}{x}e^{-x^2/2}(1 + o(1)).$$

Combining the estimates gives

$$F(n) = \frac{e^{\varrho+\varrho^2/2}}{\sqrt{2\pi}\varrho^n}\left(\frac{1}{\sqrt{\varrho+2\varrho^2}}(1 + O(\varrho^{-1/2})) + O(e^{-\sqrt{\varrho}/2})\right)$$

$$= \frac{e^{\varrho+\varrho^2/2}}{\sqrt{2\pi}\varrho^n\sqrt{\varrho+2\varrho^2}}(1 + O(\varrho^{-1/2})).$$

It remains to substitute $\varrho = \sqrt{n+1/4} - 1/2$ into this formula.
Since $F(n) = T_n^{(2)}/n!$, we find that

$$\log T_n^{(2)} = \log n! + \varrho + \frac{\varrho^2}{2} - n\log\varrho - \frac{1}{2}\log(\varrho + 2\varrho^2) - \log\sqrt{2\pi} + O(\varrho^{-1/2}).$$

$$(5.1.5)$$

Replace $\log n!$ by Stirling's formula

$$\log n! = n\log n - n + \frac{1}{2}\log n + \log\sqrt{2\pi} + O(n^{-1}). \qquad (5.1.6)$$

It is easily seen that

$$\varrho = \sqrt{n+1/4} - \frac{1}{2} = \sqrt{n}\left(1 + \frac{1}{4n}\right)^{1/2} - \frac{1}{2}$$

$$= \sqrt{n}\left(1 + \frac{1}{8n} + O\left(\frac{1}{n^2}\right)\right) - \frac{1}{2}$$

$$= \sqrt{n} - \frac{1}{2} + \frac{1}{8\sqrt{n}} + O\left(\frac{1}{n\sqrt{n}}\right)$$

$$= \sqrt{n}\left(1 - \frac{1}{2\sqrt{n}} + \frac{1}{8n} + O\left(\frac{1}{n^2}\right)\right), \qquad (5.1.7)$$

$$\varrho^2 = n - \sqrt{n} + \frac{1}{2} + O\left(\frac{1}{\sqrt{n}}\right). \qquad (5.1.8)$$

When we use (5.1.7), we find

$$n \log \varrho = \frac{1}{2} n \log n + n \log \left(1 - \frac{1}{2\sqrt{n}} + \frac{1}{8n} + O\left(\frac{1}{n^2}\right) \right)$$

$$= \frac{1}{2} n \log n - \frac{1}{2}\sqrt{n} + O\left(\frac{1}{\sqrt{n}}\right). \qquad (5.1.9)$$

Finally,

$$\log(\varrho + 2\varrho^2) = \log 2 + 2\log \varrho + \log\left(1 + \frac{1}{2\varrho} \right)$$

$$= \log n + \log 2 + O\left(\frac{1}{\sqrt{n}}\right). \qquad (5.1.10)$$

By substituting estimates (5.1.6)–(5.1.10) into (5.1.5), we obtain the final formula for $\log T_n^{(2)}$:

$$\log T_n^{(2)} = \frac{n}{2} \log n - \frac{n}{2} + \sqrt{n} - \frac{1}{4} - \log\sqrt{2} + O\left(n^{-1/4}\right),$$

which implies the assertion of the theorem. ∎

5.2. Equations of prime degree

According to (4.3.4), the number $a_{n,R}$ of permutations in $S_{n,R}$ can be represented in the form

$$a_{n,R} = \frac{n!}{x^n} e^{B_R(x)} \sum_{N=1}^{\infty} \frac{(B_R(x))^N}{N!} e^{-B_R(x)} \mathbf{P}\{\xi_1^{(R)} + \cdots + \xi_N^{(R)} = n\}, \qquad (5.2.1)$$

where

$$B_R(x) = \sum_{k \in R} \frac{x^k}{k}, \qquad (5.2.2)$$

and $\xi_1^{(R)}, \ldots, \xi_N^{(R)}$ are independent identically distributed random variables,

$$\mathbf{P}\{\xi_1^{(R)} = k\} = \frac{x^k}{k B_R(x)}, \qquad k \in R, \qquad (5.2.3)$$

and the positive parameter x can be chosen arbitrarily from the domain of convergence of the series in (5.2.2).

If p is a prime number, then the number $T_n^{(p)}$ of solutions of equation (5.1.2) is $a_{n,R}$, where $R = \{1, p\}$. Therefore

$$B_R(x) = x + \frac{x^p}{p},$$

and by (5.2.1),

$$T_n^{(p)} = \frac{n!}{x^n} e^{x+x^p/p} \sum_{N=1}^{\infty} \frac{(x+x^p/p)^N}{N!} e^{-x-x^p/p} \, \mathbf{P}\{\zeta_N = n\}, \qquad (5.2.4)$$

where $\zeta_N = \xi_1 + \cdots + \xi_N$, $\xi_1, \ldots, \xi_N$ are independent identically distributed random variables and

$$\mathbf{P}\{\xi_1 = 1\} = \frac{px}{px+x^p}, \qquad \mathbf{P}\{\xi_1 = p\} = \frac{x^p}{px+x^p}. \qquad (5.2.5)$$

Thus, to find the asymptotics of $T_n^{(p)}$, it suffices to choose an appropriate value of x and to prove a local limit theorem for the sum $\zeta_N = \xi_1 + \cdots + \xi_N$. The summation of independent random variables taking two values is a simple problem that is solved by the de Moivre–Laplace theorem. Therefore the approach based on the representation (5.2.4) seems more suitable here than the saddle-point method.

We begin by applying this approach to the proof of Theorem 5.1.3.

Proof of Theorem 5.1.3. If $R = \{1, 2\}$, then obviously

$$\mathbf{P}\{\xi_1 = 1\} = \frac{x}{B(x)} = \frac{2}{2+x}, \qquad \mathbf{P}\{\xi_1 = 2\} = \frac{x^2}{2B(x)} = \frac{x}{2+x},$$

where $B(x) = B_R(x) = x + x^2/2$, and $\mathbf{E}\zeta_N = N\mathbf{E}\xi_1 = N(x+x^2)/B(x)$.

In the main part of the sum in (5.2.4), the parameter N takes values close to $B(x)$; therefore we choose x such that

$$x + x^2 = n.$$

Hence,

$$x = \sqrt{n+1/4} - \frac{1}{2}, \qquad B(x) = x + \frac{x^2}{2} = \frac{n}{2} + \frac{1}{2}\sqrt{n+1/4} - \frac{1}{4},$$

$$\mathbf{E}\xi_1 = \frac{x+x^2}{B(x)} = \frac{n}{B(x)}, \qquad \mathbf{D}\xi_1 = \frac{x^3}{2B^2(x)},$$

and $\mathbf{D}\xi_1 = 2n^{-1/2}(1 + o(1))$ as $n \to \infty$ (where $\mathbf{D}$ denotes the variance).

Let

$$u = \frac{2(N - B(x))}{\sqrt{B(x)\mathbf{D}\xi_1}}, \qquad A = \sqrt{2\log n},$$

and divide the sum from (5.2.4) into two parts so that

$$T_n^{(2)} = \frac{n! \, e^{B(x)}}{x^n} (S_1 + S_2), \qquad (5.2.6)$$

where

$$S_1 = \sum_{N:|u|\leq A} \frac{B^N(x)}{N!} e^{-B(x)} \mathbf{P}\{\zeta_N = n\},$$

$$S_2 = \sum_{N:|u|> A} \frac{B^N(x)}{N!} e^{-B(x)} \mathbf{P}\{\zeta_N = n\}.$$

In the first sum,

$$\frac{|N - B(x)|}{\sqrt{B(x)}} \leq \frac{A\sqrt{\mathbf{D}\xi_1}}{2} = \frac{\sqrt{\log n}}{n^{1/4}}(1 + o(1)),$$

and by using the normal approximation to the Poisson distribution, we obtain, as $n \to \infty$,

$$\frac{B^N(x)}{N!} e^{-B(x)} = \frac{1}{\sqrt{2\pi B(x)}}(1 + o(1))$$

uniformly in the integers N such that $|u| \leq A$.

The sum $\zeta_N - N$ has the binomial distribution with N trials and the probability of success $p(x) = x/(2 + x)$. If $|u| \leq A$, then $N = B(x)(1 + o(1))$, and

$$Np(x)(1 - p(x)) = \frac{2xN}{(2 + x)^2} = \sqrt{n}(1 + o(1))$$

as $n \to \infty$. Therefore the normal approximation to the binomial distribution is valid. For $|u| \leq A = \sqrt{2 \log n}$,

$$\frac{n - N\mathbf{E}\xi_1}{\sqrt{N\mathbf{D}\xi_1}} = \frac{n(B(x) - N)}{B(x)\sqrt{N\mathbf{D}\xi_1}} = -u(1 + O(n^{-1/2})).$$

Therefore, by the de Moivre–Laplace theorem,

$$\mathbf{P}\{\zeta_N = n\} = \frac{1}{\sqrt{2\pi N\mathbf{D}\xi_1}} e^{-u^2/2}(1 + o(1))$$

uniformly in the integers N such that $|u| \leq A$.

The behavior of the functions $\varphi_1(N) = B^N(x)e^{-B(x)}/N!$ and $\varphi_2(N) = \mathbf{P}\{\zeta_N = n\}$ is represented approximately in Figure 5.2.1.

The sum S_1 can be estimated as follows:

$$S_1 = \sum_{N:|u|\leq A} \frac{B^N(x)}{N!} e^{-B(x)} \mathbf{P}\{\zeta_N = n\}$$

$$= \sum_{N:|u|\leq A} \frac{1}{\sqrt{2\pi B(x)}\sqrt{2\pi N\mathbf{D}\xi_1}} e^{-u^2/2}(1 + o(1))$$

$$= \frac{1}{2\sqrt{2\pi B(x)}\sqrt{2\pi}} \sum_{N:|u|\leq A} \frac{2}{\sqrt{B(x)\mathbf{D}\xi_1}} e^{-u^2/2}(1 + o(1)).$$

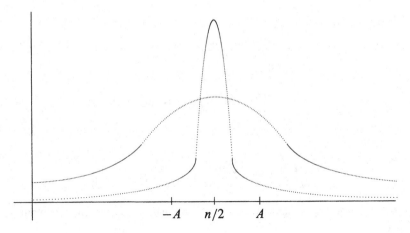

Figure 5.2.1. The graphs of $\varphi_1(N)$ and $\varphi_2(N)$

The last sum is an integral sum of the function $e^{-u^2/2}$ with step $2(B(x)\mathbf{D}\xi_1)^{-1/2}$, so as $n \to \infty$,

$$S_1 = \frac{1}{2\sqrt{2\pi B(x)}} \frac{1}{\sqrt{2\pi}} \int_{-\infty}^{\infty} e^{-u^2/2} du(1 + o(1)) = \frac{1}{2\sqrt{2\pi B(x)}}(1 + o(1)).$$

By virtue of monotonicity, for $|u| > A$,

$$\mathbf{P}\{\zeta_N = n\} \leq \frac{1}{\sqrt{2\pi B(x)\mathbf{D}\xi_1}} e^{-A^2/2}(1 + o(1)),$$

and there exists a constant c such that

$$\mathbf{P}\{\zeta_N = n\} \leq cn^{-1/4}e^{-A^2/2} \leq cn^{-5/4}.$$

Therefore

$$S_2 = \sum_{N:|u|>A} \frac{B^N(x)}{N!} e^{-B(x)} \mathbf{P}\{\zeta_N = n\} \leq cn^{-5/4}.$$

Thus

$$S = S_1 + S_2 = S_1(1 + o(1)) = \frac{1}{2\sqrt{2\pi B(x)}}(1 + o(1)),$$

and by substituting this estimate into (5.2.6), we obtain

$$T_n^{(2)} = \frac{n!\,e^{B(x)}}{2x^n \sqrt{2\pi B(x)}}(1 + o(1)).$$

It remains to substitute

$$x = \sqrt{n + 1/4} - \frac{1}{2}, \qquad B(x) = \frac{n}{2} + \frac{1}{2}\sqrt{n + 1/4} - \frac{1}{4}$$

into the formula. It is easily seen that

$$e^{B(x)} = e^{n/2+\sqrt{n}/2-1/4}(1+o(1)),$$

$$x^n = n^{n/2}e^{-\sqrt{n}/2}(1+o(1)).$$

Therefore

$$T_n^{(2)} = \frac{n^n\sqrt{n}e^{-n}\sqrt{2\pi}e^{n/2+\sqrt{n}/2-1/4}}{n^{n/2}e^{-\sqrt{n}/2}2\sqrt{2\pi}\sqrt{n/2}}$$

$$= e^{-1/4}2^{-1/2}\left(\frac{n}{e}\right)^{n/2}e^{\sqrt{n}}(1+o(1)),$$

and Theorem 5.1.3 with the remainder term of the form $1+o(1)$ is proved. ∎

We now turn to the case where p is a fixed prime number, $p \geq 3$, and consider the number $T_n^{(p)}$ of solutions of equation (5.1.2).

Theorem 5.2.1. *If* $n \to \infty$ *and* p *is prime,* $p \geq 3$, *then*

$$T_n^{(p)} = \left(\frac{n}{e}\right)^{n(1-1/p)}p^{-1/2}e^{n^{1/p}}(1+o(1)).$$

Proof. The proof is almost the same as the proof of Theorem 5.1.3 given above and is also based on relation (5.2.4). For $R = \{1, p\}$,

$$B(x) = B_R(x) = x + x^p/p,$$

and the independent random variables $\xi_1, \ldots, \xi_N$ in (5.2.4) have the distribution

$$\mathbf{P}\{\xi_1 = 1\} = \frac{x}{B(x)} = \frac{px}{px + x^p}, \qquad \mathbf{P}\{\xi_1 = p\} = \frac{px^p}{B(x)} = \frac{x^p}{px + x^p}.$$

We choose the parameter x such that

$$x + x^p = n. \tag{5.2.7}$$

Then

$$x = n^{1/p} - \frac{1}{p}n^{-1+2/p} + O(n^{-2+2/p}),$$

$$B(x) = x + x^p/p = \frac{n}{p} + \frac{p-1}{p}n^{1/p} + O(n^{-1+2/p}),$$

$$p(x) = \frac{x^p}{px + x^p} = 1 - pn^{-1+1/p} + O(n^{-2+2/p}),$$

$$\mathbf{E}\xi_1 = n/B(x), \qquad \mathbf{D}\xi_1 = (p-1)^2pn^{-1+1/p}(1+o(1)).$$

Let

$$u = \frac{p(N - B(x))}{\sqrt{B(x)\mathbf{D}\xi_1}}, \qquad A = \sqrt{2\log n},$$

and divide the sum in (5.2.4) into two parts so that

$$T_n^{(p)} = \frac{n!\,e^{B(x)}}{x^n}(S_1 + S_2),$$

where

$$S_1 = \sum_{N:|u|\leq A} \frac{B^N(x)}{N!} e^{-B(x)} \mathbf{P}\{\zeta_N = n\},$$

$$S_2 = \sum_{N:|u|> A} \frac{B^N(x)}{N!} e^{-B(x)} \mathbf{P}\{\zeta_N = n\}.$$

In the first sum, $N = B(x)(1 + o(\sqrt{B(x)}))$ and

$$\frac{B^N(x)}{N!} e^{-B(x)} = \frac{1}{\sqrt{2\pi B(x)}}(1 + o(1))$$

uniformly in the integers N such that $|u| \leq A$.

Let $\xi_i^* = (\xi_i - 1)/(p - 1)$, $i = 1, \ldots, N$. The sum

$$\zeta_N^* = \xi_1^* + \cdots + \xi_N^*$$

has the binomial distribution with N trials and the probability of success

$$p(x) = x^p/(px + x^p) = 1 - pn^{-1+1/p} + O(n^{-2+2/p})$$

as $n \to \infty$. It is clear that

$$\mathbf{P}\{\zeta_N = n\} = \mathbf{P}\{\zeta_N^* = (n - N)/(p - 1)\},$$

and if $(n - N)/(p-1)$ is not an integer, then $\mathbf{P}\{\zeta_N = n\} = 0$. Since $\mathbf{E}\xi_1 = n/B(x)$, $B(x) = n/p(1 + o(1))$, and

$$\frac{(n - N)/(p - 1) - N\mathbf{E}\xi_1^*}{\sqrt{N\mathbf{D}\xi_1^*}} = \frac{n - N\mathbf{E}\xi_1}{\sqrt{N\mathbf{D}\xi_1}} = \frac{n(B(x) - N)}{B(x)\sqrt{N\mathbf{D}\xi_1}}$$

$$= -\frac{nu}{p\sqrt{B(x)N}} = -u(1 + o(1))$$

as $n \to \infty$ and $|u| \leq A$, by using the de Moivre–Laplace theorem, we obtain

$$\mathbf{P}\{\zeta_N = n\} = \mathbf{P}\{\zeta_N^* = (n - N)/(p - 1)\} = \frac{1}{\sqrt{2\pi N\mathbf{D}\xi_1^*}} e^{-u^2/2}(1 + o(1))$$

uniformly in the integers N such that $(n - N)/(p - 1)$ is an integer and $|u| \leq A$.

Therefore

$$
S_1 = \sum_{N:|u|\le A} \frac{B^N(x)}{N!} e^{-B(x)} \mathbf{P}\{\zeta_N = n\}
$$

$$
= \sum_{N:|u|\le A} \frac{p-1}{\sqrt{2\pi B(x)}\sqrt{2\pi N \mathbf{D}\xi_1}} e^{-u^2/2}(1+o(1))
$$

$$
= \frac{p-1}{p\sqrt{2\pi B(x)}} \frac{1}{\sqrt{2\pi}} \sum_{N:|u|\le A} \frac{p}{\sqrt{2\pi B(x)\mathbf{D}\xi_1}} e^{-u^2/2}(1+o(1)),
$$

where the summation is over the integers N such that $(n-N)/(p-1)$ is an integer. The last sum is an integral sum of the function $e^{-u^2/2}$ with step $p(B(x)\mathbf{D}\xi_1)^{-1/2}$. Since the summation is over N such that $(n-N)/(p-1)$ is an integer, that is, only each $(p-1)$th term is included in the sum, we obtain

$$
\frac{p-1}{\sqrt{2\pi}} \sum_{N:|u|\le A} \frac{p}{\sqrt{2\pi B(x)\mathbf{D}\xi_1}} e^{-u^2/2} \to \frac{1}{\sqrt{2\pi}} \int_{-\infty}^{\infty} e^{-u^2/2} du = 1.
$$

Therefore, as $n \to \infty$,

$$
S_1 = \frac{1}{p\sqrt{2\pi B(x)}}(1+o(1)).
$$

For $|u| > A$,

$$
\mathbf{P}\{\zeta_N = n\} \le \frac{p-1}{\sqrt{2\pi B(x)\mathbf{D}\xi_1}} e^{-A^2/2}(1+o(1)),
$$

and there exists a constant c such that

$$
\mathbf{P}\{\zeta_N = n\} \le cn^{-1-1/(2p)},
$$

and $S_2 \le cn^{-1-1/(2p)}$. Thus

$$
S = S_1 + S_2 = S_1(1+o(1)) = \frac{1}{p\sqrt{2\pi B(x)}}(1+o(1)),
$$

and by substituting this estimate into (5.2.6), we obtain

$$
T_n^{(p)} = \frac{n!\,e^{B(x)}}{px^n\sqrt{2\pi B(x)}}(1+o(1)). \tag{5.2.8}
$$

It is easily seen that

$$
e^{B(x)} = e^{n/p+(p-1)n^{1/p}/p}(1+o(1)),
$$

$$
x^n = n^{n/p}e^{-n^{1/p}/p}(1+o(1)).
$$

When we substitute these expressions into (5.2.8), we obtain the assertion of Theorem 5.2.1. ∎

A slight refinement of the estimates used in the proof of Theorem 5.2.1 allows us to show that the assertion of the theorem is valid if p tends to infinity slowly, as specified below, where we prove a more general result.

Theorem 5.2.2. *If p is prime and n, $p \to \infty$ in such a way that $p/n \to 0$, then*

$$T_n^{(p)} = \left(\frac{n}{e}\right)^{n(1-1/p)} p^{1/2} \sum_{k=0}^{\infty} \frac{(n^{1/p})^{m+kp}}{(m+kp)!}(1+o(1)); \qquad (5.2.9)$$

in particular, if $p^{-2}n^{1/p} \to \infty$, then

$$T_n^{(p)} = \left(\frac{n}{e}\right)^{n(1-1/p)} p^{-1/2} e^{n^{1/p}}(1+o(1)), \qquad (5.2.10)$$

and if $p^{-1}n^{1/p} \to 0$, then

$$T_n^{(p)} = \left(\frac{n}{e}\right)^{n(1-1/p)} p^{1/2} \frac{n^{m/p}}{m!}(1+o(1)), \qquad (5.2.11)$$

where $m = n - p[n/p]$, and $[c]$ is the integer part of c.

Proof. The proof is similar to the proof of Theorem 5.2.1, but now we need to trace the effect of the parameter p in the remainder terms of the asymptotic formulas and to use a representation in terms of the Poisson probabilities instead of the representation (5.2.4).

It follows from the equation $x + x^p = n$ that under the conditions of the theorem,

$$x = n^{1/p} - \frac{n^{2/p}}{np} + O\left(\frac{n^{3/p}}{n^2 p^2}\right), \qquad (5.2.12)$$

$$B = B(x) = \frac{n}{p} + \frac{(p-1)n^{1/p}}{p} + O\left(\frac{n^{2/p}}{np}\right). \qquad (5.2.13)$$

Therefore it is easy to confirm that

$$p(x) = \mathbf{P}\{\xi_1 = p\} = \frac{x^p}{px + x^p} = 1 - pn^{-1+1/p} + O\left(n^{-2+2/p}\right).$$

The random variable $(\zeta_N - N)/(p-1)$ can be represented in the form

$$\frac{\zeta_N - N}{p-1} = N - \eta_N,$$

where η_N has the binomial distribution with N trials and probability of success

$$q = q(x) = 1 - p(x) = pn^{-1+1/p}\left(1 + O\left(n^{-1+1/p}\right)\right). \qquad (5.2.14)$$

Therefore it is not difficult to see that for $n = m + p[n/p]$, the probability

$P\{\zeta_N = n\}$ is nonzero if

$$N = [n/p] + m + k(p-1), \quad 0 \leq k \leq [n/p],$$

and for such N,

$$P\{\zeta_N = n\} = P\{\eta_N = l\},$$

where $l = m + kp$. Thus, the representation (5.2.4) takes the form

$$T_n^{(p)} = \frac{n!}{x^n} \sum_{N=1}^{\infty} \frac{B^N}{N!} e^{-B} P\{\zeta_N = n\}$$

$$= \frac{n!}{x^n} \sum_{k=0}^{[n/p]} \frac{B^N}{N!} e^{-B} \binom{N}{l} q^l (1-q)^{N-l}.$$

This results in the representation

$$T_n^{(p)} = \frac{n!}{x^n} \sum_{k=0}^{[n/p]} \frac{(Bq)^l}{l!} e^{-Bq} \frac{(B(1-q))^{N-l}}{(N-l)!} e^{-B(1-q)}, \qquad (5.2.15)$$

where $l = m + pk$, $N = [n/p] + m + k(p-1)$, $m = n - p[n/p]$; and to obtain the basic assertion of the theorem, we must sum the products of two Poisson probabilities. Let

$$s = \sum_{k=0}^{[n/p]} \frac{(Bq)^{m+pk}}{(m+pk)!} e^{-Bq}, \qquad a = \left(n^{1/p} \sqrt{p/n}\right)^{1/3},$$

and divide s into two parts,

$$s_1 = \sum_{k:|(N-B)b^{-1/2}|\leq a} \frac{(Bq)^{m+pk}}{(m+pk)!} e^{-Bq},$$

$$s_2 = \sum_{k:|(N-B)b^{-1/2}|>a} \frac{(Bq)^{m+pk}}{(m+pk)!} e^{-Bq}.$$

Note that $a \to 0$ under the conditions of the theorem, and the normal approximation to the second multiplier

$$\frac{(B(1-q))^{N-l}}{(N-l)!} e^{-B(1-q)} = \frac{1}{\sqrt{2\pi B}} (1 + o(1)) \qquad (5.2.16)$$

is valid for all l, N such that $|(N - B)B^{-1/2}| \leq a$, and outside this region,

$$\frac{(B(1-q))^{N-l}}{(N-l)!} e^{-B(1-q)} \leq \frac{c}{\sqrt{2\pi B}}, \qquad (5.2.17)$$

where c is a constant.

It remains to show that $s_2 = o(s_1)$ and

$$s_1 = \sum_{k=0}^{\infty} \frac{(n^{1/p})^{m+pk}}{(m+pk)!} e^{-n^{1/p}} (1+o(1)).\tag{5.2.18}$$

For the sake of brevity, we let $b = Bq$. It follows from (5.2.13) and (5.2.14) that under the conditions of the theorem,

$$b = n^{1/p} \left(1 + O\left(pn^{-1+1/p}\right)\right).\tag{5.2.19}$$

It is clear that

$$s_1 \geq \frac{b^{[b]+p}}{([b]+p)!} e^{-b},$$

since at least one of the summands with l from the interval $([b], [b]+p)$ is included in the sum s_1.

On the other hand, the summation over $N \geq B + a\sqrt{B}$ is the summation over l, with $l = m + pk$ such that $l \geq b + a\sqrt{B} + o(\sqrt{B})$. Let $l_0 = b + a\sqrt{B} + o(\sqrt{B})$. Then,

$$s_2 \leq \frac{b^{l_0}}{l_0!} e^{-b} \left(1 + \frac{b}{l_0} + \frac{b^2}{l_0^2} + \cdots\right) \leq \frac{b^{l_0} e^{-b} l_0}{l_0!(l_0 - b)} \leq \frac{c b^{l_0}}{l_0!},$$

since $b/l_0 \to 0$. Therefore,

$$\frac{s_2}{s_1} \leq \frac{c b^{l_0 - [b] - p}}{l_0(l_0 - 1) \cdots ([b] + p + 1)}$$

$$\leq \frac{c}{(1 + (l_0 - b)/b) \cdots (1 + ([b] - b + p + 1)/b)}$$

$$\leq \frac{c_1 b^3}{(l_0 - b)^3} \leq \frac{c_2 b^3}{(a\sqrt{B})^3} \leq \frac{c_3 n^{3/p} p^{3/2}}{a^3 n^{3/2}},$$

where $c_1, c_2,$ and c_3 are constants. By the choice of a, the last bound tends to zero.

This estimate, (5.2.16), (5.2.17), and (5.2.19) imply (5.2.18). Assertion (5.2.9) follows from (5.2.15), (5.2.16), (5.2.17), and (5.2.18).

If $p^{-2} n^{1/p} \to \infty$, then by using the normal approximation, we obtain

$$\sum_{k=0}^{\infty} \frac{(n^{1/p})^{m+pk}}{(m+pk)!} e^{-n^{1/p}} = \frac{1}{p}(1+o(1)).$$

This yields assertion (5.2.10) of the theorem.

Assertion (5.2.11) follows from the fact that if $p^{-1} n^{1/p} \to 0$, then

$$\sum_{k=0}^{\infty} \frac{(n^{1/p})^{m+pk}}{(m+pk)!} = \frac{n^{m/p}}{m!}(1+o(1)).$$

5.3. Equations of compound degree

In this section, we consider the number $T_n^{(d)}$ of solutions of the equation

$$X^d = e, \tag{5.3.1}$$

where d is a natural number, e is the identity permutation, and X is an unknown element of the symmetric group S_n. The cases where d is a prime number were considered in the previous sections. Let d be a compound number and let $1 = d_0 < d_1 < \cdots < d_r = d$ be all different divisors of d. A permutation X is a solution of equation (5.3.1) if and only if the lengths of cycles of X belong to the set $\{d_0, \ldots, d_r\}$. Therefore $T_n^{(d)}$ is equal to the number $a_{n,R}$ of permutations in $S_{n,R}$, where $R = \{d_0, \ldots, d_r\}$. The following is a generalization of Theorems 5.1.3 and 5.2.1.

Theorem 5.3.1. *If $n \to \infty$ and d is a fixed number, $d \geq 2$, then*

$$T_n^{(d)} = \left(\frac{n}{e}\right)^n n^{-n/d} \frac{1}{\sqrt{d}} \exp\left\{\sum_{j|d} \frac{n^{j/d}}{j}\right\} (1 + o(1))$$

if d is odd, and

$$T_n^{(d)} = \left(\frac{n}{e}\right)^n n^{-n/d} \frac{1}{\sqrt{d}} \exp\left\{\sum_{j|d} \frac{n^{j/d}}{j} - \frac{1}{2d}\right\} (1 + o(1))$$

if d is even.

Note that the summation in the above formulas is over the divisors j of the number d, and if we put $d = 2$ and $d = p$, we obtain Theorem 5.1.3 and 5.2.1, respectively.

Proof. Let $1 = d_0 < d_1 < \cdots < d_r = d$ be all the divisors of d, $R = \{d_0, \ldots, d_r\}$,

$$B(x) = B_R(x) = \sum_{k \in R} \frac{x^k}{k},$$

and let $\xi_1, \ldots, \xi_N$ be independent identically distributed random variables,

$$\mathbf{P}\{\xi_1 = k\} = \frac{x^k}{kB(x)}, \quad k \in R, \tag{5.3.2}$$

where the positive parameter x can be chosen arbitrarily. Since d is compound, $r > 2$.

Put $\zeta_N = \xi_1 + \cdots + \xi_N$. It is clear that

$$\mathbf{E}\zeta_N = N\mathbf{E}\xi_1 = (x + x^{d_1} + \cdots + x^{d_{r-1}} + x^d)/B(x).$$

We choose the parameter x such that

$$x + x^{d_1} + \cdots + x^{d_{r-1}} + x^d = n, \qquad (5.3.3)$$

and in what follows, we consider the random variables $\xi_1, \ldots, \xi_N$ with distribution (5.3.2), where x is the solution of this equation.

By iteration, it is not difficult to determine that

$$x^d = n - n^{d_{r-1}/d} - \cdots - n^{1/d} + o(1) \qquad (5.3.4)$$

if d is odd, and

$$x^d = n - n^{d_{r-1}/d} - \cdots - n^{1/d} + 1/2 + o(1) \qquad (5.3.5)$$

if d is even.

Since $T_n^{(d)} = a_{n,R}$, where $R = \{1, d_1, \ldots, d_{r-1}, d\}$, we can use the representation (5.2.1) and obtain

$$T_n^{(d)} = \frac{n!}{x^n} e^{B(x)} \sum_{N=1}^{\infty} \frac{B^N(x)}{N!} e^{-B(x)} \mathbf{P}\{\zeta_N = n\}. \qquad (5.3.6)$$

Therefore, to obtain the assertions of Theorem 5.3.1, it is sufficient to find the asymptotics of $\mathbf{P}\{\zeta_N = n\}$.

It is not difficult to see that

$$\mathbf{E}\xi_1 = \frac{n}{B(x)},$$

$$\mathbf{D}\xi_1 = \frac{B(x)\left(x + d_1 x^{d_1} + \cdots + d x^d\right) - n^2}{B^2(x)},$$

$$B(x) = x + \frac{x^{d_1}}{d_1} + \cdots + \frac{x^d}{d} = \sum_{j|d} \frac{x^j}{j},$$

where the summation is over the integers j, which are the divisors of d. In view of (5.3.4) and (5.3.5),

$$B(x) = \sum_{j|d} \frac{n^{j/d}}{j}(1 + o(1)), \qquad (5.3.7)$$

as $n \to \infty$. By estimating the second and third central moments of ξ_1 and using the characteristic function of ζ_N, we can prove that the distribution of the random variable $(\zeta_N - N\mathbf{E}\xi_1)/\sqrt{N\mathbf{D}\xi_1}$ converges to the normal law with parameters $(0, 1)$ as $N\mathbf{D}\xi_1 \to \infty$. If h is the maximal step of the lattice containing the set R, then the local limit theorem is valid on this lattice. We omit the proof of this local theorem.

The remaining part of the proof of Theorem 5.3.1 repeats the corresponding part of the proof of Theorem 5.1.3 from Section 5.2. We put

$$v = \frac{n - N\mathbf{E}\xi_1}{\sqrt{N\mathbf{D}\xi_1}}, \qquad u = \frac{d(N - B(x))}{\sqrt{B(x)\mathbf{D}\xi_1}}, \qquad A = 2\sqrt{2\log n},$$

and divide the sum from (5.3.6) into two parts so that

$$T_n^{(s)} = \frac{n!}{x^n} e^{B(x)}(S_1 + S_2),$$

where

$$S_1 = \sum_{N:|u|\leq A} \frac{B^N(x)}{N!} e^{-B(x)} \mathbf{P}\{\zeta_N = n\},$$

$$S_2 = \sum_{N:|u|> A} \frac{B^N(x)}{N!} e^{-B(x)} \mathbf{P}\{\zeta_N = n\}.$$

It is easy to see that $N = B(x)(1 + o(1))$ for $|u| \leq A = 2\sqrt{2\log n}$ and

$$v = \frac{n(B(x) - N)}{B(x)\sqrt{N\mathbf{D}\xi_1}} = -u(1 + O(n^{-1/2})), \qquad (5.3.8)$$

and by the local limit theorem,

$$\mathbf{P}\{\zeta_N = n\} = \frac{h}{\sqrt{2\pi N\mathbf{D}\xi_1}} e^{-u^2/2}(1 + o(1))$$

uniformly in the integers N such that $|u| \leq A$ and $(n - N)/h$ are integers. Recall that h is the maximal span of the distribution of ξ_1.

As in the proof of Theorem 5.1.3, Section 5.2, we obtain

$$S_1 = \frac{1}{d\sqrt{2\pi B(x)}} \frac{1}{\sqrt{2\pi}} \sum_{N:|u|\leq A} \frac{dh}{\sqrt{B(x)\mathbf{D}\xi_1}} e^{-u^2/2}(1 + o(1)).$$

The last sum is an integral sum of the function $e^{-u^2/2}$, with step $d(B(x)\mathbf{D}\xi_1)^{-1/2}$, and the summation is over N such that $(n - N)/h$ are integers, that is, only each hth term is included in the sum. Since h and d are relatively prime, we see that

$$\frac{1}{\sqrt{2\pi}} \sum_{N:|u|\leq A} \frac{hd}{\sqrt{B(x)\mathbf{D}\xi_1}} e^{-u^2/2} \to \frac{1}{\sqrt{2\pi}} \int_{-\infty}^{\infty} e^{-u^2/2} du = 1,$$

and

$$S_1 = \frac{1}{d\sqrt{2\pi B(x)}}(1 + o(1)).$$

In estimating S_2, it will not be possible now to use the monotonicity of the tails of the function $\varphi_2(N) = \mathbf{P}\{\zeta_N = n\}$ as we did in the proof of Theorem 5.1.3 in

Section 5.2 (see Figure 5.2.1). By (5.3.8), in the first sum, $|v| \leq \sqrt{2 \log n}$ for a sufficiently large n. Therefore, in the second sum,

$$P\{\zeta_N = n\} \leq \sum_{n:|v|>\sqrt{2\log n}} P\{\zeta_N = n\}.$$

By the integral limit theorem,

$$\sum_{n:|v|>\sqrt{2\log n}} P\{\zeta_N = n\} = \frac{2}{\sqrt{2\pi}} \int_{\sqrt{2\log n}}^{\infty} e^{-z^2/2} dz (1 + o(1)),$$

and there exists a constant c such that, in the second sum,

$$P\{\zeta_N = n\} \leq cn^{-1}.$$

Thus, $S_1 + S_2 = S_1(1 + o(1))$, and we obtain

$$T_n^{(d)} = \frac{n! \, e^{B(x)}}{x^n d \sqrt{2\pi B(x)}} (1 + o(1)). \tag{5.3.9}$$

This implies the assertions of the theorem because

$$e^{B(x)} = \exp\left\{ \sum_{j|d} \frac{x^j}{j} \right\},$$

and x^n can be represented in the cases of odd and even d as follows.

Let d be odd, then according to (5.3.4),

$$x^n = n^{n/d} e^{-(n^{d_r-1/d} + \cdots + n^{1/d})/d} (1 + o(1)).$$

For $1 \leq j < d$,

$$x^j = n^{j/d} + o(1),$$

and for $j = d$,

$$x^d = n - n^{d_r-1/d} - \cdots - n^{1/d} = o(1).$$

Thus

$$e^{B(x)} = \exp\left\{ \sum_{j|d} \frac{n^{j/d}}{j} - \frac{1}{d}\left(n^{d_r-1/d} + \cdots + n^{1/d}\right) + o(1) \right\},$$

and

$$x^{-n} e^{B(x)} = n^{-n/d} \exp\left\{ \sum_{j|d} \frac{n^{j/d}}{j} \right\} (1 + o(1)).$$

When we substitute the last expression into (5.3.9), we obtain the first assertion of the theorem.

If d is even, we note that $2d_{r-1} = d$ and use (5.3.5) to obtain

$$x^n = n^{n/d} e^{-(n^{d_r-1/d} + \cdots + n^{1/d} - 1/2)/d - 1/(2d)} (1 + o(1)).$$

For $1 \le j < d_{r-1}$,

$$x^j = n^{j/d} + o(1);$$

for $j = d$,

$$x^d = n - n^{d_r-1/d} - \cdots - n^{1/d} + 1/2 + o(1);$$

and for $j = d_{r-1}$,

$$x^{d_{r-1}} = n^{d_r-1/d} - d_{r-1}/d + o(1).$$

Thus

$$e^{B(x)} = \exp\left\{\sum_{j|d} \frac{n^{j/d}}{j} - \left(n^{d_r-1/d} + \cdots + n^{1/d} - 1/2\right)/d + o(1)\right\},$$

$$x^{-n} e^{B(x)} = x^{-n/d} \exp\left\{\sum_{j|d} \frac{n^{j/d}}{j} - \frac{1}{2d}\right\} (1 + o(1)).$$

The substitution of the last expression into (5.3.9) gives us the second assertion of the theorem. ∎

5.4. Notes and references

The study of equations of the form $X^d = e$ in the symmetric group S_n is directly related to one of the significant characteristics of the elements of S_n: the order of permutations. By the order $O_n(s)$ of a permutation $s \in S_n$, we mean the least positive integer k such that s^k is the identity permutation. The orders of elements in S_n vary from 1 to the maximal value $G(n)$ over all $s \in S_n$. E. Landau [95] shows that

$$\lim_{n \to \infty} \frac{\log G(n)}{\sqrt{n \log n}} = 1.$$

In spite of such a wide range of $\log O_n(s)$, the typical values of $\log O_n(s)$ are considerably less than $\log G(n)$ and are concentrated near $2^{-1} \log^2 n$. Let O_n be the order of a random permutation from S_n with uniform distribution. The following assertion is well known.

Theorem 5.4.1. *For any fixed x,*

$$\lim_{n \to \infty} P\left\{(\log O_n - 2^{-1} \log^2 n)/\sqrt{3^{-1} \log^3 n}\right\} = \frac{1}{\sqrt{2\pi}} \int_{-\infty}^{x} e^{-u^2/2} du.$$

The asymptotic normality of $\log O_n$ was first proved by P. Erdős and P. Turan [39]. Other proofs of Theorem 5.4.1 can be found in [106, 18, 27]. All the proofs are rather cumbersome and involve many analytical difficulties. From our point of view, the simplest proof, but still not a sufficiently simple one, is suggested in [78], where the approach based on the generalized scheme is used.

It seems to us that investigating the numbers of solutions of equations of the form $X^d = e$ could provide the basis for the study of the local behavior of O_n. Indeed, if p is prime, then $T_n^{(p)}$ is just the number of permutations $s \in S_n$ whose order $O_n(s) = p$. Since the leading term of the asymptotics of the number $T_n^{(d)}$ for a compound d is $(n/e)^{n(1-1/d)}$, almost all permutations counted by $T_n^{(d)}$ probably have the order d. It would be of considerable interest to find the asymptotics of the local probabilities $\mathbf{P}\{O_n = d\}$ for d that lie in a neighborhood of $\exp\{2^{-1} \log^2 n\}$ and to see whether the integral limit theorem follows from these results in spite of the fact that the behavior of the probabilities $\mathbf{P}\{O_n = d\}$ is likely to be rather complicated. By virtue of the irregularity of the behavior of $\mathbf{P}\{O_n = d\}$, this problem is not usually as trivial as is obtaining the integral limit theorem from the local theorem because now we have to obtain the local theorem for d of a specified form and, in addition, we have to know how many d of such a form exist.

Theorems 5.1.1 and 5.1.2 for $R = \{1, 2\}$ and Theorem 5.1.3 were proved in [32]. Theorem 5.1.2 for $R = \{1, p\}, p \geq 2$, was proved in [61], and for an arbitrary R in [33].

Theorem 5.1.3 was proved in [103], where the result of Theorem 5.2.1 was also presented. Assertion (5.2.9) of Theorem 5.2.2 was proved by the saddle-point method in [144].

Theorem 5.3.1 was proved in [108, 145, 150] independently and almost simultaneously.

The approach based on the generalized scheme of allocation, presented in Chapter 5 of this book, was first published in [82], where the proof of Theorem 5.1.3 was realized with the help of this approach. The proof of Theorem 5.3.1 in Section 5.3 follows A. V. Kolchin [68], who, in addition, extended this theorem to the case $d \to \infty$ such that $d \ln \ln n / \ln n \to 0$.

The general conditions of existence of a solution of the equation $X^d = a$, where a is a fixed permutation and X is an unknown permutation from S_n, are given in [102].

The system of equations

$$X_1^{m_1} = X_2^{m_2} = \cdots = X_k^{m_k} = e,$$

where $k \geq 2, m_1, \ldots, m_k$ are fixed natural numbers, $X_1, \ldots, X_k \in S_n$, and e is the identity permutation in S_n, is considered in [110]. The asymptotic representation of the number of solutions $X = (X_1, \ldots, X_k)$ such that $X_i X_j = X_j X_i$ for all $i \neq j$ is found.

BIBLIOGRAPHY

[1] Sh. M. Agadzhanyan. On a general method of estimating the number of graphs from given classes. *Avtomatika*, (1):10–21, 1981. In Russian.

[2] Sh. M. Agadzhanyan. The asymptotic formulae for the number of *m*-component graphs. *Avtomatika*, (4):27–33, 1986. In Russian.

[3] D. J. Aldous. Exchangability and related topics. *Lecture Notes in Math.*, 1117:1–198, 1985.

[4] D. J. Aldous. Brownian bridge asymptotics for random mappings. *Adv. Appl. Probab.*, 24:763–764, 1992.

[5] J. Arney and E. A. Bender. Random mappings with constraints on coalescence. *Pacific J. Math.*, 103:269–294, 1982.

[6] R. A. Arratia. Independent process approximation for random combinatorial structures. *Adv. Appl. Probab.*, 24:764–765, 1992.

[7] R. Arratia and S. Tavaré. Limit theorems for combinatorial structures via discrete process approximations. *Random Structures and Algorithms*, 3:321–345, 1992.

[8] G. N. Bagaev. Distribution of the number of vertices in a component of an indecomposable mapping. *Belorussian Acad. Sci. Dokl.*, 21(12):1061–1063, 1977. In Russian.

[9] G. N. Bagaev. Limit distributions of metric characteristics of an indecomposable random mapping. In *Combinatorial and Asymptotic Analysis*, pp. 55–61. Krasnoyarsk Univ., Krasnoyarsk, 1977. In Russian.

[10] G. N. Bagaev and E. F. Dmitriev. Enumeration of connected labelled bipartite graphs. *Belorussian Acad. Sci. Dokl.*, 28:1061–1063, 1984. In Russian.

[11] G. V. Balakin. On random matrices. *Theory Probab. Appl.*, 12:346–353, 1967. In Russian.

[12] G. V. Balakin. The distribution of random matrices over a finite field. *Theory Probab. Appl.*, 13:631–641, 1968. In Russian.

241

[13] G. V. Balakin, V. I. Khokhlov, and V. F. Kolchin. Hypercycles in a random hypergraph. *Discrete Math. Appl.*, 2:563–570, 1992.

[14] A. D. Barbour. Refined approximations for the Ewens sampling formula. *Adv. Appl. Probab.*, 24:765, 1992.

[15] A. D. Barbour. Refined approximations for the Ewens sampling formula. *Random Structures and Algorithms*, 3:267–276, 1992.

[16] E. A. Bender, E. R. Canfield, and B. D. McKay. The asymptotic number of labeled connected graphs with a given number of vertices and edges. *Random Structures and Algorithms*, 1:127–170, 1990.

[17] E. A. Bender, E. R. Canfield, and B. D. McKay. Asymptotic properties of labeled connected graphs. *Random Structures and Algorithms*, 3:183–202, 1992.

[18] M. R. Best. The distribution of some variables on a symmetric group. *Nederl. Akad. Wetensch. Indag. Math. Proc.*, 73:385–402, 1970.

[19] L. Bieberbach. *Analytische Fortsetzung*. Springer-Verlag, Berlin, 1955.

[20] B. Bollobas. The evolution of random graphs. *Trans. Amer. Math. Soc.*, 286:257–274, 1984.

[21] B. Bollobas. *Random Graphs*. Academic Press, London, 1985.

[22] Yu. V. Bolotnikov. Convergence to the Gaussian and Poisson processes of the variable $\mu_r(n, n)$ in the classical occupancy problem. *Theory Probab. Appl.*, 13:39–50, 1968. In Russian.

[23] Yu. V. Bolotnikov. Convergence to the Gaussian process of the number of empty cells in the classical occupancy problem. *Math. Notes*, 4:97–103, 1968. In Russian.

[24] Yu. V. Bolotnikov. Limit processes in a non-equiprobable scheme of allocating particles into cells. *Theory Probab. Appl.*, 13:534–542, 1968. In Russian.

[25] Yu. V. Bolotnikov. On some classes of random variables on cycles of permutations. *Math. USSR Sb.*, 36:87–99, 1980.

[26] Yu. V. Bolotnikov, V. N. Sachkov, and V. E. Tarakanov. Asymptotic normality of some variables connected with the cyclic structure of random permutations. *Math. USSR Sb.*, 28:107–117, 1976.

[27] J. D. Bovey. An approximate probability distribution for the order of elements of the symmetric group. *Bull. London Math. Soc.*, 12:41–46, 1980.

[28] V. E. Britikov. Limit theorems on the maximum size of trees in a random forest of non-rooted trees. In *Probability Problems of Discrete Mathematics*, pp. 84–91. MIEM, Moscow, 1987. In Russian.

[29] V. E. Britikov. The asymptotic number of forests from unrooted trees. *Math. Notes*, 43:387–394, 1988.

[30] V. E. Britikov. The limit behaviour of the number of trees of a given size in a random forest of nonrooted trees. In *Stochastic Processes and Applications*, pp. 36–41. MIEM, Moscow, 1988. In Russian.

[31] I. A. Cheplyukova. Emergence of the giant tree in a random forst. *Discrete Math. Appl.*, 8(1):17–34, 1998.

[32] S. Chowla, I. N. Herstein, and K. Moore. On recursions connected with symmetric groups. *Canad. J. Math.*, 3:328–334, 1951.

[33] S. Chowla, I. N. Herstein, and W. R. Scott. The solution of $x^d = 1$ in symmetric groups. *Norske Vid. Selsk.*, 25:29–31, 1952.

[34] J. M. DeLaurentis and B. G. Pittel. Random permutations and Brownian motion. *Pacific J. Math.*, 119:287–301, 1985.

[35] P. J. Donnelly. Labellings, size-biased permutations and the gem distribution. *Adv. Appl. Probab.*, 24:766, 1992.

[36] P. J. Donnelly, W. J. Ewens, and S. Padmadisastra. Functionals of random mappings: Exact and asymptotic results. *Adv. Appl. Probab.*, 23:437–455, 1991.

[37] P. Erdős and A. Rényi. On the evolution of random graphs. *Publ. Math. Inst. Hungarian Acad. Sci., Ser. A*, 5(1–2):17–61, 1960.

[38] P. Erdős and A. Rényi. On random matrices. *Magyar Tud. Akad. Mat. Kutató Int. Közl.*, 8:455–461, 1963.

[39] P. Erdős and P. Turan. On some problems of statistical group theory. iii. *Acta Math. Acad. Hungar.*, 18(3–4):309–320, 1967.

[40] W. J. Ewens. The sampling theory of selectively neutral alleles. *Theoret. Pop. Biol.*, 3:87–112, 1972.

[41] W. J. Ewens. Sampling properties of random mappings. *Adv. Appl. Probab.*, 24:773, 1992.

[42] M. V. Fedoryuk. *Saddle Point Method*. Nauka, Moscow, 1977. In Russian.

[43] W. Feller. *An Introduction to Probability Theory and Its Applications*, vol. 2. Wiley, New York, 1966.

[44] P. Flajolet. The average height of binary trees and other simple trees. *Journal of Computer and System Sciences*, 25:171–213, 1982.

[45] P. Flajolet. Random tree models in the analysis of algorithms. In P.-J. Courtois and G. Latouche, editors, *Performance'87*, pp. 171–187. North-Holland, Amsterdam, 1988.

[46] P. Flajolet, D. E. Knuth, and B. Pittel. The first cycles in an evolving graph. *Discrete Math.*, 75:167–215, 1989.

[47] P. Flajolet and A. M. Odlyzko. Random mapping statistics. In J.-J. Quisquarter and J. Vandewalle, editors, *Advances in Cryptology, Lecture Notes in Computer Science, Vol. 434*, pp. 329–354. Springer-Verlag, Berlin, 1990.

[48] P. Flajolet and M. Soria. Gaussian limiting distributions for the number of components in combinatorial structures. *J. Combinatorial Theory, Series A*, 53:165–182, 1990.

[49] B. V. Gnedenko and A. N. Kolmogorov. *Limit Distributions for Sums of Independent Random Variables*. Addison-Wesley, Reading, MA, 1949.

[50] S. W. Golomb. *Shift Register Sequences*. Aegean Park Press, Laguna Hills, CA, 1982.

[51] V. L. Goncharov. On the distribution of cycles in permutations. *Soviet Math. Dokl.*, 35(9):299–301, 1942. In Russian.

[52] V. L. Goncharov. On the alternation of events in a sequence of Bernoulli trials. *Soviet Math. Dokl.*, 36(9):295–297, 1943. In Russian.

[53] V. L. Goncharov. On the field of combinatorics. *Soviet Math. Izv., Ser. Math.*, 8:3–48, 1944. In Russian.

[54] A. A. Grusho. Random mappings with bounded multiplicity. *Theory Probab. Appl.*, 17:416–425, 1972.

[55] A. A. Grusho. Distribution of the height of mappings of bounded multiplicity. In *Asymptotic and Enumerative Problems of Combinatorial Analysis*, pp. 7–18. Krasnoyarsk Univ., Krasnoyarsk, 1976. In Russian.

[56] J. C. Hansen. Order statistics for random combinatorial structures. *Adv. Appl. Probab.*, 24:774, 1992.

[57] B. Harris. Probability distributions related to random mappings. *Ann. Math. Statist.*, 31:1045–1062, 1960.

[58] C. C. Heyde. A contribution to the theory of large deviations for sums of independent random variables. *Z. Wahrscheinlichkeitstheorie und verw. Gebiete*, 7:303–308, 1967.

[59] W. Hoeffding. Probability inequalities for sums of bounded random variables. *J. Amer. Statist. Assoc.*, 58(301):13–30, 1963.

[60] I. A. Ibragimov and Yu. V. Linnik. *Independent and Stationary Related Variables*. Nauka, Moscow, 1965. In Russian.

[61] E. Jacobstal. Sur le nombre d'éléments du group symmetric S_n dont l'ordre est un nombre premier. *Norske Vid. Selsk.*, 21:49–51, 1949.

[62] S. Janson. Multicyclic components in a random graph process. *Random Structures and Algorithms*, 4:71–84, 1993.

[63] S. Janson, D. E. Knuth, T. Łuczak, and B. Pittel. The birth of the giant component. *Random Structures and Algorithms*, 4:233–358, 1993.

[64] I. B. Kalugin. The number of cyclic points and the height of a random mapping with constraints on multiplicities of the vertices. In *Abstracts of the All-Union Conference Probab. Methods in Discrete Math.*, pp. 35–36. Karelian Branch of the USSR Acad. Sci., Petrozavodsk, 1983. In Russian.

[65] V. I. Khokhlov. On the structure of a non-uniformly distributed random graph. *Adv. Appl. Probab.*, 24:775, 1992.

[66] V. I. Khokhlov and V. F. Kolchin. On the structure of a random graph with nonuniform distribution. In *New Trends in Probab. and Statist.*, pp. 445–456. VSP/Mokslas, Utrecht, 1991.

[67] J. F. C. Kingman. The population structure associated with the Ewens sampling formula. *Theoret. Pop. Biol.*, 11:274–284, 1977.

[68] A. V. Kolchin. Equations in unknown permutations. *Discrete Math. Appl.*, 4:59–71, 1994.

[69] V. F. Kolchin. A class of limit theorems for conditional distributions. *Litovsk. Mat. Sb.*, 8:53–63, 1968. In Russian.

[70] V. F. Kolchin. On the limiting behavior of extreme order statistics in a polynomial scheme. *Theory Probab. Appl.*, 14:458–469, 1969.

[71] V. F. Kolchin. A problem of allocating particles into cells and cycles of random permutations. *Theory Probab. Appl.*, 16:74–90, 1971.

[72] V. F. Kolchin. A problem of the allocation of particles in cells and random mappings. *Theory Probab. Appl.*, 21:48–63, 1976.

[73] V. F. Kolchin. Branching processes, random trees, and a generalized scheme of arrangements of particles. *Math. Notes*, 21:386–394, 1977.

[74] V. F. Kolchin. Moment of degeneration of a branching process. *Math. Notes*, 24:954–961, 1978.

[75] V. F. Kolchin. Branching processes and random trees. In *Cybernetics, Combinatorial Analysis and Graph Theory*, pp. 85–97. Nauka, Moscow, 1980. In Russian.

[76] V. F. Kolchin. *Asymptotic Methods of Probability Theory*. MIEM, Moscow, 1984. In Russian.

[77] V. F. Kolchin. On the behavior of a random graph near a critical point. *Theory Probab. Appl.*, 31:439–451, 1986.

[78] V. F. Kolchin. *Random Mappings*. Optimization Software, New York, 1986.

[79] V. F. Kolchin. *Systems of Random Equations*. MIEM, Moscow, 1988. In Russian.

[80] V. F. Kolchin. On the number of permutations with constraints on their cycle lengths. *Discrete Math. Appl.*, 1:179–194, 1991.

[81] V. F. Kolchin. Cycles in random graphs and hypergraphs. *Adv. Appl. Probab.*, 24:768, 1992.

[82] V. F. Kolchin. The number of permutations with cycle lengths from a fixed set. In *Random Graphs'89*, pp. 139–149. Wiley, New York, 1992.

[83] V. F. Kolchin. Consistency of a system of random congruences. *Discrete Math. Appl.*, 3:103–113, 1993.

[84] V. F. Kolchin. A classification problem in the presence of measurement errors. *Discrete Math. Appl.*, 4:19–30, 1994.

[85] V. F. Kolchin. Random graphs and systems of linear equations in finite fields. *Random Structures and Algorithms*, 5:135–146, 1994.

[86] V. F. Kolchin. Systems of random linear equations with small number of non-zero coefficients in finite fields. In *Probabilistic Methods in Discrete Mathematics*, pp. 295–304. VSP, Utrecht, 1997.

[87] V. F. Kolchin and V. I. Khokhlov. An allocation problem and moments of the binomial distribution. In *Probab. Problems of Discrete Math.*, pp. 16–21. MIEM, Moscow, 1987. In Russian.

[88] V. F. Kolchin and V. I. Khokhlov. On the number of cycles in a random non-equiprobable graph. *Discrete Math. Appl.*, 2:109–118, 1992.

[89] V. F. Kolchin and V. I. Khokhlov. A threshold effect for systems of random equations of a special form. *Discrete Math. Appl.*, 5:425–436, 1995.

[90] V. F. Kolchin, B. A. Sevastyanov, and V. P. Chistyakov. *Random Allocations*. Wiley, New York, 1978.

[91] I. N. Kovalenko. A limit theorem for determinants in the class of Boolean functions. *Soviet Math. Dokl.*, 161:517–519, 1965. In Russian.

[92] I. N. Kovalenko. On the limit distribution of the number of solutions of a random system of linear equations in the class of Boolean functions. *Theory Probab. Appl.*, 12:51–61, 1967. In Russian.

[93] I. N. Kovalenko, A. A. Levitskaya, and M. N. Savchuk. *Selected Problems of Probabilistic Combinatorics*. Naukova Dumka, Kiev, 1986. In Russian.

[94] J. B. Kruskal. The expected number of components under a random mapping function. *Amer. Math. Monthly*, 61:392–397, 1954.

[95] E. Landau. *Handbuch der Lehre von der Verteilung der Primzahlen*, vol. 1. Teubner, Berlin, 1909.

[96] A. A. Levitskaya. Theorems on invariance of the limit behaviour of the number of solutions of a system of random linear equations over a finite ring. *Cybernetics*, (2):140–141, 1978. In Russian.

[97] A. A. Levitskaya. Theorems on invariance for the systems of random linear equations over an arbitrary finite ring. *Soviet Math. Dokl.*, 263:289–291, 1982. In Russian.

[98] A. A. Levitskaya. The probability of consistency of a system of random linear equations over a finite ring. *Theory Probab. Appl.*, 30:339–350, 1985. In Russian.

[99] T. Łuczak. Component behaviour near the critical point of the random graph process. *Random Structures and Algorithms*, 1:287–310, 1990.

[100] T. Łuczak. Cycles in a random graph near the critical point. *Random Structures and Algorithms*, 2:421–439, 1991.

[101] T. Łuczak and B. Pittel. Components of random forests. *Comb. Probab. and Comput.*, 1:35–52, 1992.

[102] M. P. Mineev and A. I. Pavlov. On the number of permutations of a special form. *Math. USSR Sb.*, 99:468–476, 1976. In Russian.

[103] L. Moser and M. Wyman. On the solution of $x^d = 1$ in symmetric groups. *Canad. J. Math.*, 7:159–168, 1955.

[104] L. R. Mutafchiev. Local limit theorems for sums of power series distributed random variables and for the number of components in labelled relational structures. *Random Structures and Algorithms*, 3:403–426, 1992.

[105] E. Palmer. *Graphical Evolution*. Wiley, New York, 1985.

[106] A. I. Pavlov. On the limit distribution of the number of cycles and the logarithm of the order of a class of permutations. *Math. USSR Sb.*, 42:539–567, 1982.

[107] A. I. Pavlov. On the number of cycles and the cycle structure of permutations from some classes. *Math. USSR Sb.*, 46:536–556, 1984.

[108] A. I. Pavlov. On the permutations with cycle lengths from a fixed set. *Theory Probab. Appl.*, 31:618–619, 1986. In Russian.

[109] A. I. Pavlov. Local limit theorems for the number of components of random substitutions and mappings. *Theory Probab. Appl.*, 33:196–200, 1988. In Russian.

[110] A. I. Pavlov. The number and cycle structure of solutions of a system of equations in substitutions. *Discrete Math. Appl.*, 1:195–218, 1991.

[111] Yu. L. Pavlov. The asymptotic distribution of maximum tree size in a random forest. *Theory Probab. Appl.*, 22:509–520, 1977.

[112] Yu. L. Pavlov. Limit theorems for the number of trees of a given size in a random forest. *Math. USSR Sb.*, 32:335–345, 1977.

[113] Yu. L. Pavlov. A case of limit distribution of the maximum size of a tree in a random forest. *Math. Notes*, 25:387–392, 1979.

[114] Yu. L. Pavlov. Limit distributions of some characteristics of random mappings with a single cycle. In *Math. Problems of Modelling Complex Systems*, pp. 48–55. Karelian Branch of the USSR Acad. Sci., Petrozavodsk, 1979. In Russian.

[115] Yu. L. Pavlov. Limit theorems for a characteristic of a random mapping. *Theory Probab. Appl.*, 27:829–834, 1981.

[116] Yu. L. Pavlov. Limit distributions of the height of a random forest. *Theory Probab. Appl.*, 28:471–480, 1983.

[117] Yu. L. Pavlov. On the random mappings with constraints on the number of cycles. In *Proc. Steklov Inst. Math.*, pp. 131–142. Nauka, Moscow, 1986.

[118] Yu. L. Pavlov. Some properties of plane planted trees. In *Abstr. All-Union Conference on Discrete Math. and its Appl. to Modelling of Complex Systems*, p. 14. Irkutsk State Univ., Irkutsk, 1991. In Russian.

[119] Yu. L. Pavlov. Some properties of planar planted trees. *Discrete Math. Appl.*, 3:97–102, 1993.

[120] Yu. L. Pavlov. The limit distributions of the maximum size of a tree in a random forest. *Discrete Math. Appl.*, 5:301–316, 1995.

[121] Yu. L. Pavlov. Limit distributions of the number of trees of a given size in a random forest. *Discrete Math. Appl.*, 6:117–133, 1996.

[122] V. V. Petrov. *Sums of Independent Random Variables*. Springer-Verlag, New York, 1975.

[123] B. Pittel. On tree census and the giant component in sparse random graphs. *Random Structures and Algorithms*, 1:311–342, 1990.

[124] G. Pólya and G. Szegő. *Aufgaben und Lehrsätze aus der Analysis*. Springer-Verlag, Berlin, 1925.

[125] Yu. V. Prokhorov. Asymptotic behaviour of the binomial distribution. *Uspekhi Matem. Nauk*, 8(3):135–142, 1953. In Russian.

[126] J. Riordan. *Combinatorial Identities*. Wiley, New York, 1968.

[127] A. Ruciński and N. C. Wormald. Random graph processes with degree restrictions. *Combinatorics, Probability and Computing*, 1:169–180, 1992.

[128] V. N. Sachkov. Mappings of a finite set with restraints on contours and height. *Theory Probab. Appl.*, 17:640–656, 1972.

[129] V. N. Sachkov. Random mappings with bounded height. *Theory Probab. Appl.*, 18:120–130, 1973.

[130] V. N. Sachkov. *Probability Methods in Combinatorial Analysis*. Nauka, Moscow, 1978. In Russian.

[131] A. I. Saltykov. The number of components in a random bipartite graph. *Discrete Math. Appl.*, 5:515–523, 1995.

[132] B. A. Sevastyanov. Convergence of the number of empty cells in the classical allocation problems to Gaussian and Poisson processes. *Theory Probab. Appl.*, 12:144–154, 1967. In Russian.

[133] V. E. Stepanov. On the probability of connectedness of a random graph $g_m(t)$. *Theory Probab. Appl.*, 15:55–67, 1970.

[134] V. E. Stepanov. Phase transition in random graphs. *Theory Probab. Appl.*, 15:187–203, 1970.

[135] V. E. Stepanov. Structure of random graphs $g_n(x \mid h)$. *Theory Probab. Appl.*, 17:227–242, 1972.

[136] L. Takacs. On the height and widths of random rooted trees. *Adv. Appl. Probab.*, 24:771, 1992.

[137] S. G. Tkachuk. Local limit theorems on large deviations in the case of stable limit laws. *Izvestiya of Uzbek Academy of Sciences*, (2):30–33, 1973. In Russian.

[138] V. A. Vatutin. Branching processes with final types of particles and random trees. *Adv. Appl. Probab.*, 24:771, 1992.

[139] A. M. Vershik and A. A. Shmidt. Symmetric groups of high degree. *Soviet Math. Dokl.*, 13:1190–1194, 1972.

[140] A. M. Vershik and A. A. Shmidt. Limit measures arising in the asymptotic theory of symmetric groups. i. *Theory Probab. Appl.*, 22:78–85, 1977.

[141] A. M. Vershik and A. A. Shmidt. Limit measures arising in the asymptotic theory of symmetric groups. ii. *Theory Probab. Appl.*, 23:36–49, 1978.

[142] V. A. Voblyi. Asymptotic enumeration of labelled connected sparse graphs with a given number of planted vertices. *Discrete Analysis*, 42:3–16, 1985. In Russian.

[143] V. A. Voblyi. Wright and Stepanov–Wright coefficients. *Math. Notes*, 42:969–974, 1987.

[144] L. M. Volynets. The number of solutions of an equation in the symmetric group. In *Probab. Processes and Appl.*, pp. 104–109. MIEM, Moscow, 1985. In Russian.

[145] L. M. Volynets. On the number of solution of the equation $x^s = e$ in the symmetric group. *Math. Notes*, 40:155–160, 1986. In Russian.

[146] L. M. Volynets. An estimate of the rate of convergence to the limit distribution for the number of cycles in a random substitution. In *Probab. Problems of Discrete Math.*, pp. 40–46. MIEM, Moscow, 1987. In Russian.

[147] L. M. Volynets. The generalized scheme of allocation and the distribution of the number of cycles in a random substitution. In *Abstracts of the Second All-Union Conf. Probab. Methods of Discrete Math.*, pp. 27–28. Petrozavodsk, 1988. In Russian.

[148] L. M. Volynets. The generalized scheme of allocation and the number of cycles in a random substitution. In *Probab. Problems of Discrete Math.*, pp. 131–136. MIEM, Moscow, 1988. In Russian.

[149] L. M. Volynets. An example of a nonstandard asymptotics of the number of substitutions with restrictions on the cycle lengths. In *Probab. Processes and Appl.*, pp. 85–90. MIEM, Moscow, 1989. In Russian.

[150] H. Wilf. The asymptotics of $e^{P(z)}$ and the number of elements of each order in S_n. *Bull. Amer. Math. Soc.*, 15:228–232, 1986.

[151] E. M. Wright. The number of connected sparsely edged graphs. iii. *J. Graph Theory*, 4:393–407, 1980.

[152] E. M. Wright. The number of connected sparsely edged graphs. iv. *J. Graph Theory*, 7:219–229, 1983.

[153] A. L. Yakymiv. On the distribution of the number of cycles in random a-substitutions. In *Abstracts of the Second All-Union Conference Probab. Methods in Discrete Math.*, p. 111. Karelian Branch of the USSR Acad. Sci., Petrozavodsk, 1988. In Russian.

[154] A. L. Yakymiv. Substitutions with cycle lengths from a fixed set. *Discrete Math. Appl.*, 1:105–116, 1991.

[155] A. L. Yakymiv. Some classes of substitutions with cycle lengths from a given set. *Discrete Math. Appl.*, 3:213–220, 1993.

[156] N. Zierler. Linear recurring sequences. *J. Soc. Ind. Appl. Math.*, 7:31–48, 1959.

INDEX